The Global Thermohaline Paleocirculation

Elena V. Ivanova

The Global Thermohaline Paleocirculation

Springer

Dr. Elena V. Ivanova
Shirshov Institute of Oceanology
Russian Academy of Sciences
36 Nakhimovsky Prospekt
117997 Moscow
Russia
e_v_ivanova@ocean.ru

Translated by Ivar Murdmaa and Anna Stepanova.

ISBN 978-90-481-2414-5 e-ISBN 978-90-481-2415-2
DOI 10.1007/978-90-481-2415-2
Springer Dordrecht Heidelberg London New York

Library of Congress Control Number: 2009928018

The Russian language edition of the book Globalnaya termohalinnaya paleotcirculatcia was first published by Nauchnij Mir, Moscow, Russia, 2006.

Printed on acid-free paper

Springer is part of Springer Science+Business Media (www.springer.com)

Memory of Sergey Lappo

Preface

Since 2006, when the Russian edition of *The Global Thermohaline Paleocirculation* was issued by *Scientific World*, several books and hundreds of relevant international papers have been published in the domain of paleoceanography and paleoclimatology. The uprising interest in this subject encouraged me to revise and translate the monograph initially destined to summarize my own research over the last years and to compile the overview of the up-to-date knowledge on the past ocean circulation to my Russian colleagues who still have limited access to the international publications.

Although this English edition is extensively revised, I was not pretending to consider all currently debatable topics in such large and rapidly evolving field as the study of past ocean circulation. Very important subjects of the climate research like greenhouse gases, orbital theory of climate, and others are only mentioned briefly in this book since the major goal of this work was to consider more carefully the data and ideas on the global paleocirculation from my perspective. Several important areas of the World Ocean where I have not worked are omitted or shortly discussed in the last chapter on the global teleconnections. I am aware of these disadvantages in the book and regret about not being able to provide a complete review of all aspects of current research and debates on paleocirculation in the same detail. Instead, I hope the book will enable interested readers to review the recent and classical scientific publications cited here and learn about the achievements in modern and past circulation studies in Russia, as the reference list contains more than a hundred publications in Russian. In total, the list of literature in the English editions contains 140 new references.

Chapter 1 presents a short review of ideas led up to the concept of the global conveyor belt and the basic knowledge on the present global thermohaline circulation (THC) and ocean–atmosphere interaction. It also includes (Section 1.3) the overview of the major tectonic and paleoceanographic events that affected the interoceanic exchange during the Cenozoic and caused the onset of the modern-like pattern of the global oceanic overturning which substantially enlarged the English edition.

The data and ideas on the orbital and millennial-scale paleocirculation variability over the last climatic cycle and the Holocene in the North Atlantic where the global conveyor originates and in several key regions of the globe are discussed in

Chapters 3, 4, 5, and 6. I imply the last climatic cycle as the time interval between Terminations II and I (Broecker and van Donk 1970) including the Eemian interglacial and the last continental glaciation of the Northern Hemisphere. Methods of the study are summarized in Chapter 2.

Chapter 3 scrutinizes the up-to-date knowledge on the modes of the Atlantic meridional overturning circulation over the specific intervals of the last interglacial-to-glacial cycle and the Holocene including the last interglacial, last glacial maximum, Dansgaard–Oeschger cycles, Heinrich events, Younger Dryas, and the 8 cal. ka event. The English edition also contains some new data and modeling results. Chapters 4, 5, and 6 and Section 7.6 present the paleoceanographic reconstructions in the western Eurasian seas, Indian Ocean, South China Sea, and eastern equatorial Pacific obtained within the framework of several international projects in which I have had a pleasure to participate over the last 15 years. The English edition of the book contains some of the latest unpublished data and figures.

In the western Eurasian Arctic Seas, paleoenvironmental changes are mainly controlled by variations in the Atlantic water input and river discharge, thus by the interaction between the Atlantic and Arctic oceanic and atmospheric circulation. These marginal basins greatly contribute to the freshwater inflow from the Arctic into the North Atlantic and thus influence the global thermohaline circulation. As it is shown in Chapter 4, this influence was even more pronounced during the last deglaciation. Unfortunately, unlike low latitudes, the chronological frame of our research in the Barents and Kara seas is limited to the last 15–20 ka, the interval which can be recovered by gravity cores. Generally high sedimentation rates in the marginal seas provide records for high-resolution paleoreconstructions as compared to the most part of the World Ocean. In particular, we succeeded to recover and investigate the unique High-Arctic archive of the Last Millennium from the Russian Gavan' Fjord, Novaya Zemlya which is discussed in Section 4.5.4.

Paleoenvironments of the northern Indian Ocean and the South China Sea have been strongly affected by the variability in seasonal monsoonal circulation and in interoceanic heat, salt, and nutrient exchange via the return branch of the global thermohaline circulation. The links between the Indian and East Asian monsoons are still strongly debatable, as well as the intensity of the return route of the conveyor during the glacials. In Chapters 5 and 6, I tried to demonstrate that the return branch of the global thermohaline circulation and both Indian and East Asian monsoons were generally stronger during the interglacials and interstadials as compared to the glacials, and that the re-arrangement of the upper limb occurred over the terminations.

The last Chapter 7 discusses the contemporary ideas about the climatic teleconnections and plausible mechanisms with the special emphasis on the links provided by the thermohaline circulation. In particular, I consider the pole-to-pole and equator-to-pole linkages manifested in the Antarctic, Eurasian Arctic, northern Pacific, and low latitudes Indo-Pacific.

Moscow, Russia Elena V. Ivanova

Acknowledgments

During the last 15 years, I had a luxury to carry out the joint paleoceanographic research with Jean-Claude Duplessy, Ivar Murdmaa, Michael Sarnthein, Ralf Shiebel, Luejiang Wang†, Christoph Hemleben, Luc Beaufort, Uwe Pflaumann, Laurent Labeyrie, Leonid Polyak, Martine Paterne, Elsa Cortijo, Gerhard Schmidel, and Laurence Vidal who have especially influenced my thoughts and work.

Bjørg Risebrobakken, Michal Kucera, Claire Waelbroeck, Elena Krylova, Guillaume Leduc, Thibault de Garidel, Joel Guiot, Françoise Chalié, Frauke Rostek, my PhD students Natalia Chistyakova, and Ekaterina Ovsepyan generously helped me sharing their materials or providing the expertise on specific subjects.

I have benefited greatly from discussions with colleagues who are more deeply immersed in the particular aspects of modern oceanography, paleoceanography, micropaleontology, climatology, biology, and atmospheric sciences. Completing this work became possible thanks to the suggestions and advices of Sergey Lappo† (the director of Shirshov Institute of Oceanology RAS in 1995–2005), Max Barash, Natalia Blyum†, Natalia Os'kina, Tatyana Khusid, Valentin Burkov†, Vitaliy Stepanov†, Sergey Korsun, Mikhail Koshlyakov, Victor Neiman, Anatoliy Shcherbinin, Vyacheslas Bol'shakov, Michail Levitan, Leonid Galerkin, Yuri Romanov, Boris Filyushkin, Vladimir Vedernikov, Leonid Stefantsev, Irina Burmistrova, Ekaterina Taldenkova, Nikita Kucheruk, Alexander Vetrov, Nick Shackleton†, Eystein Jansen, David Anderson, Igor Belkin, Morten Hald, Heloise Leclerc, Kate Darling, Laeticia Pichavin, Laetitia Licari, Joseph Ortiz, Marie-France Loutre, Michael Schulz, Stefan Rahmstorf, John Murry, Evguenia Kandiano, Kazuyo Tachikawa, Steve Clemens, and Hans-Stefan Niebler.

Ray Bradley, Rainer Zahn, Jean-Claude Duplessy, Andrey Ganopolski, and Ralf Schiebel encouraged me to publish the monograph in English.

Meetings and discussions with the colleagues from IMAGES (The International Marine Global Change Study) and SCOR/IMAGES working group PACE (Reconstruction of Past Ocean Circulation) were always very stimulating while I was a Russian representative for IMAGES over 10 years and an expert of PACE over 3 years.

I am deeply grateful to all above-mentioned colleagues, as well as to V.P. Kazakova, T.N. Alekseeva, and A.N. Rudakova for grain-size analysis. The crews and members of scientific parties on the research vessels *Academic Sergey Vavilov*,

Boris Petrov, *Professor Shtokman,* and *Marion Dufresne* are warmly thanked for their help in obtaining the materials during marine expeditions.

I appreciate very much the work of my colleagues Ivar Murdmaa and Anna Stepanova to translate the Russian version of the book into English. Special thanks to Andrey Ganopolski, Mikhal Kucera, Sergey Gulev, and Andrey Lapenis, who critically read the drafts of several sections and offered me the advices on the necessary corrections.

The warmest thanks go to Galina Alekhina who made the time consuming editing of figures and to my son Andrey who helped a lot with preparing the reference list and especially supporting me with his great sense of humor.

The radiocarbon dates (AMS-^{14}C) used in this work were obtained at the Laboratoire des Sciences du Climat et de l'Environnement (LSCE CNRS-CEA, France), Leibniz Laboratory of Kiel University (Germany), laboratories of the University of Arizona and «Geochron» (USA), Poznań Radiocarbon Laboratory (Poland). Oxygen and carbon isotopes measurements were carried out at LSCE and CEREGE (France), Leibniz Laboratory (Kiel, Germany), Ecole Politechnique de Zurich (ETH, Switzerland), Oceanographical Institute of Woods Hole (WHOI, USA).

The research of the author was partly supported by RFBR-CNRS (grants RFBR 98-05-22029-CNRS_a and RFBR 06-05-22000-CNRS_a), RAS-CNRS (PECO-16348), RFBR (grants 97-05-68698-a and 06-05-64803-a), the Russian Federal Programme 'World Ocean". French Ministry of Foreign Affairs, German Geoscience Foundation (DFG), German Foundation of Academic Exchange Programs (DAAD), University of Tübingen (Germany), International Program START, Bjerknes Centre for Climate Research, University of Bergen (Norway), Smithsonian Institute (USA).

Contents

Introduction

Owing to extensive efforts of the international paleoceanographic community, a crucial role of the global thermohaline circulation in past climatic teleconnections on orbital and suborbital time scales is well recognized today. The marine archives record not only major but even short-term climatic events controlled by both global and regional mechanisms. Therefore, reconstruction of the thermohaline paleocirculation gives us a key to understand the history of the global climate system and thus contributes to long-term forecast of future climate changes.

According to the generally accepted orbital theory of climate proposed by J. Croll and developed by M. Milankovich, the cyclic alternation of glacials and interglacials during the Pleistocene, accompanied by the growth and collapse of the continental ice sheets in the Northern Hemisphere, was mainly controlled by the orbitally driven oscillations in summer insolation at high latitudes (notably 65°N). Many scientists still support the Milankovich theory accounting for the new data on the deglaciations (e.g., Kawamura et al. 2007) However, an increasing number of researchers now believe that the worldwide changes in annual (e.g., Imbrie 1982; Bol'shakov 2003, 2008) or annual mean (e.g., Loutre et al. 2004) insolation also need to be considered. The insolation is known to depend on the cyclic variations in three parameters of the Earth's orbit: eccentricity (elongation degree of the elliptic Earth's orbit), obliquity (inclination of the equator on the ecliptic), and precession (motion of the Earth's axis relative to the fixed stars) (Broecker and van Donk 1970; Shackleton and Opdyke 1973; Hays et al. 1976; Berger, 1977; Imbrie 1982; Imbrie et al. 1992; Bol'shakov 2003). Indeed, combination of the "external" orbital parameters defines the changes in solar energy received at the top of the atmosphere (Berger 1977; Monin and Shishkov 1979; Bol'shakov 2003; Loutre et al. 2004).

The orbitally driven climatic oscillations are modulated by the interplay of several "internal" (that is pertained to the planet Earth itself) factors and feedbacks. The most important internal factors are: topography of continents, distribution of land and sea on the planet surface, extension of ice sheets and permafrost, greenhouse gases content in the atmosphere, and winds pattern. The global thermohaline circulation occupies a crucial position among the internal factors due to its capacity to sustain the stability of the global climate and to transfer the abrupt regional changes to the remote areas. Indeed, the heat and moisture transport to high latitudes by the thermohaline circulation exerted strong control on the extension of the Northern Hemisphere ice sheets via positive and negative feedbacks.

The heat flux is generally directed from the ocean to the atmosphere, first of all, in the energy-active areas, as the overall heat inertia of the ocean considerably exceeds that of the atmosphere. The ocean largely controls the intensity of highs and lows in atmospheric pressure. These centers of action may influence surface ocean currents through the changes in winds stress curl (Lacombe 1972; Lappo 1984; Lappo et al. 1990; Marchuk 1983; Marchuk and Sarkisyan 1988). Global long-term stability of the atmosphere is also maintained by the ocean (Lappo et al. 1990). Besides, the ocean accumulates excess carbon dioxide from the atmosphere and contains 50 times more CO_2 than the atmosphere (Gordon 1989) As the ocean is capable to absorb the greenhouse gases from the atmosphere, it may damp the global warming.

It was determined already in the 1970s that oxygen isotope records, which primarily reflect variations in the global ice sheets volume, also demonstrate slow growth followed by fast collapse and melting of ice sheets (Broecker and van Donk 1970; Shackleton and Opdyke 1973) and the resulting variations in the sea level (Labeyrie et al. 1987; Fairbanks 1989; Waelbroeck et al. 2002). This allowed developing the standard oxygen isotope record (SPECMAP) on planktic (Imbrie et al. 1984) and benthic (Martinson et al. 1987) foraminifers. Since that time SPECMAP has been widely used in paleoceanography for the age control of the proxy time series and remote correlations. The new benthic oxygen isotope stack (Lisiecki and Raymo 2005) has a temporal resolution of 1–3 thousand years for the last 3 million years and about 5,000 years for the previous 2 million years. This achievement significantly improved the perspective of worldwide correlation of paleoceanographic and climatic events.

N. Shackleton (2000) ascertained that in-phase changes in the ocean deep-water temperature, Antarctic air temperature at Vostok site and atmospheric carbon dioxide content preceded changes in the global ice volume at terminations. This finding suggests that ocean together with atmospheric CO_2 amplify the 100-ka eccentricity cycle and thus questions the hypothesis by J. Croll about the determining role of ice sheet dynamics in reinforcing these cycles. Instead, N. Shackleton's observation confirms the importance of the oceanic overturning and greenhouse gases in triggering of the glacial–interglacial oscillations. On the contrary, W. Ruddiman (2006a, b) holds that the atmospheric CO_2 content is directly controlled by the global ice volume and that the positive feedback between these factors explains the nearly in-phase response of both hemispheres to glacial terminations. In turn, O. Sorokhtin and S. Ushakov (2003) argue that changes of the carbon dioxide content in the atmosphere were the consequence, but not the cause of the global temperature changes. Thus, confusion still persists over the significance of greenhouse gases, overturning circulation, and other factors in triggering the climate change.

Over the last years, variations in the deep ocean temperature and salinity, in particular their regional distribution pattern, have drawn increasing attention of paleoceanographers as one of the plausible mechanisms of global climate change.

W. Broecker et al. (1990) proposed that the abrupt climate changes could be triggered by the "salt oscillator" in the Atlantic. As shown by J.-C. Duplessy and co-authors (2002), regional differences in deep-ocean temperatures are linked to the intensity of NADW and AABW formation, thus to a persistence of the thermohaline circulation. Recent reconstructions by J. Adkins and colleagues (2002, 2005) demonstrated that through the last glacial Antarctic bottom waters were much cooler and saltier than the bottom waters in the rest of the World Ocean due to intensive freezing and brines release during sea-ice formation. The authors hypothesized that the ocean was more stratified during glaciations than at present and the stratification was destabilized only temporarily owing to the storage and catastrophic release of the geothermal heat in the deep water layers via the thermobaric effect (that is the reduction of the very dense deep water buoyancy). Further development of these ideas can significantly influence the existing understanding of the North Atlantic and Southern Ocean impacts on the variations of the global thermohaline circulation and global climate.

The early Southern Hemisphere and deep-water warming at the very beginning of Termination I due to increased austral spring insolation and sea ice albedo feedbacks are recently suggested by L. Stott and co-authors (2007) after the precise dating and correlation of several proxy time series from the western tropical Pacific, Southern Ocean, and Antarctica. The ~1,000 years lead of the deep-water temperature rise as compared to increase in the atmospheric CO_2 concentration and tropical ocean warming may be interpreted as an imprint of the Southern Ocean overturning in the further triggering of the deglacial climate change, in particular via the subsequent rise in atmospheric carbon dioxide. However, recently obtained high-resolution *Mg/Ca* and oxygen isotope records derived from the same oceanic fossils (mainly planktic foraminifers) showed a very early deglacial warming of the ocean surface waters in the tropics at 20–19 cal. ka BP that preceded melting of the Northern Hemisphere ice sheets by 3–5 thousand years (Lea et al. 2000, 2006; Nűrnberg and Müller 2000; Spero and Lea 2002; Rosenthal et al. 2003). These conflicting data contribute valuable insights into the importance of low latitudes in triggering global climate changes.

Within the last 10–20 years, the special emphasis in paleoclimatic and paleogeographic research was shifted from the investigation of orbital scale glacial–interglacial oscillations to the study of Heinrich events, suborbital Dansgaard–Oeschger cycles, and even centennial–decadal variations. It was shown that the short-term fluctuations can be recognized at least within the last 3.2 million years of the geological history (Raymo et al. 1998; McManus et al. 1999; Lisiecki and Raymo 2005). Such high-resolution records narrow the gap between the study of climatic and geological phenomena, thus contributing to the development of numerical modeling of natural processes and forecasting of future climate changes.

Abrupt high-amplitude climatic oscillations were especially well expressed in the Greenland ice cores, where the temperature changes reached 10–16°C within a few decades (Dansgaard et al. 1993; GRIP project members 1993; Landais et al. 2005). The high-resolution multiproxy study of Greenland ice cores showed that the remarkable variations of environmental parameters cannot only reflect local temperature changes, but in fact they document the climatic variability of the vast North Atlantic region linked to the mode of the thermohaline circulation (Lappo 1984; Broecker 1992; Duplessy 1996; Rahmstorf 2002, 2003; Rasmussen and Thomsen 2004). The thermohaline circulation in turn is very sensitive even to relatively minor additional freshwater input into this region (Fichefet et al. 1994; Ganopolski et al. 1998; Seidov and Haupt 1999; Ganopolski and Rahmstorf 2001; Knutti et al. 2004; Piotrowski et al. 2004). Thus, variations in environmental parameters recorded in marine sediments and ice cores reflect abrupt climatic changes, as well as instability of the thermohaline circulation (e.g., Zahn 1999; Vidal and Arz 2004). By now, millennial and centennial-scale variations are also documented in the Antarctic (Blunier et al. 1998, 1999; Blunier and Brook 2001, Shackleton 2001; Lal et al. 2004). However, the theory of decadal–centennial climatic cycles has not been developed yet.

In the ocean, millennial-decadal-scale events, such as changes of sea surface temperature (SST), paleoproductivity, mixed-layer and thermocline depths, and intensity of the oxygen minimum zone were revealed in many remote regions. Examples came from: the Santa Barbara Basin in California (Heusser and Sirocko 1997; Hendy and Kennett 1999, 2000; Hendy et al. 2002, 2004), Arabian Sea (Sirocko et al. 1993, 1996; Schulz et al. 1998; Schulte and Müller 2001), South China Sea (Wang et al. 1999a, b; Wang and Sarnthein 1999; Hong et al. 2005), north-western Pacific (Kiefer et al. 2001; Kiefer and Kienast 2005), eastern equatorial Pacific (Leduc et al. 2007), Iberian Margin and Mediterranean (Moreno et al. 2005; Martrat et al. 2004, 2007; Voelker et al. 2006) and elsewhere. Besides, thorough analysis of the surface and bottom water proxies demonstrate a close coupling between changes in the upper and lower limbs of the thermohaline circulation during short-term events (Curry and Oppo 1997). Moreover, abrupt changes, comparable to Heinrich events and Dansgaard–Oeschger cycles, were recorded on land (e.g., Gasse 2000; Wang et al. 2005; Cruz et al. 2005).

Thus, one of the primary goals of modern paleogeography and paleoceanography is to understand the still ambiguous mechanisms of climatic teleconnections. In this context, the importance of modeling simulations must be emphasized. Some models and data series demonstrated the transfer of the short-term climatic signals between remote regions through the atmosphere, and not through the ocean (e.g., Mikolaewich et al. 1997; Sirce et al. 2006). The role of the ocean in these processes still remains poorly studied (Zahn 1999; Vidal and Arz 2004). However, increasing number of proxy time series and modeling results shows that the thermohaline circulation likely provides a link between remote areas on the centennial–millennial time scale (e.g., EPICA members 2006; Rahmstorf 2006).

Whereas glacial–interglacial cycles are known to be roughly simultaneous on a global scale, suborbital, and centennial-scale oscillations cannot be unilateral because of existing feedbacks and non-linear oceanic overturning response to the local perturbations. Slowing down of the thermohaline circulation and weakening of the heat transfer to the northern North Atlantic leads to the cooling in this region.

However, another consequence of the same process could be warming in the tropics and South Atlantic (Crowley 1992; Stocker et al. 1992; Zahn 1999; Ganopolski and Rahmstorf 2001; Stocker 2002). According to several authors (e.g., Kiefer et al. 2001; Kiefer and Kienast 2005), the northwestern Pacific shows concurrent opposite climatic trends to those of Greenland and the North Atlantic, since it is located on the other end of the global conveyor. During Dansgaard–Oeschger stadials, the reducing of NADW production was most likely accompanied by the weakening of deep-water upwelling in the northern Pacific that could contribute to warming in this region. Thus, the regional events superimpose on the global ones complicating paleoenvironmental reconstructions and identification of natural mechanisms of global changes.

It is necessary to emphasize the importance of properly developed time scales for distant correlation of paleoceanographic events. Increase in the resolution and accuracy of datings for reference ice cores, as well as marine and terrestrial paleoclimatic archives led to the development of the hypothesis of "bipolar seesaw" (e.g., Blunier et al. 1998; Broecker 1997, 1998, 1999, 2000; Stocker 1998; Toggweiler 2005), postulating the antiphase relationship between high northern and southern latitudes over the millennial-scale climate variability. The abrupt perturbations of the thermohaline circulation are believed to trigger changes in the cross-equatorial heat transport and hence in sea surface and air temperature. Although this hypothesis still does not satisfactorily explain some empirical data and modeling results (e.g., Seidov et al. 2005), it has been recently supported by the new persuasive time series from the Antarctic (EPICA community members 2006) and some other areas (e.g., Martrat et al. 2007, see also Chapter 7).

I would like to conclude with the statement about the importance of the international cooperative efforts in reconstructing the past ocean circulation as one of the fundamental constituents of our knowledge on the global climate change. The contribution of the IMAGES and DSDP/ODP/IODP programs is highly important for the development of modern paleoceanography and paleoclimatology. The results obtained within the framework of these programs are widely used in this book.

Chapter 1
The Global Thermohaline Circulation and the Main Stages of Its Development During the Cenozoic

Abstract The concept of the global conveyor suggested in the 1980s evolved in the generally accepted theory of the global thermohaline circulation (THC). One of the major mechanisms driving the THC is the deep convection in the Subpolar North Atlantic due to the high density of the surface waters. However, significant amount of deep, bottom, and intermediate waters are formed in several other areas of the ocean, notably in the Antarctic, providing the large-scale overturning. The large oceanic heat transport, ocean–atmosphere interaction, heat and moisture release and uptake in particular energy-active areas exert a strong control on the global climate. Along with the millennial and longer-term variations linked to the THC, the atmospheric circulation modes are of crucial importance for the interannual and decadal-scale variability. The development of the modern THC pattern during the Neogene resulted from several dramatic tectonic and paleogeographic events which contributed to the reduction of interoceanic exchange in low latitudes and its strengthening in high southern latitudes.

Keywords Conveyor belt · Atlantic meridional overturning · Return branch · Heat transfer · North Atlantic Oscillation · El-Niño–Southern Oscillation · Heat transfer · Interoceanic exchange · Ocean–atmosphere interaction · Antarctica · Central American Seaway · Indonesian Passage · Northern Hemisphere glaciation · Tectonic events

1.1 Modern Thermohaline Circulation and the Global Conveyor Belt

Already in the 1960s, geochemists from the Shirshov Institute of Oceanology of the USSR Academy of Sciences found out that phosphorus and silica contents increase southward from minimum values in the North Atlantic to greater values in Antarctic waters, and northward from the South Pacific to the Aleutian Islands. Maximum concentrations of phosphorus were recorded in the northeastern Pacific. The above-mentioned distribution pattern of biogenic elements can be traced in surface as well as in deep waters (2,000–3,000 m). Finally, it was concluded that the

E.V. Ivanova, *The Global Thermohaline Paleocirculation*,
DOI 10.1007/978-90-481-2415-2_1,

global process of intensive accumulation of nutrients ends in the Pacific (Bruevich 1966; Sapozhnikov and Chernyaev 1967). Similarly, the concentration of free carbon dioxide and dissolved carbonates increases in deep waters (about 4,000 m) from the Atlantic toward the Pacific and further, from the south to the north. The area occupied by carbonate sediments is correspondingly reduced, whereas carbonate-free red clay is more widespread (Lisitzin 1974). However, one has to keep in mind that such a comparison of the spatial sediment distribution needs a topographic correction. The area of the ocean floor deeper than carbonate compensation depth (CCD) is more extensive in the Pacific (especially in the North Pacific) than in the Indian and Atlantic oceans. Alkalinity and the dissolved oxygen content in deep waters gradually decrease from the Atlantic toward the Pacific and from the south to the north in the latter. On the basis of these data, S.V. Bruevich (1966) came to the conclusion about "aging" of the World Ocean waters from the Atlantic through the Antarctic and further toward the North Pacific. By "aging" he implied the process of organic matter mineralization starting in the surface water and continuing throughout its sinking and transport from the Atlantic to the Pacific. This process is accompanied by the decline in the dissolved oxygen and organic matter content and by the increase in the content of nutrients and free carbon dioxide. This conclusion, based on the analysis of a large set of geochemical data, represents one of the most convincing proofs of the existence of the thermohaline circulation that starts in the North Atlantic and ends in the North Pacific. A comprehensive review of ideas and relevant schematics led up to the concept of the thermohaline circulation is recently published by P. Richardson (2008).

In the Atlantic, the average annual heat flux across the equator is directed to the north, while in the Pacific and Indian oceans – to the south. Correspondingly, positive, as compared to middle latitude values, air and surface water temperature anomalies dominate in the North Atlantic (Lappo 1984; Hastenrath 1991). Heat transfer across the equator in the Atlantic is estimated at 0.5×10^{15} W. Surface water temperature (SST) in the North Atlantic is higher than the average World Ocean's by 5°C, and the air temperature above it by 9°C. Besides, the Atlantic on average is 0.3°C warmer and 0.3 psu saltier than the Pacific, and the surface water salinity is 1 psu higher than the oceanic average (Lappo 1984, 1995). Difference in temperature and salinity of these oceans is linked to the interoceanic circulation pattern, predetermined by the dynamic topography of the World Ocean, in particular by the higher level of the Pacific in the mid-latitudinal area, elevated almost by 100 cm compared to the Atlantic (according to the average-multiannual map of V.A. Burkov (1980)). S.S. Lappo (1984) argued that this elevation is supposed to cause slow motion of warm surface Pacific waters from a higher toward a lower water level from the north to the south, toward the Indian Ocean, and further around the southern extremity of the African continent into the North Atlantic. Dense North Atlantic Deep Water (NADW) moves southward to the Antarctic and, interacting with its cold bottom water (AABW), continues its route toward the North Pacific. This phenomenon is well illustrated by the maps of deep-water circulation (e.g., Stepanov 1974). The NADW and AABW spread in the World Ocean primarily via the deep western boundary currents (Stommel 1958; Talley 1999).

High salinity and density of surface waters in the North Atlantic (Color Plate 1.1) is generally explained by the excess of evaporation vs. precipitation. Increased surface water salinity promotes intensive evaporation and further transport of moisture from the North Atlantic to the continent by the trade winds owing to the relatively small cross-sectional dimension of the ocean. The North Atlantic represents the most important energy-active zone with negative heat balance. In this region, heat and moisture fluxes to the atmosphere considerably exceed the value of those above the central parts of the oceans. Interoceanic deep-water circulation starts here, initiating the transfer of heat and salt to the North Pacific, where negative anomalies of these parameters are observed in the surface water layer (Lappo 1984, 1995).

In 1991, W. Broecker developed a theory of the global ocean conveyor belt (Color Plate 1.2) to explain the interoceanic transport of heat and salt. Analysis of the hydrochemical data given above shows that the conveyor also transfers nutrients as well as dissolved oxygen. The length of the conveyor is about 30,000–40,000 km. According to different estimates, the overall oceanic mixing time ranges from 200–300 (Lappo 1984; Anisimov et al. 2002) to 2,000 years (Seidov and Haupt 1997). Today, the conveyor theory is generally accepted as a simplified and generalized concept of a two-layer global thermohaline circulation, which allows estimating the heat, water, and salt transport in the World Ocean. Owing to the exchange between the ocean and the atmosphere, marine waters become enriched with carbon dioxide that contains unstable ^{14}C isotope. The decrease in the radiocarbon content from the North Atlantic toward the North Pacific (Fig. 1.1) is considered to be a robust

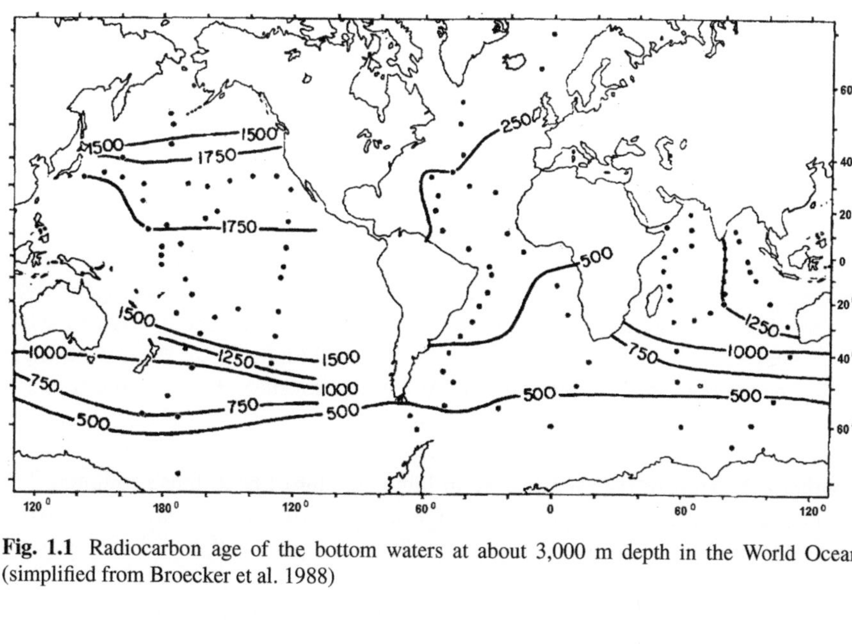

Fig. 1.1 Radiocarbon age of the bottom waters at about 3,000 m depth in the World Ocean (simplified from Broecker et al. 1988)

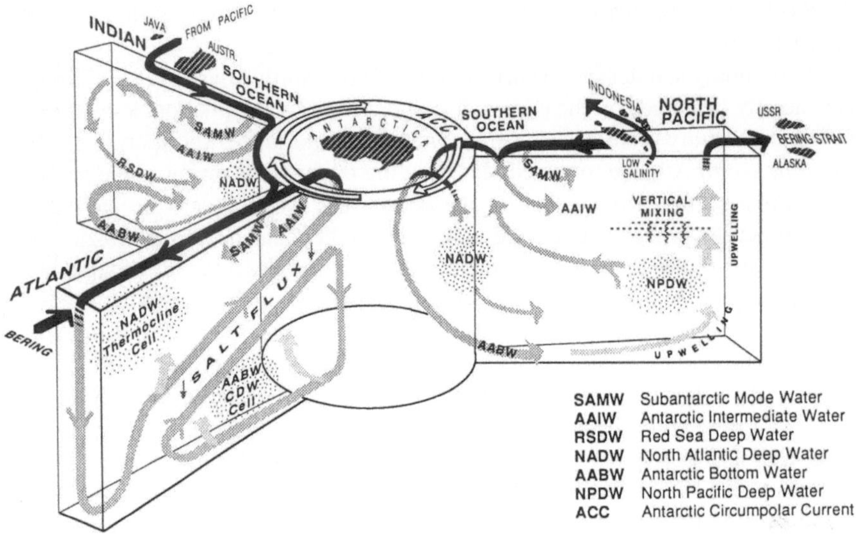

Fig. 1.2 Map view of the major pathways for links between the North Atlantic Deep Water (NADW) and Antarctic Bottom Water (AABW) in the global conveyor (Gordon 1989, courtesy by A. Gordon)

evidence of the water "aging" within the conveyor (Broecker et al. 1988; Lappo 1995). Much longer global oceanic overturning as compared to the duration of the exchange processes in the atmosphere, along with considerably higher heat capacity of sea water vs. that of the air, defines the crucial role of the ocean in the global long-term climatic oscillations.

Slightly different version of the conveyor theory, including the Arctic Ocean, was proposed by A. Gordon (1989) (Fig. 1.2). He argued that the deepest convection occurs in the North Atlantic and around the Antarctica where the Antarctic Bottom Water (AABW) with a temperature below –1°C and salinity about 34.7 psu is formed on the continental slope. In this area, surface water freezes and sinks owing to the strong cooling of the entire polar region and strong offshore katabatic winds. AABW moves northward in all oceans and slowly rises and mixes with less dense deep waters and after that returns to the Southern Ocean as circumpolar deep water. This relatively warm deep water has a temperature of about 1°C and salinity around 34.8 psu; it rises and transforms into cold (about –1.85°C surface water of the Antarctic Circumpolar Current (ACC) and warmer (3–5°C, salinity about 3–4.3 psu) Antarctic Intermediate Water (AAIW). In 1–2 years these water masses contribute to the formation of the AABW. Such relatively fast overturning in the Southern Ocean results in the release of huge amount of heat, thus promoting fast ice melting in spring and hindering the formation of the thick winter sea ice cover like that found in the Arctic. At the same time ice separates the cold winter atmosphere from the ocean preventing it from the stronger cooling (Gordon 2001). About 10–15 Sv (1 Sv = $10^6\,m^3\,s^{-1}$) of AABW is formed annually (MacDonald and

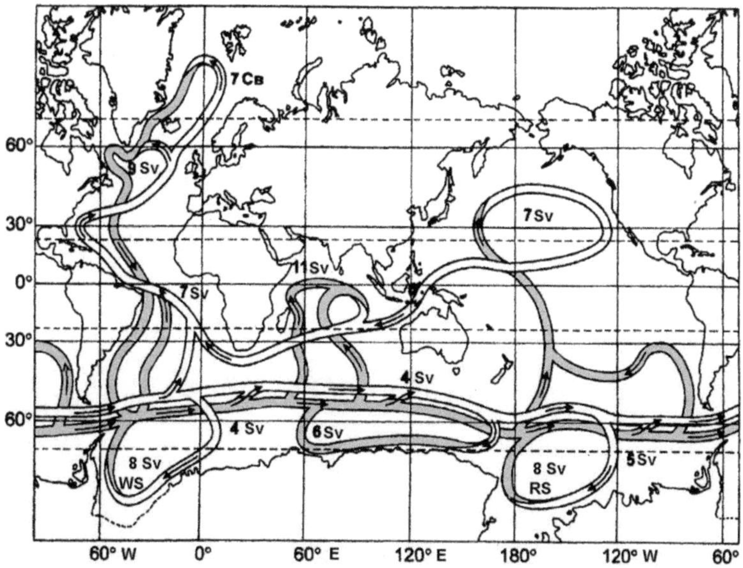

Fig. 1.3 The global oceanic conveyor. The deep-water component of the thermohaline circulation (*lower limb*) is shaded and the surface component (*upper limb*) is shown by open double line. *Arrows* indicate direction of the water transport. Vertical water discharge in direct and return branches of circulation cells is shown in sverdrups (Sv). Deep-water upwelling into the thermocline that occurs throughout the Antarctic Circumpolar Current with a total discharge value 13 Sv is arbitrarily shown for three regions with discharge values of 4 and 5 Sv (modified from Koshlyakov et al. 2001). WS = Weddell Sea, RS = Ross Sea. Printed with permission from AGU

Wunsch 1996). M. Koshlyakov and co-authors (2001) argue that the Ross and Weddell Seas are to be considered as very important convective cells (Fig. 1.3). Stability in the rate of deep-water formation is of crucial importance for the climate variability in the future.

In the North Pacific, a deep convection is hampered by low surface water salinity. Unlike the Atlantic, zonal circulation driven by trade winds prevails over meridional surface currents in the Pacific (Lappo 1984). Since deep waters of the North Pacific are low in oxygen and rich in nutrients, they differ strongly from NADW. They move southward, merge with circumpolar deep water, and further flow into the Atlantic. North Pacific deep water (NPDW) is formed as a result of northward transport of NADW and especially AABW, which rises and mixes here with low-salinity subsurface waters. Because of the great dimension of the Pacific even weak vertical mixing can lead to the formation of 10 Sv of NPDW (Gordon 1989). This process does not represent sinking of 10 Sv of surface waters, but more likely is a slow diffusion of surface water parameters; that is why the advection of atmospheric gases into the deep waters does not occur here, unlike the Atlantic. Part of NPDW reaches the surface and flows into the Arctic through the Bering Strait, whereas another part penetrates into the Indian Ocean through the Indonesian Seaway. Variations in this interoceanic water exchange must exert a significant influence on the global climate (Gordon 1989, 2001; Clarke et al. 2001).

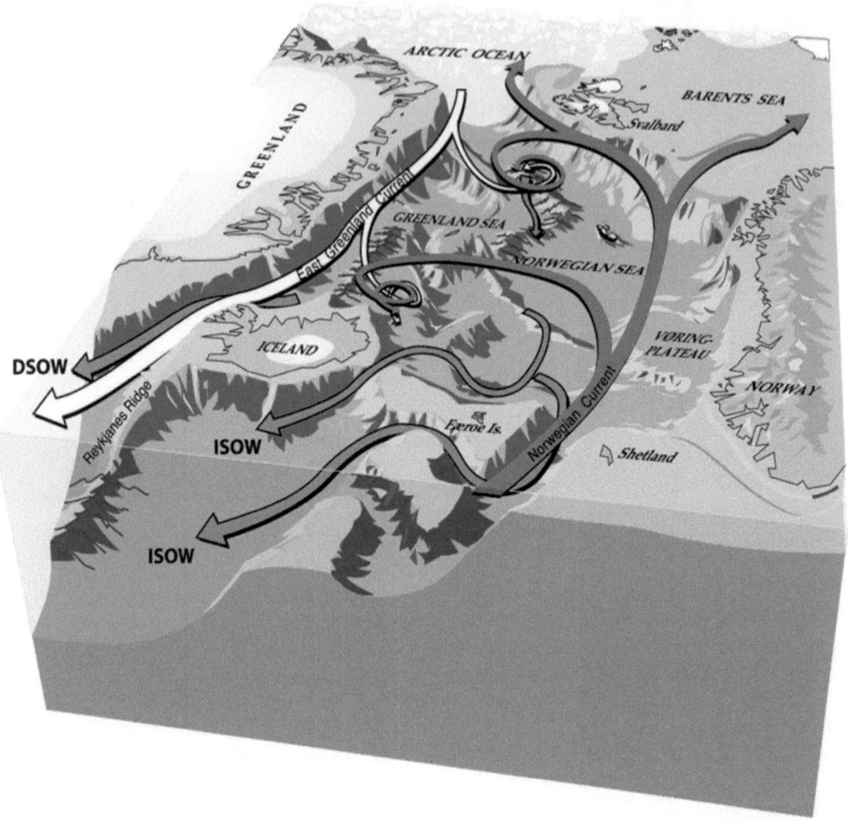

Fig. 1.4 Physiography and circulation of the North Atlantic (drawing by Masaoki Adachi, University of Bergen, Norway, for free public use). DSOW = Denmark Strait Overflow Water, ISOW = Iceland–Scotland Overflow Water

Early publications on the global overturning (Stommel 1961; Lappo 1984; Broecker et al. 1985; Gordon 1986; Broecker and Denton 1989; Broecker 1991; Color Plate 1.2) considered only the two-layer thermohaline circulation, but later the importance of intermediate waters, in particular Mediterranean water, in the global processes was recognized (Color Plate 1.3–1.5; Zahn 1997; Gordon 2001; Clarke et al. 2001). According to the up-to-date concept (Gordon et al. 1992; Broecker 1992, 1999; Dobrolyubov 2004; Lappo 2004; Rahmstorf 2002, 2003, 2006), the global conveyor and the THC originate in the North Atlantic (Color Plate 1.3; Fig. 1.4), where the most intensive sinking of relatively warm highly saline and, therefore, dense surface waters occurs down to depths of more than 2 km. It is generally accepted that the Atlantic Meridional Overturning Circulation (AMOC) represents the most important "driver" of the modern THC, and some authors consider the AMOC and THC to be synonymous (IPCC 2007). However, the THC is not confined to the meridional transport but is associated with zonal overturning cells (Togweiller and Samuels 1995; Rahmstorf 2006). The size of deep convection cells is about 200 km (Clarke et al. 2001). The estimates of annual NADW formation vary from 13 to 22 Sv (Broecker 1991; MacDonald 1998; Gordon 1989; MacDonald and Wunsch 1996; Anisimov et al. 2002; Talley et al. 2003) whereas the global overturning exceeds 30 Sv.

At present, NADW formation mainly occurs in the Norwegian–Greenland Basin due to the strong heat release into the atmosphere, whereas the waters southward of Iceland are too warm and thus not dense enough to sink. The Arctic Ocean significantly contributes to NADW formation since its relatively warm and saline deep waters, after passing the Fram Strait, mix with colder and freshened deep waters of the Greenland Sea at depths exceeding 1,700 m, as well as with the North Atlantic waters south of Iceland (Aagaard et al. 1985, 1991; Aagaard and Carmack 1989). Low salinity of the Greenland Sea waters is caused by the export of the freshwater sea ice and freshened surface waters from the Arctic. These waters come from vast Arctic shelves where the river run-off is very strong (Treshnikov 1985, see also Chapter 4). Besides, part of the 0.8–1 Sv of freshened waters from the North Pacific flowing to the Arctic through the Bering Strait also enters the Atlantic (Aagaard and Carmack 1989; Gordon 1986, 2001). Intermediate and deep waters of the Labrador Sea are formed as a result of intensive winter cooling and represent an important part of NADW (Weaver 1995; Dobrolyubov 2004). The Labrador Sea waters merge with deep waters from the Norwegian–Greenland Basin and contribute 2–4 Sv to the AMOC. The main area of NADW formation is separated from the rest of the ocean by the Greenland–Iceland–Scotland system of ridges, forming a sill of only 500–800 m water depth. After crossing the sill, NADW flows into the cascading up to several kilometers long and 100-m thick with intensive mixing and entrainment of the surrounding water masses. The mechanism of this process is not yet well understood (Clarke et al. 2001). NADW with a temperature of about 3°C and salinity of more than 34.9 psu moves southward at the depths of 2–4 km. NADW distribution can be detected down to the equatorial Atlantic from the $\delta^{18}O$ isoline of +1‰ (Kroopnick 1985; Duplessy et al. 1991a, 2002; Sarnthein et al. 1994b). Recent studies demonstrated that the intensity of the AMOC strongly depends on the deep-water influx from the Arctic and is inversely correlated to the rate of the Labrador Water formation (Dobrolyubov 2004).

To the south of Azores Islands, NADW merges with AABW. Then, the deep waters of the conveyor move eastward, to the South Pacific where they turn to the north, toward the Aleutian Islands. Supposedly, deep waters rise to the surface in the North Pacific due to Ekman divergence and mixing to complete the global overturning (Broecker and Peng 1982; Koshlyakov et al. 2001). According to the alternative point of view the upwelling in the northwestern Pacific can be caused by deep convection in the Sea of Japan and Sea of Okhotsk (Lappo 1984). The return surface flow (so-called warm water route) passes through the Western Pacific Warm Pool and the Indonesian Seaway, crosses the Indian Ocean in the westward direction, flows around the Cape of Good Hope, turns to the North, and ends in the area of NADW formation (Color Plates 1.2, 1.5, Fig. 1.3).

Return Indonesian Throughflow (ITF) controls and moderates the surface water temperature in the Indo-Pacific Warm Pool. ITF is mainly fed by surface and intermediate North Pacific waters and to a less extent by South Pacific waters, flowing

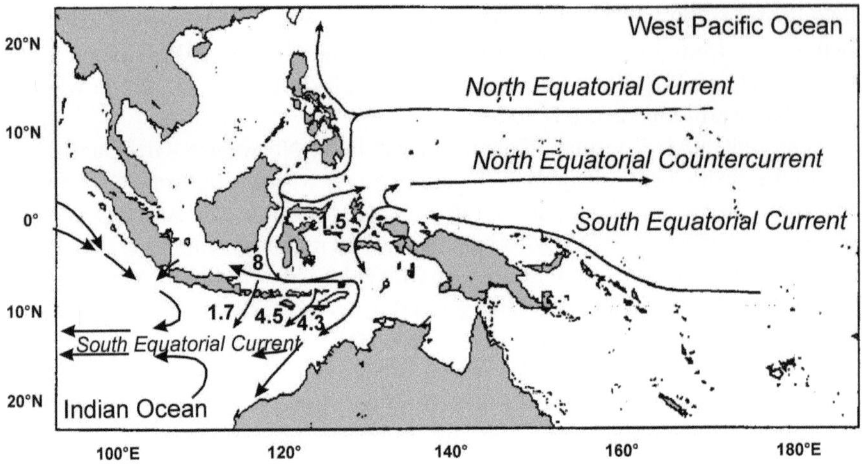

Fig. 1.5 Surface circulation via the Indonesian Throughflow (after M. Moore et al. 2000 with additions). Estimates of total volume transport (in Sv = 10^6 m^3 s^{-1}), after (Gordon 2005). Printed with permission from AGU

on the thermocline depths through the Makassar and Lombok Straits, and then through the Banda and Flores seas (Fig. 1.5). According to M. Anisimov and co-authors (2002), the present-day intensity of the conveyor's return branch in the northern Pacific reaches 12 Sv provided the thickness of the layer is about 400 m. In the Indonesian Straits, the return flow merges with the weaker flow of the eastern deep branch of the conveyor that moves northward from the Antarctic to the southeastern Pacific and upwells to the surface in the South Equatorial Current. This upwelling creates the so-called equatorial called tongue in the eastern equatorial Pacific. Intensive mixing and heat and moisture exchange with the atmosphere take place in the return flow. About 10 Sv is transported through the upper 400 m of the Timor Sea in the strait between Java and Australia. The estimates of water transport from the Pacific into the Indian Ocean in low latitudes range from 9–10 to 20 Sv (Gordon and Fine 1996; Gordon 2001; Seidov and Haupt 1999; Shriver and Hurlburt 1997; Anisimov et al. 2002). The strengthening of the return flow of warm freshened waters through the Indonesian Seaway influences the ocean–atmosphere interaction, and consequently the development of monsoons and El Niño. In turn, seasonal oscillations of the return surface flow can depend on the intensity of the monsoon winds on the southern and northern borders of the Makassar Strait (MacDonald and Wunsch 1996; Gordon and Fine 1996; Gordon 2001).

Hence, the deeper parts of the oceans are mainly filled by the waters formed as a result of deep convection in high latitudes. Whereas these waters advance toward the equator, the surface waters move poleward to compensate the downwelling in convection cells according to the continuity equation (Clarke et al. 2001). As a result, export of NADW into the other oceans is balanced by the water inflow from the Pacific and Indian oceans at shallower depths, above 1–2 km (Lynch-Stieglitz et al. 2007), primarily in low latitudes. The second return flow from the southern Pacific into the Atlantic through the Drake Passage (so-called cold water route) is generally considered to be less important for the global overturning. However, it has been recently estimated to bring about 25 Sv (Anisimov et al. 2002). Transport via the Bering Strait, as mentioned above, does not exceed 1 Sv (Aagaard and Carmack 1989). Convection down to the thermocline and intermediate water depths also takes place in the Mediterranean, Arabian, and Red seas, as well as in the Sea of Okhotsk (Talley 1999; Gordon 2001). Intermediate waters play a very important role in the thermohaline circulation. Partly they have North Atlantic origin. Upon reaching the latitude of Antarctic Circumpolar Current, NADW feeds the Antarctic Intermediate Water and rises up to the thermocline level in all oceans. NADW reaches the Indonesian Gateway and then moves back with the return flow and its continuation – Agulhas Linkage to the Atlantic (Gordon 2001).

The Agulhas Current transports up to 70–100 Sv (Bryden and Beal 2001; Gordon 2001). The main flow does not enter the Atlantic, but turns back into the Indian Ocean at the latitude of about 40–37°C as so-called Agulhas Retroflexion. From the 25 Sv carried by the extension of the Agulhas Current in the South Atlantic, the Benguela Current, 15–20 Sv is brought from the Indian Ocean in the upper 1,500 m layer (Gordon et al. 1992; Anisimov et al. 2002). Another 9 Sv from the Indian Ocean together with the AAIW feeds the Northern Brazilian Coastal Current at the depth of the upper and lower thermoclines. This current flows to the North Atlantic, while the rest of the water mainly circulates in the subtropical gyre of the South Atlantic and returns into the Indian Ocean. Agulhas Current brings warm and salty water into the South Atlantic, since salinity in the return flow increases after crossing the Indian Ocean. The transport is made by means of the anticyclonic eddies, rings, and filaments. A flow more than 500-m thick in the Agulhas Current confirms the presence of Red Sea waters participating in the transformation of Antarctic intermediate and bottom waters passing through the Southern Indian Ocean to join the current. In fact, water transport from the Indian Ocean into the Atlantic maintains the return flow toward the north and hence the Atlantic Meridional Overturning Circulation (AMOC). Otherwise, the transport over the Bering Strait would have become very important (Gordon 2001).

The estimates of NADW transport are highly variable. From the 15–20 Sv of NADW that reaches the Southern Ocean, 5–6 Sv enters the Indian Ocean and 6–15 Sv flows into the Pacific. The return surface flow through the South Atlantic northward contains 16–20 Sv, and the inflow of AABW into the Atlantic reaches 8 Sv (Lappo 1995; Rahmstorf 1995; Seidov and Haupt 1999; Schulz et al. 2001; Anisimov et al. 2002). Besides regions of NADW and AABW formation, energy-active convective cells of anticyclonic gyres play a very important role in supplying the deep branch of the THC, particularly in the South Atlantic (Lappo 1984; Lappo et al. 1990; Koshlyakov et al. 2001). The THC is a broader concept than the oceanic conveyor, since it also includes currents that are not considered by the conveyor theory. For example, Togweiller and Samuels (1995) consider the westerly winds-driven deep-water upwelling in the Southern Ocean and corresponding northward Ekman transport as the major driving force of the global overturning. The latest models of the THC display not only surface and deep, but also bottom, and sometimes even intermediate waters. One of the most often-used schemes by S. Ramstorf (2002) is shown here in Color Plate 1.5. Presently, efforts to develop the general numerical model of the global (thermohaline and wind-induced) circulation are being undertaken (MacDonald and Wunsch 1996; MacDonald 1998). However, the combined modeling of the three very important components of the oceanic circulation, such as thermohaline, wind –induced, and eddy induced, is very complicated and hardly possible at the moment (Clarke et al. 2001).

Over the three main deep convection areas of the World Ocean, the Norwegian–Greenland Sea, Ross and Weddell seas and Labrador Sea, the air temperatures are ~10°C higher than the latitudinal mean (Rahmstorf 2006). The coupled ocean–atmosphere models demonstrate the Northern Hemisphere cooling and Southern Hemisphere warming provided the NADW formation is switched off and the cross-equatorial heat transport is correspondingly reduced. The maximum cooling occurs near the sea ice margin in the Nordic Seas because of the positive albedo feedback. The NADW shutdown is followed by the southward shift of the thermal equator and ITCZ (intertropical convergence zone) and by rather rapid rise of the sea level in the northern Atlantic (Rahmstorf 2006).

Meridional transfer of heat and salt with the global conveyor (especially in the North Atlantic) could have been strongly variable in the past (Gordon 1989, 2001; Seidov and Haupt 1999; Boyle and Keigwin 1987; Broecker et al. 1985; Duplessy et al. 1988; Oppo and Lehman 1993; Sarnthein et al. 1994b, 1995, 2000b; Duplessy 1996, see Chapter 3). Results of numerical modeling show that relatively small variations in the inflow of fresh waters to the North Atlantic can lead to significant changes in the thermohaline circulation (Aagaard and Carmack 1989; Gordon 1989; Seidov and Haupt 1999; Fichefet et al. 1994; Ganopolski and Rahmstorf 2001; Manabe and Stouffer 1995, 1997; Ramstorf 1995) and that the THC could have been unstable in the wide range of variable parameters of the surface layer (Seidov and Haupt 1999). According to several authors, the thermohaline circulation in the North Atlantic declined by 30–65% during the Last Glacial Maximum (Sarnthein et al. 1994b; Winguth et al. 1999; Schulz et al. 2001), but according to other estimates (Yu and Harrison 1996) it was as intense as it is now. Reconstructions of the early stage of the deglaciation 17–14.7 cal. ka are also contradictory (Schulz et al. 2001). Linkage between northern parts of the Pacific and Atlantic oceans through the global conveyor could have been interrupted during the terminations and glacial inceptions (Seidov and Haupt 1999; Gordon et al. 1992; Broecker and Denton 1989). Unlike the sub-Arctic Pacific, the mid and low latitudes are very sensitive to cooling and freshening of the North Atlantic. Important changes occurred in the South Atlantic, southeastern parts of the Indian and Pacific oceans, as well as to the east of Australia (Seidov and Haupt 1999).

In the 1970s, the Great Salinity Anomaly presumably caused by the enhanced freshening of the Subpolar North Atlantic was recorded. Numerous measurements made during those years showed that salinity decreased by 0.1 psu in 1968 in the upper 200 m layer east of Greenland. The anomaly of a lower salinity started to propagate toward Newfoundland, and by 1977 it approached the Subpolar Front (Dickson et al. 1988; Lappo 1995; Belkin et al. 1998). The anomaly led to the substantial reduction of the convection depth in the Greenland, Irminger and Labrador seas limiting the upper layer to 500 m, as the temperature of the maximum density was close to the freezing point in 1989 when the salinity was around 34 psu. Extremely high sea ice extent was observed in the Labrador Sea due to the conservation of cold in the upper water layer and significant diminishing of meridional heat transfer through the Subpolar Front. At the same time the heat flux across the equator nearly doubled, from 0.5×10^{15}W up to 1×10^{15}W (Lappo 1995); the heat release into the atmosphere also increased on the Subpolar Front. Particularly, this resulted in mild and humid winters in Central Russia and in increase of Volga River runoff in the 1980s. Investigation of the Great Salinity Anomaly revealed the variable intensity of circulation at different depths in the source area of the global conveyor belt and the possibility of existence of different modes of meridional heat transfer.

1.2 Atmospheric Circulation Modes and Their Linkage to Thermohaline Circulation

Among the most important atmospheric circulation modes affecting global climate and the thermohaline circulation are the following: the North Atlantic Oscillation (NAO), the El-Niño–Southern Oscillation (ENSO), the Arctic Oscillation (AO), and the Pacific Decadal Oscillation (PDO) (Walker 1924; Lappo et al. 1990). The combined influence of NAO and ENSO defines up to 50–60% of the processes in the atmosphere (Wallace et al. 1995). It is also noteworthy, that the ocean–atmosphere interaction to a high degree depends on the distribution of sea ice, in particular in the North Atlantic (Sarnthein et al. 2003a; Piotrowski et al. 2005).

1.2.1 NAO, AO, and PDO

Higher SSTs on the west of the North Atlantic subtropical gyre compared to the North Pacific favor the intensive evaporation, thus increasing salinity and enhancing deep-water formation in the North Atlantic. In addition, they promote the transport of moisture, condensed above the warm Atlantic, by westerly winds (Broecker et al. 1985; Gordon 2001). "The beginning" and "the end" of the global conveyor, according to its geographical location, correspond to the Icelandic Low and the Aleutian Low, which interact through the atmosphere and "work" in the antiphase mode. According to several authors (e.g., Seidov and Haupt 1999), the link between the North Atlantic and the northern Pacific is mostly maintained through the atmosphere.

The North Atlantic Oscillation is controlled by the fluctuations in the atmospheric pressure gradient on the sea level between the Icelandic Low and the Azores High (Walker 1924; Rodwell et al. 1999). If the Icelandic Low and the Azores High become more pronounced, meaning the meridional pressure gradient is increasing, the NAO index is known to be positive (Color Plate 1.6). Westerly winds become stronger and the cyclones they bring are more frequent in high latitudes of the

North Atlantic. As a result the precipitation increases in the northern Europe and the Barents Sea region (see Section 4.5.4), while the near-surface continental temperatures above the southeastern United States and northern Europe including the northern part of Russia concurrently rise. In contrast, arid climate predominates in southern Europe; the Mediterranean and northern Africa temperatures are lower to the south of Mediterranean. The amount of sand brought from the Sahara to the North Atlantic and Caribbean Basin rises (Hurrell and van Loon 1997; Jung et al. 2003). More of Atlantic waters flow into the Arctic. The influence of Siberian river run-off can be traced to the west, and more ice is produced and drifted out than during the negative phase of the NAO index. The Canarian upwelling intensifies. The correlation coefficient between the NAO and the dominant frequency mode of cyclones in the North Atlantic equals 0.86 (Gulev et al. 2001).

Strengthening of the NAO can be caused by the acceleration of the rotation of the atmosphere or due to the processes occurring in the stratosphere. It was discovered, in particular, that the NAO mode changed about 25 years ago: between 1960 and 1972 variations of the NAO were simultaneous to the PDO, and between 1972 and 1995 they were almost in antiphase (Gulev et al. 2001). The positive NAO index promotes the intensified ocean baroclinicity that is the activation of the Gulf Stream and the increase of heat release and volume of fresh cold water convection in the Labrador Sea down to depths of 1.5–2 km (Gulev et al. 2001; Jung et al. 2003). The stronger intermediate water formation in the Labrador Sea in turn invokes the weakening of the AMOC by 30–40% (Dobrolyubov 2004). The NAO also influences the convection in the Iceland and Sargasso seas (Zahn 1999). However, there is also a feedback. Modeling results demonstrated that surface temperatures of the North Atlantic affect the NAO variability on interannual and interdecadal time scales, causing changes in temperature, atmospheric precipitation, and storms above Europe (Rodwell et al. 1999). The intensity of the NAO is also sensitive to the "see-saw" effect between the Icelandic and Aleutian Lows, as well as to the Southern Oscillation effect reaching up to the North Atlantic (Honda and Nakamura 2001).

The Arctic Oscillation (AO) is closely linked to the NAO. It represents the stratospheric vortex above the Arctic, which defines to a high degree the climatic changes in the region. When the anticyclonic gyre in the Beaufort Sea is strong (negative AO index), the intensity of water exchange with the North Atlantic diminishes and less sea ice drifts there from the Arctic Basin. The opposite situation occurs when the Icelandic Low deepens and the NAO/AO index is positive. The AO can result from long-term (on the synoptic timescale) temperature changes in northern parts of the Atlantic and Pacific oceans in the dipole over the Rocky Mountains (Thompson and Wallace 1998; Proshutinsky and Johnson 1997).

The PDO is defined as variations of the difference in sea level pressure between Hawaii High and Aleutian Low. Positive PDO index is correlated with the strengthening of these centers of the atmospheric pressure, as well as with the increase of the meridional pressure gradient between them. The PDO is characterized by strong westerly winds and frequent cyclones in high latitudes of the North Pacific, and by weakening of the winds at mid and tropical latitudes. The correlation coefficient between the PDO and the frequency of cyclones in the northern Pacific is 0.68 (Gulev et al. 2001).

1.2.2 ENSO

The El Niño-–Southern Oscillation (ENSO) represents the interannual variability in the winds intensity, precipitation, sea level, SST, and thermocline tilt. The southern oscillation index (SOI) is controlled by the atmospheric pressure gradient in the Southern Hemisphere, between Darwin (12°C, 131°E, Australia) and Tahiti (17°C, 150°W). ENSO variability seems to be mainly triggered by oscillations in the Walker circulation that is by the trade winds-driven westward air and surface water transport along the equator. In turn, the intensity of the Walker circulation is controlled by the surface water temperature gradient between the warm western and the cooler eastern part of the Equatorial Pacific (Bjerknes 1969; Lappo et al. 1990; Neelin et al. 1998; Wallace et al. 1998; Wyrtki 1982). The distribution of surface water temperatures generates a thermocline tilt in the tropical Pacific from the east to the west (Fig. 1.6). As a result, the Western Pacific Warm Pool (WPWP, also called Indo-Pacific Warm Pool) is formed with the thickened surface water layer, while on the east the Ekman upwelling and shallow thermocline promote the formation of the so-called Equatorial cold tongue (Lappo et al. 1990; Neelin et al. 1998). In normal conditions, the strong atmospheric convection in the WPWP causes heavy rains (Fig. 1.6).

When the SST rises in the eastern equatorial Pacific (EEP), the meridional temperature gradient decreases and the Walker circulation (easterlies) and trade winds weaken in the equatorial area. The thermocline in the equatorial region along the American coast deepens and the South equatorial current weakens. These changes result in a positive feedback initiating further surface water temperature rise in the EEP. Interaction between the ocean and the atmosphere becomes unstable giving rise to ENSO (Bjerknes 1969; Neelin et al. 1998). El Niño corresponds to the negative phase of the southern oscillation that is to the abnormal pressure rise in the Darwin area and decrease of the Subtropical High above Tahiti in the southeastern Pacific.

Observation and modeling results show that ENSO is controlled by the heat and moisture exchange between the ocean and the atmosphere in the equatorial Pacific (Bjerknes 1969; Alexander et al. 2002; Wallace et al. 1998; Wyrtki 1982). Already Wyrtki (1975) concluded that the El Niño phenomenon does not merely reflect local wind-induced changes at the South American coast that is the weakening of the upwelling, but evidence the remote surface water temperature response to the fast weakening of the eastern transfer in the equatorial Pacific. ENSO originates as a self-maintaining cycle of 2–10 years duration in which surface water temperature anomalies in the Pacific initiate the strengthening or weakening of the trade winds. The latter in turn triggers changes of oceanic circulation that later result in surface temperature changes (Bjerknes 1969; Neelin et al. 1998; Clarke et al. 2001; Trenberth et al. 1998). Since the oceanic response is in general slower than that of the atmosphere, the ocean maintains ENSO with alternation of warm and cold phases (Bjerknes 1969; Neelin et al. 1998). Investigations, carried out by the TOGA (Tropical Ocean and Global Atmosphere Program), suggested a possible explanation of this alternation. It was called the memory paradigm of the subsurface ocean, which is the adaptation (regulation) of the near-surface ocean. The "memory" for

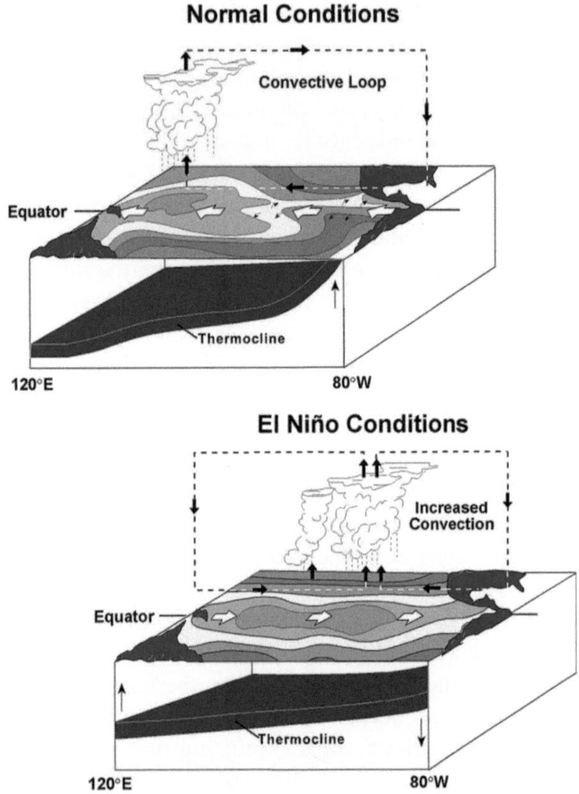

Fig. 1.6 Schematic of normal and El Niño conditions in the equatorial Pacific (McPhaden et al. 1998). Printed with permission from AGU

ENSOs is provided by the passage of the Rossby waves from the region affected by winds to the western boundary of the ocean and by the return of the signal to the east as reflected Kelvin waves (Neelin et al. 1998). According to several authors, changes in tropical convection in the WPWP associated with ENSO events affect the global atmospheric circulation (Clarke et al. 2001). Thus, ENSO can cause large-scale teleconnections (see Section 7.4). According to V.I. Byshev and co-authors (2001a), the influence of strong El Nino of 1997/1998 was observed even in the Barents Sea.

Warming in the eastern equatorial Pacific during the El Niño as a rule leads to the cooling in the centers of northern and southern parts of the ocean as well as to the strengthening of the Aleutian Low during the boreal winter and spring (Lappo et al. 1990; Alexander et al. 2002). The amount of atmospheric precipitation exceeds the normal rate in the center of the usually arid equatorial Pacific and is lower in the Indonesia and in the northern Brazil area. This implies not only an eastward shift of the humid area at the equator, but also the compression of the arid zone in its eastern part. The intensification of the westerly winds near the equator in combination with the shift of Walker circulation to the east is accompanied by the strengthening of Hadley cell in the Pacific. The westerlies at mid-latitudes (20°–40°N and C) of the Pacific intensify (Wallace et al. 1998). The shift of the Aleutian Low to the east changes the properties of the flow from the Pacific to the Arctic through the Bering Strait (Gordon 2001).

Observations and modeling experiments show that the return flow of the conveyor through the Indonesian Straits is changing in phase with ENSO: it intensifies during La Niña events and weakens during El Niño events (Gordon 2001). The transport through the Indonesian Passage diminishes when the thermocline rises in the WPWP and the baroclinic component of the horizontal pressure gradient between the northwestern Australia and Java decreases. Therefore, the upwelling off Java intensifies. Thus, variations in El Niño are transported from the Pacific to the Indian Ocean via the thermohaline circulation. In the tropical area of the Indian Ocean the average annual SSTs reflect the link with the interdecadal ENSO variability (Wallace et al. 1998). During El Niño (when SOI is negative) SSTs in the tropical ocean slightly increase, while the monsoon rains on the Arabian Sea coast generally weaken in the next year's summer. When SOI is positive, cloudiness increases above the Indian Ocean and its waters are warmed more slowly leading to the intensification of the summer monsoon. In that way, interannual fluctuations of temperature and precipitation in the tropical Indian Ocean represent a remote response to ENSO-induced variations in the Pacific (Lappo et al. 1990; Alexander et al. 2002; Webster et al. 1998; Wallace et al. 1998).

The increase of surface water temperature during El Niño, although not as significant as in the equatorial Pacific, is observed in the northern equatorial Atlantic (Alexander et al. 2002; Wallace et al. 1998). There is evidence that the spatial anomalies of the surface temperature and atmospheric precipitation in the equatorial Atlantic are similar to those linked to ENSO in the equatorial Pacific (Wallace et al. 1998). In turn, temperature anomalies in the tropical Atlantic influence the monsoon rains in the western Africa adjacent to Sahara.

Modeling results obtained by Alexander and co-authors (2002) showed that the major part of the dominant low-frequency (more than 10 years) variability of the SST in the northern Pacific is defined by the influence of the tropics. The atmosphere serves as a bridge connecting the Equatorial Pacific to its northern and southern parts, as well as to the Atlantic and Indian oceans. This teleconnection will result not only in changes in surface water and near-surface air temperature but also in changes in the atmospheric precipitation, salinity, mixed-layer depth, winds, and currents. These linkages existed in the recent geological past and are discussed in the following chapters.

1.3 An Overview of the Global Thermohaline Circulation Development over the Course of the Cenozoic

Global cooling in the Neogene is thought to be mainly initiated by the uplift of the Tibetan Plateau, which started already 55 Ma as a result of the collision of the Indian and Asian plates. In the early Miocene, about 20 Ma, the Tibetan Plateau became high enough and its elevation accelerated. The next step in increasing elevation and erosion rates in the Himalayas and Tibet (at least in its southwestern part) is recorded in the middle Miocene, about 10 Ma (Klootwijk et al. 1992; Ruddiman 1998). According to several studies (Ruddiman and Kutzbach 1989; Prell and Kutzbach 1992), the uplift of the Tibetan Plateau resulted in the onset and intensification of the Asian Monsoon. Weathering of the plateau amplified by strong monsoon rains sequestered carbon from the atmosphere leading to global cooling. Some authors consider this hypothesis to be the best explanation of global cooling in the Neogene and possibly in the Cenozoic as a whole (Ruddiman 1997). Recently M. Allen and H. Armstrong (2008) suggested the initial collision of Arabian and Eurasian plates in the late Eocene has to be considered as a complimentary mechanism of global cooling and reduction of the atmospheric CO_2 content. It is generally accepted that orbitally forced variations in climate (10^4–10^6-year cyclicity) were superimposed on the progressive cooling induced by the tectonic and paleogeographic changes occurring on 10^5–10^7-year time scale (e.g., Zachos et al. 2001).

The most critical event in the onset of THC during the Cenozoic must have been the development of the mature ACC after the opening of the Drake Passage and Tasmanian Gateway. Before the circumpolar exchange commenced and the Antarctic ice sheet was formed due to the increasing isolation of the continent, the deep ocean had been much warmer (Gordon 1989; Shackleton and Kennett 1975; Kennett 1982). The "haline" circulation was replaced by the thermohaline most likely at the Paleogene/Neogene boundary (Krasheninnikov et al. 1999). Prior to the early Oligocene, the Indian Ocean waters had flown to the Pacific via the Tethys and Central American gateways (Fig. 1.7; Allen and Armstrong 2008 and references therein). The opening of Drake Passage started in the early Eocene (Eagles et al. 2006) and ended about 22 Ma when the current velocity in the Australian–Antarctic Gateway had reached its maximum value (Pfuhl et al. 2001). Increased Pacific Water flow via the Drake Passage to the Atlantic is believed to be indicated by positive shifts (to more radiogenic values) in the neodymium isotope record of ODP Site 1090 from the Agulhas Ridge around the early/late Oligocene boundary and the Oligocene/Miocene boundary (Scher and Martin 2008). The further long-term decreasing trend in the record (to less radiogenic values) through the Oligocene–Miocene documented in the three records from the Agulhas, Walvis, and Ninetyeast Ridges suggests the increased export of the Northern Component Water to the Southern Ocean. Thus, according to several authors (e.g., Scher and Martin 2008 and references therein) the late Paleogene circulation with the major convection cell(s) and AABW formation in the Southern Ocean gradually evolved into the modern bipolar mode of deep-water formation. However, several authors believe the NADW formation and modern-like Atlantic circulation started in the late Eocene, at about 35 Ma (Via and Thomas 2006; Allen and Armstrong 2008; Fig. 1.7). The initial NADW penetration into the western Tethys seems to have occurred as early as about the Eocene/Oligocene transition (Barbieri et al. 2003).

M. Allen and H. Armstrong (2008) argue that the closure of Tethys ocean gateway started at about the same time (35 Ma) whereas according to several previous publications (e.g., Adams et al. 1983; Krasheninnikov et al. 1999) the free circum–equatorial exchange ceased only in the middle Miocene as a result of the closure

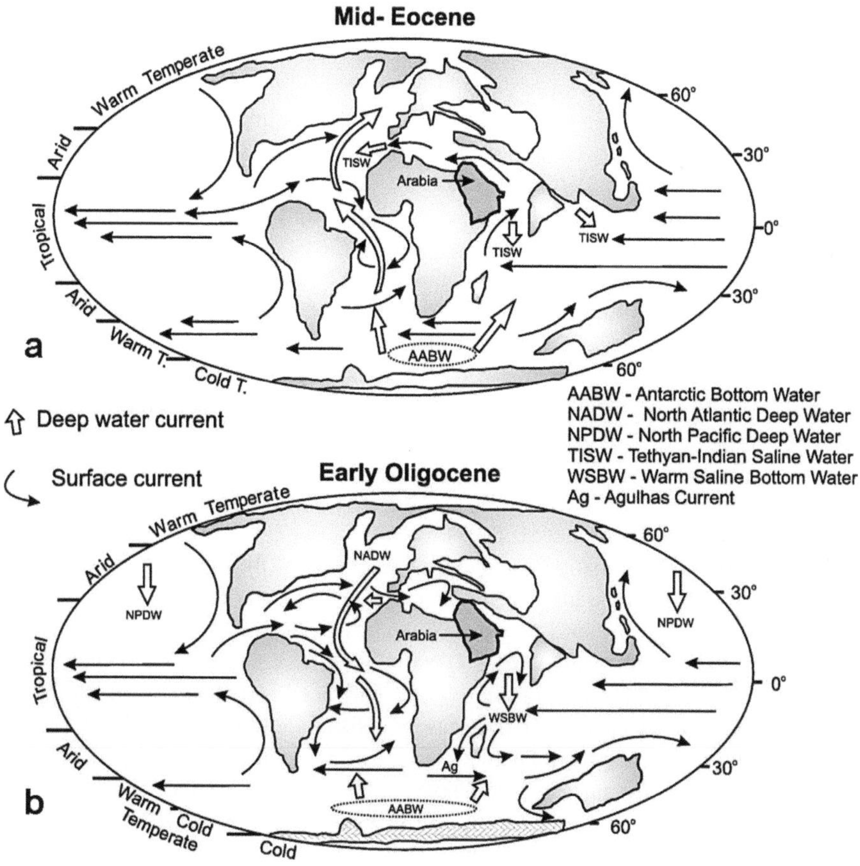

Fig. 1.7 Paleogeographic and oceanographic reconstructions before and after the demise of the Tethys Ocean gateway (after Allen and Armstrong 2008, with permission of Elsevier). (**a**) Eocene period, with westerly transport of warm Indian Ocean water into the Atlantic via Tethys. (**b**) Oligocene, with connection between the Indian and Atlantic oceans impeded by the Arabia–Eurasia collision zone

of the eastern Tethys and formation of the Mediterranean Sea. Some authors suggested that at the end of Oligocene–early Miocene, when the connection between the Indian Ocean and remnant basins of the Tethys still existed on the northwest, in the area of the modern Red Sea and Persian Gulf, warm and saline waters of the Tethys served as a source of deep water that moved southward and upwelled in the sub-Antarctic. At the same time, waters from the Indian Ocean could penetrate the Tethys basins in the surface layer (Sancetta 1978; Zachariasse 1992). The closure of Tethys must have significantly affected the water exchange in the low-middle latitudes and the intermediate deep-water formation in the Indian Ocean. Summarizing the data from several publications M. Allen and H. Armstrong (2008) concluded that the present-day major sources of deep waters in the middle to high latitudes, notably proto-NADW, AABW, and NPDW, already existed in the early Oligocene (Fig. 1.7).

According to the members of DSDP Leg 45 (Rea et al. 1995; Haug et al. 1995), the most significant geological evidence of the change in the THC pattern in the beginning of middle Miocene, about 17–16 million years ago, was the shift of the main area of silica accumulation from the Atlantic to the Pacific, which ceased about 12–11 Ma. However, most of the studies published up to date demonstrate that the modern-like oceanic conveyor developed only in the beginning of the late Miocene. Prior to this time, the deep-water formation in the North Atlantic was still sporadic and unstable. Benthic foraminiferal assemblages and relatively low carbon isotope values monitor weak near-bottom circulation in the northeastern Atlantic in the early Miocene and its further intensification in the middle Miocene (Berggren and Schnitker 1983; Miller et al. 1989; Krasheninnikov et al. 1999). Tectonic processes in the North Atlantic led to the onset of a stable linkage to the Norwegian–Greenland Basin and formation of NADW, favored by the global cooling of the late Miocene (Kennett et al. 1985; Krasheninnikov and Basov 1986; Blyum et al. 1988; Ivanova et al. 1989; Krasheninnikov et al. 1999).

One of the most dramatic paleogeographic events of the Neogene that strongly affected the THC was the uplift of the Isthmus of Panama that resulted in the progressive closure of the Central American Seaway between 8.3 and 3.5 Ma (Haug and Tiedemann 1998; Jain and Collins 2007; Fig. 1.7). According to modeling experiments, a significant amount of deep water had been also forming in the North Atlantic prior to the closure of the seaway (Nisancioglu et al. 2003). A part of proto-NADW could reach the Pacific, provided the depth of the passage exceeded 1,000 m, whereas the surface water layer was characterized by the presence of the geostrophic flow from the Pacific to the Atlantic due to the difference in the atmospheric pressure across the seaway (Nisancioglu et al. 2003). The similarity of planktic foraminiferal assemblages in the early, middle, and late Miocene ascertain interoceanic exchange in the surface layer between the Atlantic and Pacific in low latitudes (Ivanova et al. 1986; Blyum et al. 1988). The planktic oxygen isotope records show close to zero difference between the Caribbean and eastern equatorial Pacific interpreted by G. Haug and co-authors (2001) in terms of very similar salinity distribution on both sides of the Central American Seaway prior to its shoaling.

Since the end of the middle Miocene, the Atlantic deep waters could not enter the Pacific via the Central American Seaway due to its shoaling to intermediate depth of $\sim$1,000 m. Thus, the modern pattern of deep-water circulation evolved (Nisancioglu et al. 2003). When the sill had reached the water depth of 130 m the salinity increased by $\sim$1 psu. As a result, the Atlantic–Pacific salinity contrast was established at 4.2 Ma (Haug et al. 2001). According to the global climate–ocean ecosystem model study (Schneider and Schmittner 2006), the biological productivity reduced in the North Atlantic and increased in the eastern equatorial Pacific due to the restricted outflow of the nutrient-rich subsurface Pacific waters. This conclusion is supported by the micropaleontological and geochemical data from

ODP Site 999, which indicated the gradual decrease in paleoproductivity and ventilation improvement in the Caribbean during the seaway constriction (Jain and Collins 2007).

The restriction of the Indo-Pacific connection in low latitudes sometime between 10 and 8 Ma resulted in the pronounced changes of the Pacific circulation, notably in the development of the equatorial undercurrent and appearance of trans-equatorial planktic foraminiferal assemblage instead of previously existing western and eastern tropical assemblages (Kennett et al. 1985). In the late Miocene, planktic foraminiferal assemblages to the west and to the east of the Sunda Islands differed markedly indicating the plainly distinguishable tropical water masses in the eastern Indian Ocean and western Pacific (Blyum et al. 1988).

The interoceanic exchange in low latitudes was over after the closure of the Central American Seaway, while the meridional flow of NADW into the South Atlantic and meridional heat transfer to the North Atlantic intensified (Haug and Tiedemann 1998). The meridional transport was supported by the inflow of Mediterranean waters as a result of the opening of the Strait of Gibraltar after the Messinian crisis (Haq et al. 1987) at ~5.3 Ma.

Closure of the Isthmus of Panama led to the lysocline deepening in the Cariaco Basin about 4.6 million years ago (ODP Site 665), which is indicated by the improvement of bottom-water ventilation and preservation of carbonate sediments in the Caribbean. This is interpreted as a strengthening of the less-aggressive upper NADW (Haug and Tiedemann 1998). The lower NADW was also reinforced, as demonstrated by the data on the ocean ventilation and carbonate preservation in the equatorial Atlantic from the water depths exceeding 3,000 m at ODP/DSDP Sites 659, 665, 925–929. Oxygen isotope composition of planktic foraminifers points to the increase of surface water salinity around 4.6 Ma. Conversely, the preservation of carbonate sediments in the Pacific diminished owing to the increase of geochemical isolation of the two basins. As indicated by the distribution of shallow-water microfossils, the closure of the Isthmus of Panama was almost complete by 3.6 Ma. At that time, the first ventilation maximum is documented in the Caribbean Sea and in the deep Atlantic (more than 3,000 m) by similar high values on the carbon isotope records of sites 999 and 659, respectively. This maximum reflects the further strengthening of NADW after 3.6 Ma (Haug and Tiedemann 1998).

Correspondence of the increasing opal accumulation rate in the sub-Arctic Pacific (Site 882) and general trend toward heavier epibenthic $\delta^{13}C$ values from the ODP Site 999 indicating the improvement of ventilation in the Atlantic and Caribbean between 5.3 and 2.73 Ma points to the strengthening of the deep conveyor branch (Fig. 1.8). This intensification favored the nutrient entrainment into the surface water layer of the sub-Arctic Pacific, as well as the increase in the opal content in the sediments. It also led to the strengthening of the return flow of the conveyor in the surface and intermediate layers. In the Pliocene, the upwelling in the northern Pacific could have been stronger than now owing to the absence of the pronounced halocline (Haug et al. 1999). The closure of the Isthmus of Panama amplified the deep-water formation in the Labrador Sea.

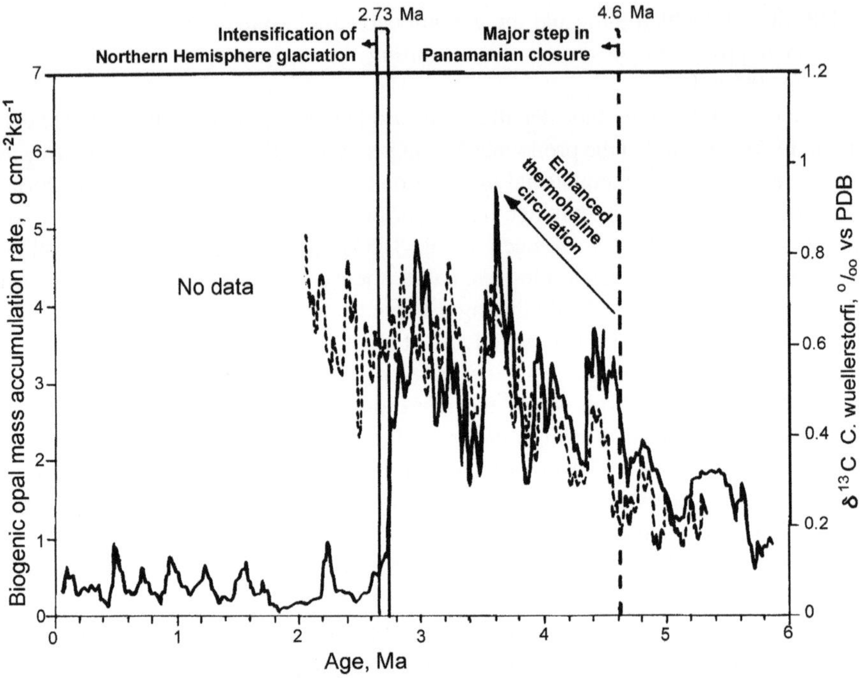

Fig. 1.8 Mass accumulation rate of biogenic opal in the northwestern Pacific (ODP Site 882) compared to $\delta^{13}C$ record on benthic foraminifera from the Caribbean Sea (ODP Site 999) during the Pliocene and major paleogeographic events (modified from Haug et al. 1999)

The reorganization of the thermohaline circulation contributed to the early Pliocene warming (Haug and Tiedemann 1998), which was reflected in the micropaleontological and isotopic data at low and mid latitudes of the World Ocean (Blyum et al. 1987; Ivanova et al. 1989; Barash et al. 1989). However, the sea ice formation in the Arctic intensified due to the increase of Siberian river run-off and evaporation above the North Atlantic, provided the intensification of the Gulf Stream took place. This resulted in the freshening of surface Arctic waters and, hence, their freezing at higher temperatures (Haug and Tiedemann 1998).

According to the hypothesis of Cane and Molnar (2001), the passage between the Halmahera and New Guinea Islands was much wider (5 Ma). New Guinea and Australia were located 3–5° to the south from their present position. The passage between the New Guinea and Sulawesi Islands was deeper and wider, implying that the Indonesian Passage was situated to the southeast from its modern location. It was deeper and wider than the Makassar Strait. Halmahera Island, which prevents the South Pacific waters from penetrating the Indonesian Passage, uplifted due to volcanic activity and enlarged after 5 Ma. The equatorial Pacific had a colder western part and a warmer eastern part. Accordingly, the atmospheric convection was weaker above the western equatorial area, as well as the Walker circulation, trade winds, and the west to east thermocline tilt (Fig. 1.6; Cane and Molnar 2001).

Thus, the situation was similar to the modern El Niño. According to the recent data and modeling results, El Niño was continuous or almost permanent during the early Pliocene warming (e.g., Fedorov et al. 2006; Ravelo et al. 2006). The strong heat flux from the tropics to high latitudes prevented the development of Northern Hemisphere glaciation. Around 3–4 Ma, dramatic narrowing of the Indonesian Passage and its shift to the north led to the change of water inflow source: the passage now was receiving waters from the northern Pacific, instead of the southern Pacific as it used to be (Cane and Molnar 2001). The WPWP expanded northward and the inflow of Pacific waters into the Indian Ocean via the Indonesian Passage reduced substantially (Sato et al. 2008). The East Asian monsoon strengthened at about 3.1 Ma (Jian et al. 2001). The narrowing of the Indonesian Passage resulted in the decrease of the mixed layer temperature in the northern Indian Ocean, as well as in the transfer of heat from tropics to high latitudes. The latter contributed to the development of the Northern Hemisphere glaciation (Cane and Molnar 2001).

Continental glaciation of Greenland started 3.3 million years ago (Jansen et al. 2000). The intervals enriched in the ice-rafted debris (IRD) in the sediments of the Norwegian and western Barents Seas indicate the onset and intensification of the Northern Hemisphere glaciation at 2.73–2.73 million years ago (Thiede et al. 1998; Haug et al. 1999; Jansen et al. 2000). The increased moisture advection from low to high latitudes is suggested to be a necessary precondition of the ice sheets build-up (Haug and Tiedemann 1998; Elverhøi et al. 1998; Haug et al. 2001), which was triggered by changes in obliquity (e.g., Haug and Tiedemann 1998).

By that time the final closing phase of the Isthmus of Panama occurred (Fig. 1.8), concurrently with the global sea level drop due to the expansion of the ice sheets. This is reflected in the decrease of biogenic opal concentration at about 2.73 million years ago, the abrupt increase in IRD, and enrichment of planktic oxygen isotope composition at ODP Site 882 (Haug et al. 1999). The break of connection between Pacific and Atlantic in the Pliocene is evidenced by changes in planktic foraminiferal assemblages, which evolved differently in low latitudes of the two basins (Oskina et al. 1982).

The dramatic reorganization of interoceanic exchange in the Pliocene finally caused the modern differences in vertical and horizontal circulation (Gordon 2001). Maximum concentrations of biogenic opal within last 2.73 million years are linked to interglacials (Haug et al. 1999). The drop of opal concentration most likely points to the weakening of the silica-rich deep-water upwelling in the northern Pacific. However, the decrease in opal concentration cannot be explained by the diminishing inflow of nutrients with deep waters from the Atlantic, as well as by the weakening of the Ekman upwelling in the northern Pacific due to wind relaxation, since it is not supported by geological data. G. Haug and co-authors (1999) suggested that the formation of a pronounced halocline made the northern Pacific more distinct from the Antarctic than before and also caused the weakening of the upwelling. Before the complete closure of the Isthmus of Panama, the northern Pacific resembled the Antarctic with strong wind-induced upwelling and vertical mixing, high concentrations of phosphates and nitrates in the water, and opal in the sediments. It is unknown what factor triggered the formation of the halocline, but it was part of the process of the Northern Hemisphere glaciation (Haug et al. 1999).

Thus, a set of the Miocene–Pliocene paleogeographic events, including the narrowing of the Indonesian Seaway, closing of the Isthmus of Panama, and the intensification of glaciation in the Northern and Southern hemispheres, caused the strong intensification of the THC, notably of the NADW formation, since 4.7–4.6 Ma and created new feedback mechanisms that significantly affected global climate and freshwater budgets. Besides the above-mentioned increase in poleward oceanic heat and moisture transport, the decreased interoceanic exchange in low latitudes provoked changes in the ITCZ location and tropical Pacific thermocline structure, altered the direction of freshwater flow via the Bering Strait (e.g., Haug et al. 1999 and references therein). One of the important consequences of the events described herein was the onset of the modern spatial distribution pattern of major atmospheric modes (see Sections 1.2, 5.1, and 7.5).

1.4 Summary

According to the up-to-date understanding, the global thermohaline circulation represents a system of heat, salt, and nutrient transport in the ocean at different depths. The THC largely controls the energy exchange between the ocean and the atmosphere and exerts a significant influence on the climate due to its non-linear and abrupt response to external forcing.

Variations in the SST, intensity of the THC, and atmospheric high- and low-pressure cells cause the alternating pattern of ENSO – Southern Oscillation, NAO, and other atmospheric modes. The atmospheric oscillations in turn do affect the SST anomalies in the ocean via the feedback mechanisms.

The THC experienced several dramatic changes during the Cenozoic. Besides the orbital forcing, they were associated with several tectonic and paleogeographic events, notably the Tibetan uplift, opening of the Drake Passage and development of the mature ACC, narrowing of the Indonesian Seaway and closing of the Central American Seaways, as well as with Northern and Southern hemisphere glaciation.

Chapter 2
Methods and Proxies of Paleoceanographic Reconstructions

Abstract This chapter provides a brief overview of the existing traditional and upcoming micropaleontological, geochemical, and other proxies, methods, and techniques of paleoceanographic reconstructions referred to in the monograph. Special emphasis is given to the foraminiferal transfer functions and stable isotopes as the concert of these techniques enables to obtain the quantitative estimates of the basic parameters of the thermohaline circulation, notably temperature and salinity. The carbon isotope composition of planktic and epibenthic species and several conservative and non-conservative water mass tracers are mentioned as the tools to discriminate the advection and propagation of the waters from different sources.

Keywords Micropaleontological · Geochemical · Proxies · Methods · Techniques · Transfer functions · Stable isotopes · Water mass tracers

Dozens of proxies and techniques currently applied in paleoceanographic reconstructions are extensively reviewed in special monographs (e.g., Fischer and Wefer 1999; Murray 2006; Hillaire-Marcel and de Vernal 2007) and discussed in several papers (e.g., Lynch-Stieglitz 2003; Lynch-Stieglitz et al. 1999, 2007). In this chapter, I only briefly refer to the methods used in my research and to some alternative techniques. Besides, special methods applied for regional reconstructions are indicated in corresponding chapters. The prospects of existing proxies and methods were carefully reviewed by the PACE working group (SCOR/IMAGES WG 123 – Reconstruction of Past Ocean Circulation, 2005–2007), in which I took part. The final PACE synthesis is available at the web site (http://www.scorint.org/Publications/WG123-Final-Report.pdf).

Variations in oxygen isotope composition ($\delta^{18}O$) of planktic and benthic foraminifera have been widely used in stratigraphy and correlation of deep-sea cores during more than 50 years, as this parameter reflects changes in the global ice volume (e.g., Emiliani 1955; Shackleton 1974; Labeyrie et al. 1987; Waelbroeck et al. 2002). Although the oxygen isotope composition of planktic foraminifera mainly displays the global ice volume and ambient water temperature (e.g., Emiliani 1955; Shackleton 1974), it was ascertained already in the 1960s that the $\delta^{18}O$ values also correlate with sea-surface salinity changes (e.g., Craig and Gordon 1965; Ivanova

E.V. Ivanova, *The Global Thermohaline Paleocirculation*,
DOI 10.1007/978-90-481-2415-2_2, © Springer Science+Business Media B.V. 2009

et al. 1985; Duplessy et al. 1991b, 1992), since both parameters depend on the hydrological cycle (precipitation + runoff – evaporation). In the Arctic, variations in planktic and benthic oxygen isotope ratio during the last glaciation also monitor the Atlantic water input (e.g., Duplessy et al. 2001, Lubinski et al. 2001; Risebrobakken et al. submitted, see Chapter 4) whereas in the North Atlantic they enable to trace the freshwater discharge in the surface water layer and the transfer of this signal to the intermediate depth during the brines rejection (see Chapter 3 and Chapter 4). Thus, in fact the oxygen and carbon isotope data provide the valuable tools to reconstruct the shifts in location and depth of the deep convection during the abrupt climate changes (Vidal et al. 1997, 1998; Dokken and Jansen 1999; see Chapter 3).

Along with the "traditional" paleoceanographic proxies considered below, several new geochemical tools are currently being developed, based on the use of different conservative (stable isotopes and trace metals) and non-conservative (^{14}C) water mass tracers. Widely used and new promising paleonutrient proxies include *Cd/Ca* ratio, carbon isotopes ($\delta^{13}C$), *Zn/Ca,* and *Ba/Ca* ratios (e.g., Boyle and Keigwin 1987; Duplessy et al. 1988; Zahn and Stüber 2002; Lynch-Stieglitz 2003; Martínez-Méndez et al. 2007). Geochemical tracers allow for constraining the properties of different water masses in their formation regions and further use of radiocarbon measurements in deep-sea corals and benthic foraminifera to evaluate rates of ocean circulation (e.g., Duplessy et al. 1989; Adkins et al. 1997; Keigwin 2004; Robinson et al. 2005). The carbon isotope composition of planktic and epibenthic species are widely used to discriminate the advection and propagation of the waters of different origin (e.g., Vincent and Berger 1981; Duplessy et al. 1988; Labeyrie et al. 1992, 1999; Vidal et al. 1997; Wang et al. 1999a; Elliot et al. 2002; Lynch-Stieglitz et al. 1999, 2007; see Chapter 3, Chapter 6, and Chapter 7). Benthic foraminiferal *Nd* isotopes offer a valuable tool to gain water mass end members and their mixing patterns (Piotrowski et al. 2004, 2005; Scher and Martin 2008; see also Section 1.5).

The vertical and lateral gradients of seawater density can be inferred from the combined oxygen isotope measurements on several planktic foraminiferal species calcifying at different depth (e.g., Farrell et al. 1995). The vertical density gradient displays the stratification of the upper water column, whereas the lateral gradient reflects the currents pattern. The intermediate water density can be deduced from the combined oxygen isotope and *Mg/Ca* or *Sr/Ca* (temperature proxy) measurements on benthic foraminifera (e.g., Skinner and Shackleton 2005). The insight into the density stratification of the deep ocean and modes of deep-water formation can also be derived from measuring the diffusive profile of the chlorinity and $\delta^{18}O$ of pore waters in sediments combined with the $\delta^{18}O$ data on the benthic foraminifera (e.g., Adkins et al. 2002).

Along with the traditional reconstructions of the surface and bottom water masses using indicative planktic and benthic foraminiferal species, respectively (e.g., Corliss 1979; Bauch et al. 2001; Ujiié 2003; Peeters et al. 2004), past changes in the strength of the bottom currents can be inferred from the regional variations in rock magnetic properties (nature, size, and concentration of magnetic particles) of sediment drifts along the path of the main bottom water masses (e.g., Kissel et al. 1999). Size sorting of marine mud is also used for the paleocurrent reconstruction (e.g., Bianchi and McCave 1999). $^{231}Pa/^{230}Th$ ratio in sediments provides a complementary measure of the residence time of deep water (e.g., McManus et al. 2004).

Forward modeling (with tracer simulations compared to proxy data) and inverse numerical modeling (proxy data and models combined quantitatively from the outset) enable to reconstruct variations in oceanic paleocirculation (e.g., Duplessy et al. 1988; Ganopolski et al. 1998; McManus et al. 2004; see Chapters 1,Chapter 3, and Chapter 7).

The paleoceanographic reconstructions presented in this book are primarily based on the detailed quantitative study of planktic and benthic foraminiferal assemblages, as well as on the oxygen and carbon isotope measurements. It was determined long time ago, that foraminiferal abundance in sediments is mostly defined by the relationship between their production and dissolution of carbonate tests in the water column and on the sea floor (Berger 1968, 1971; Belyaeva and Burmistrova 1984; Murdmaa and Ivanova 1985, 1987). In some regions, such as the Arctic Seas, the dilution by terrigenous material plays a very important role, along with the washout of sediments from elevated areas (Chapter 4).

In the open ocean, where the influence of these factors is negligible, the total planktic foraminiferal abundance in the sediment above the lysocline is considered to roughly reflect the surface water bioproductivity. Together with Luc Beaufort, we propose a qualitative foraminiferal productivity index for the low latitudes (FIP, see Section 7.6). It is based on the relative abundance of three productivity-related foraminiferal species *Globigerina bulloides* + *Globigerinita glutinata* + *Neogloboquadrina dutertrei*. The time series of this index is consistent with the coccolithophore-based primary productivity record calculated using the equation by Beaufort et al. (1997) in the same core MD02-2529 off Costa Rica (Ivanova et al. in prep.).

Unlike planktic foraminifera that mainly feed on phyto- and zooplankton (Bé 1977; Kennett 1982; Hemleben et al. 1989), benthic foraminifera generally exploit phytodetritus and bacteria (Hunt and Corliss 1993; Fenchel and Finlay 1995). Calculating accumulation rates of foraminiferal tests one can eliminate the influence of dilution and interspecific ecological competition (Nees 1997). However, such estimates can be considered sufficiently reliable only for the well-dated sediment cores with high sedimentation rates. In Section 5.5, the record of radiolarian accumulation rates is additionally used as a proxy of surface water bioproductivity.

Since the beginning of the 1970s, a wide range of different versions of planktic foraminifera-based transfer functions have been applied to reconstruct hydrological parameters. Planktic foraminifers are most commonly used for paleoceanographic reconstructions owing to their wide distribution, relatively low, but sufficient taxonomic diversity and the linkage of species distribution to hydrological parameters of water masses and surface water productivity (Barash 1964; Imbrie and Kipp 1971; Barash and Blyum 1974; Ivanova 1983, 1988; Prell 1985; Thompson 1981; Barash 1988; Barash et al. 1989, 2002; Le and Thunnell 1996; Pflaumann et al. 1996; Waelbroeck et al. 1998; Cayre et al. 1999a; Barash and Yushina 1999a, b;

Wang et al. 1999a; Ivanova et al. 2003b; Kucera et al. 2005; Barrows and Juggins 2005). The number of publications using other groups of micro- and macrofossils (radiolarians, dinocysts, coccolithophores, diatoms, ostracods, mollusks) is considerably smaller (e.g., de Vernal and Pedersen 1997; Matul et al. 2002; Beaufort et al. 1997, 2001; Martinez et al. 2003; Polyakova et al. 2005; Ivanova et al. 2007b; Taldenkova et al. 2008). Benthic foraminifera represent a signal rather complicated for interpretation, as their distribution in sediments is controlled by the interaction of several local factors, such as food supply, salinity, temperature, ventilation, and hydrodynamics of bottom waters (e.g., Saidova 1976; Feyling-Hanssen 1964; Nees 1997; Korsun et al. 1994; Wollenburg and Mackensen 1998; Wollenburg et al. 2001; Polyak et al. 2002a; Murray 2006; Ivanova et al. 2008b). In Chapter 4 alternation in the dominance of different benthic species is used for paleoceanographic reconstructions in the Arctic Seas. This approach provides evaluation of changes in the duration of ice-free period, the position of the sea-ice margin and the Polar front, as well as the intensity of Atlantic Water penetration, bottom currents activity, river runoff and the location of the mixing zone of marine and fresh waters.

Each of the techniques for calculating paleohydrological parameters has its advantages and limitations. Therefore, in the regional research considered in this book the same parameters were reconstructed by different methods depending on the specific investigation area and availability of extensive reference databases. For example, in low and middle latitudes the most complete coretop planktic foraminiferal databases comprise the census counts in grain size fraction > 150 μm, hence we used the same fraction in our paleoreconstructions (see Chapter 5, Chapter 6, and Chapter 7). However, in high latitudes, notably in the Eurasian Arctic Seas, we studied the smaller grain size fraction (>100 μm), which provides more reliable results due to a commonly small size of several dominant planktic and benthic foraminiferal species in the region (Chapter 4).

The selection of the coretop data set and suitable transfer function is also of primary importance for a reliable paleodata prospective (Prell 1985; Pflaumann et al. 1996; Waelbroeck et al. 1998, 2005; Cayre et al. 1999a; Barash and Yushina 1999a, b; Kucera et al. 2005; Barrows and Juggins 2005; Guiot and de Vernal 2007). For foraminifera-based paleotemperature reconstructions in the subtropical gyre of the Indian Ocean (Section 5.7; Ivanova et al. 1995) we used the Brown University coretop database for tropical and subtropical latitudes of the Indian–Pacific (Prell 1985) and supplemented it with the smaller data set compiled by the Laboratoire des Etudes du Climat et de l'Environnement, where this work was carried out by the author. Later, the new database of Brown University (Prell et al. 1999) supplemented by the O. Cayre and co-authors (1999a) coretop data was selected for paleoproductivity estimates in the Arabian Sea (Section 5.5; Ivanova et al. 2003b). The regional foraminiferal database for the South China Sea was built within the frame of the project "Monitor Monsoon" (Chapter 6; Pflaumann and Jian 1999; Wang et al. 1999a) and extended with the database for the adjacent part of the Pacific compiled by P. Thompson (1981). The foraminiferal database representing a part of the MARGO project (Kucera et al. 2005) comprising the coretops from the tropical Indo-Pacific was used for paleotemperature reconstructions in the equatorial Pacific (Section 7.6; Ivanova et al. in prep.). Modern temperature and salinity values were retrieved from the worldwide atlases (Levitus 1982; Levitus and Boyer 1994; WOA 1998, 2001, 2005). The modern mean-annual PP values were extracted from the electronic database (Antoine et al. 1996), which contains PP estimates derived from satellite ocean color monitoring.

Today, the direct Mg/Ca and U^k_{37}-SST measurements are widely used in paleoceanographic reconstructions along with the traditional faunal sea-surface temperature (SST) estimates (e.g., Wang et al. 1999a; Nürnberg 2000; Lea et al. 2000; Duplessy et al. 2002; Peeters et al. 2004; Barker et al. 2005; Elderfield and Ganssen 2000). To reconstruct past changes of intercorrelated physical parameters in the upper ocean (e.g., temperature and salinity), it is necessary to apply several independent methods like a combination of transfer functions or Mg/Ca (or U^k_{37}) SST measurements and oxygen isotope data. The transfer functions applied to obtain foraminifer-based SST estimates, and referred in this book, include the modern analogue technique (MAT, Prell 1985, and SIMMAX, Pflaumann et al. 1996), revised modern analogue technique (RAM, Waelbroeck et al. 1998), factor analysis (CABFAC, Imbrie and Kipp 1971) with further regression or with spline interpolation (Barash and Yushina 1999a, b) and artificial neural networks (ANN, Kucera et al. 2005).

Salinity changes were estimated using the paleotemperature equation (e.g., Shackleton 1974) subtracting the temperature and global ice volume effects (Labeyrie et al. 1987; Vogelsand et al. 2001; Waelbroeck et al. 2002) from the planktic oxygen isotope values to obtain the local oxygen isotope composition of the sea water. At this stage, Mg/Ca-SST, U^k_{37}-SST, or faunal SST could be used. Then, the regional linear correlation between salinity and local oxygen isotope value of the sea water has been applied as both parameters are linked to the hydrological cycle (e.g., Ivanova et al. 1985; Rostek et al. 1993, see Section 5.7; Wang et al. 1995b, 1999a see Chapter 6; Benway and Mix 2006; Leduc et al. 2007, see Section 7.6). The accuracy of estimates is discussed in the corresponding regional sections of the book. In several studies, changes in sea-surface salinity have been quantified on the basis of the diatom or dynoflagelate assemblages (e.g., Polyakova and Stein 2004; de Vernal and Pedersen 1997; de Vernal et al. 2005) or qualitative assessed from changes in benthic foraminiferal fauna (see Section 4.5).

In the northern Barents Sea, subsurface and bottom water paleotemperatures were calculated using the local variations in oxygen isotope value of the sea water. These variations were derived from planktic and benthic oxygen isotope measurements removing the global ice volume. The temperature change by 0.24°C was considered corresponding to 1‰ change in local oxygen isotope value of sea water (Shackleton 1974; Duplessy et al. 2001; Section 4.5.4). Analytical accuracy of the oxygen and carbon isotope measurements, obtained in four different laboratories (see Introduction) and presented here in Chapters 4, 5, 6, and 7, is generally about 0.08‰ and 0.05–0.06‰ versus the PDB standard, respectively (Duplessy et al. 2001, 2005; Wang et al. 1999a; Fretzdorff et al. 2000; Murdmaa et al. 2004, 2006; Simstich et al. 2004; Ivanova et al. 2003b; Leduc et al. 2007).

Transfer functions (factor analysis and regression) are also used to calculate variations in PP in the Arabian Sea (Section 5.5). As shown in a number of studies (Cayre et al. 1999; Cayre and Bard 1999; Waelbroeck et al. 1998), the distribution of planktic foraminifers in the northwestern Indian Ocean is much strongly affected by changes in PP as compared to changes in SST. In fact, the attempts we made with C. Waelbroeck to estimate SST and PP in the Arabian Sea using MAT and RAM resulted in very high dissimilarity coefficients (>0.25) for about a half of the entire set of more than 200 samples. Subsequently, the application of Q-mode principal component analysis and regression using the recently modified CABFAC by J. Imbrie and N. Kipp (1971) and the same data set has been more successful (Ivanova et al. 2003b, see Section Section 5.5). The latter method appeared to be a more flexible tool in this case because not all planktic species are equally linked to the PP distribution. CABFAC allowed obtaining the true PP estimates using only a part of the foraminiferal assemblage (one factor), unlike MAT and RAM which require consideration of the total assemblage except for the rare species (<2% of the fauna throughout the data set). This example additionally confirms the crucial importance of considering regional oceanography and alternative methods in paleoceanographic reconstructions. Application of several independent methods not only provides estimates of different parameters, but also increases the reliability of results. In Chapter 6, SSTs are estimated using both foraminiferal transfer functions and alkenone measurements. The similarity of the obtained trends and estimated SST supports the robustness of reconstructions.

The age models for all but one of the cores studied by the author and colleagues have been constrained by accelerator mass spectrometry radiocarbon (AMS-^{14}C) dates performed on foraminifera or mollusks. ^{14}C ages were converted to calendar years (expressed in cal. ka throughout the text) using the Calib 4.3 and 5.0.2 software (Stuiver and Reimer 1993), the marine calibration MARINE98 and MARINE04 (Hughen et al. 2004) and modern local reservoir ages (Duplessy et al. 1989; Stuiver and Braziunas 1993; Forman and Polyak 1997; Mangerud et al. 2006; see Chapter 4 and Chapter 6, Section 5.5). The depth to age conversion was generally obtained by linear interpolation between age constrained levels without further adjustment by graphical tuning of any of the proxy records. As a common practice, the reservoir age correction was applied before recalculating AMS-^{14}C dates to calendar years BP. In areas where radiocarbon dating is not possible due to the lack of carbonate fossils, changes in the strength of the geomagnetic field (e.g., low field magnetic susceptibility) can be employed to correlate the proxy records (e.g., Grousset et al. 2001).

To obtain better constraints beyond the ^{14}C age limit, the correlation of foraminiferal oxygen isotope record with the standard SPECMAP record (Imbrie et al. 1984; Martinson et al. 1987) was carried out using the "*AnalySeries*" program (Paillard and Labeyrie 1993; Paillard et al. 1996; see Section 5.5, and Section 6.2). The disappearance of pink specimens of *Globigerinoides ruber* about 128 cal. ka BP, at the boundary between marine oxygen isotope stages (MIS) 6 and 5, was used as a tie point in Core S-17666 with volcanic ash layers. The core was obtained from the fringe of the Subtropical Gyre in the Indian Ocean, in the vicinity of Reunion Island (Fretzdorff et al. 2000, see Section 5.7). Our data on the Arabian Sea (Ivanova et al. 2003b, Section 5.5) confirmed that in this region, as well as in the vicinity of Reunion Island, the disappearance level of *G. ruber* (pink) corresponds to

Termination II and not to MIS 5.5, as it was considered before (Thompson et al. 1979).

In the following chapters, the records of the paleoceanographic parameters reconstructed by the above-mentioned approaches are correlated with published geochemical, micropaleontological, and sedimentological proxy time series. For example, in the monsoon area, the major geochemical indicators include the total organic carbon (TOC) content in the sediment, barium and biogenic opal fluxes. The grain size of dust particles, as well as *Ti/Al* and *Cr/Al* weight ratios, reflect the transport of aeolian dust brought by monsoon and/or the northwestern offshore winds from Somalia and Arabia (Hagelberg and Mix 1991; Clemens et al. 1991; Murray and Prell 1991, 1992; Niitsuma et al. 1991). *Ba/Al, Cu/Al, Ni/Al, Zn/Al, V/Al, U/Al* ratios are considered to indicate bioproductivity variations driven by the monsoonal upwelling (Shimmield and Mowbray 1991). *Cd/Ca* ratio in planktic foraminifers is widely used as a nutrient tracer that is well documented to correlate with the content of dissolved phosphorus (Boyle and Keigwin 1987; Lynch-Stieglitz 2003).

To discuss the periodicity in variations of different paleoceanographic proxies with respect to the orbital and suborbital cycles the results of spectral analysis (Blackman–Tukey, Multi-Taper and Maximum Entropy Methods) from several publications have been used in Chapters 5, 6, and 7 (e.g., Clemens et al. 1991; Sirocko et al. 1996; Wang et al. 1999a; de Garidel et al. 2001; Beaufort et al. 1997, 2001). The advantage of the Multi-Taper Method provided by the *AnalySeries* software package (Paillard et al. 1996) provides the possibility to detect low-amplitude oscillations in relatively short time series with a high degree of statistical significance (Yiou et al. 1996). In turn, the phase relationships between time series of different proxies throughout the same sediment core are generally extracted by cross-spectral analysis (e.g., Clemens et al. 1991, Section 5.6).

Chapter 3
Variability of the Meridional Overturning Circulation and Paleoceanographic Events in the North Atlantic During the Last Climatic Cycle

Abstract The THC is known to exhibit pronounced variability at glacial–interglacial, millennial, and shorter timescales. Along with the orbital forcing and global feedbacks controlling the glacial–interglacial variability of the THC, millennial-to-centennial changes in its modes were strongly influenced by the ice sheets dynamics and interaction with the atmospheric oscillations. During the Last Interglaciation, the THC mode resembled its modern pattern but the climate was characterized by rather prominent variability. Recent data disprove the shutdown of the THC during the Last Glacial Maximum and point to a shift into the convection cells. However, a shutdown or at least dramatic slowdown of the AMOC is suggested for some Heinrich events as a result of massive freshwater discharge into the North Atlantic. The abrupt climate changes during the Dansgaard–Oeschger cycles, Younger Dryas, and 8.2 cal. ka cooling event are also thought to be associated with the weakening of the AMOC due to the large freshwater input into the North Atlantic. However, the origin of the above-mentioned events is different and still debatable.

Keywords Overturning · Convection · Brines · Last Interglaciation · Last Glacial Maximum · Heinrich events · Dansgaard–Oeschger cycles · Younger Dryas · 8.2 cal. ka event · Holocene

3.1 Introduction

As it was shown in Section 1.1, the North Atlantic represents the strongest energy-active area and the main region of deep-water formation, controlling the global thermohaline circulation. Therefore, the study of global and regional paleoceanographic events and their driving mechanisms is impossible without tracing them to the North Atlantic. Paleoceanographical data and modeling results show that the thermohaline circulation significantly affects the regional heat budget in the North Atlantic and the surrounding land. In turn, the Atlantic Meridional Overturning Circulation (AMOC) is very sensitive to fluctuations in freshwater input (Rahmstorf 1995, Seidov and

E.V. Ivanova, *The Global Thermohaline Paleocirculation*,
DOI 10.1007/978-90-481-2415-2_3,

Haupt 1999; Ganopolski and Rahmstorf 2001; Lohmann and Schulz 2000; Bard et al. 2000; Bard 2002; Stouffer et al. 2006; Rahmstorf et al. 2005). This chapter contains an overview of the currently available paleoceanographic scenarios associated with the different modes of the global conveyor during the last interglacial and glacial periods, Heinrich and Dansgaard–Oeschger events, and shorter events within the Holocene. The data and reconstructions summarized in the chapter were obtained by several research groups studying ODP/DSDP sites and deep-sea cores. The drilling and coring locations are shown in Fig. 3.1.

It was established that the glacial maximum and interglacial optimum are not the extremes in the range of climate modes, but are rather associated with the maximum stability of the climate and thermohaline circulation (McManus et al. 1999; Alley et al. 2001).

This conclusion was confirmed by modeling results (Ganopolski and Ramstorf 2001). The short-term variations seem to represent the temporary departure from "normal" stable environments (Broecker and Denton 1989; Broecker 2000). To trigger this transition it is sufficient to change the salinity budget in high latitudes. As the salinity budget is defined by the balance between atmospheric precipitation, evaporation, and freshwater input to the ocean, the oscillations in salt content in the glacial ocean were suggested to be a driving mechanism for these abrupt changes (Broecker et al. 1990; Paillard and Labeyrie 1994).

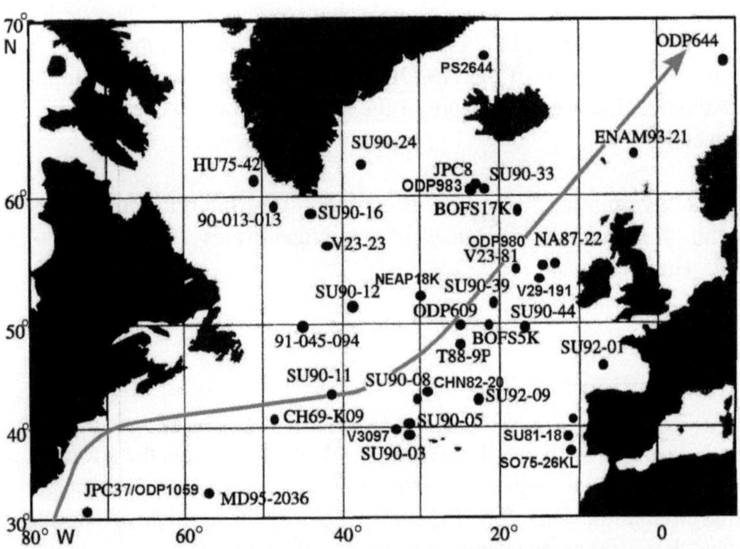

Fig. 3.1 Location of cores in the North Atlantic in which Heinrich and Dansgaard–Oeschger (DO) events have been identified (after Rasmussen et al. 1996; Dokken and Janssen 1999; Cortijo et al. 1997, 2000; Chapman et al. 2000; Oppo et al. 2001). The *arrow* shows direction of the main warm water transport by the Gulf Stream

3.2 Last Interglaciation

It was shown that the global ice volume during the last climatic cycle is comparable to recent conditions only for the time interval between 129 and 119 ka BP or 125– and 115 ka BP, defined as marine oxygen-isotope stage (MIS) 5e or 5.5 (Shackleton 1969; Shackleton et al. 2002), also called the Eemian interglaciation. At the interglacial optimum, summer SST in the North Atlantic was 2–5°C warmer than today as well as the air temperature over Arctic and Greenland (Duplessy et al. 2007 and references therein). Relatively constant *Cd/Ca* ratio (nutrient proxy), oxygen isotope values of benthic foraminifera, carbonate, and terrigenous detrital flux, estimated from ^{230}Th values, indicate that AMOC remained strong and relatively stable over several thousand years within MIS 5.5 (Zahn 1994; Adkins et al. 1997; Cortijo et al. 1999a, b; Duplessy et al. 2007; Bauch and Kandiano 2007). The main region of NADW formation was the Norwegian–Greenland Basin (Labeyrie et al. 1987; Cortijo et al. 1994; Duplessy et al. 2007) and its temperature and salinity were slightly higher (by about 0.37°±0.20°C and 0.04‰, respectively) than today due to the insolation-forced regional surface warming and increase in evaporation (Duplessy et al. 2007). Heat transfer in the Atlantic at 20ºN was estimated at 1.3×10^{15} W and reached by far higher latitudes than during the Last Glacial Maximum, reducing the area of sea ice extent in the Northern Hemisphere by 1/3 (Ganopolski and Rahmstorf 2001). High SST and salinity values reconstructed from radiolarian assemblages provided evidence of warm saline North Atlantic water input into the Labrador Sea (Matul et al. 2002). Owing to the strengthening of the Gulf Stream and the North Atlantic Current, the Subpolar Front was shifted slightly to the north compared to its modern location (Barash 1988; Zahn 1994; Cortijo et al. 1994). The number of icebergs arriving from Greenland decreased significantly due to the ice sheet retreat (Zahn 1994). AMOC transferred the warming signal to the Circumpolar Deep Water thus promoting the enhanced melting of the Antarctic ice shelves by thinning from below (Duplessy et al. 2007). J. Adkins and coauthors (1997) assumed that the interglaciation started and ended with abrupt, within less than 400 years, deep-water circulation changes that suggested the increased influence of the southern source water during the glacial inception.

However, even within stage 5.5 small fluctuations in SST were recorded in the Norwegian Sea and the North Atlantic (Cortijo et al. 1994, 1999a, b; Fronval and Jansen 1996; Duplessy 1996; Oppo et al. 2001; Fig. 3.2). Fast cooling and reduction in surface water salinity followed the minimum summer insolation at 122 ka BP north of 70°N (Cortijo et al. 1994). However, changes in the thermohaline circulation in the North Atlantic and Norwegian Sea occurred only after abrupt cooling at about 118 ka BP, when the ice sheets had already considerably increased compared to their size during the interglacial optimum (Adkins et al. 1997; Bauch et al. 2000). It was suggested that these changes reflect the response of the ocean–atmosphere system to insolation changes (Cortijo et al. 1999b).

The study of sediment cores from the North Atlantic, obtained in the area between 30 and 70°N, revealed that in the late MIS 5.5, a gradual cooling and sea

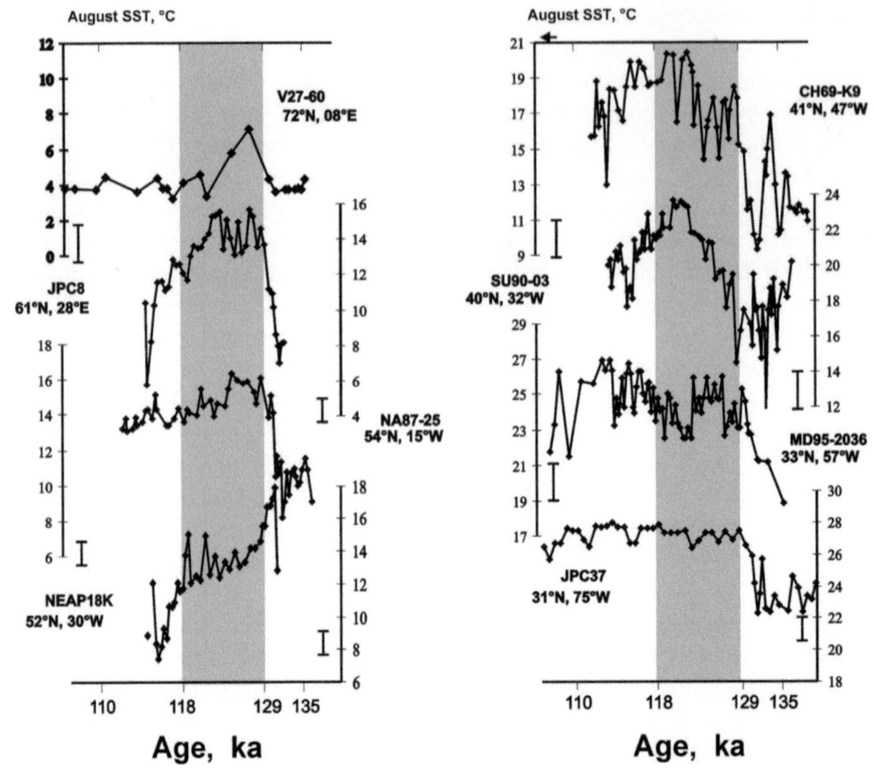

Fig. 3.2 Summer sea surface temperature (SST) variations during the interglacial optimum reconstructed by the modern analogue technique on planktic foraminiferal assemblages in cores from the North Atlantic (Cortijo et al. 1999b, courtesy by Elsa Cortijo). The *arrows* indicate modern summer SST values (from Levitus 1982)

surface salinity decrease can be traced in the north of this region, while the temperature and salinity rise is observed in the south. These observations are well correlated with the inverse trend of low-amplitude changes in average annual insolation at 50° and 31°N controlled by obliquity-driven changes in mean annual insolation (Cortijo et al. 1999a; Pahnke and Sachs 2006; Duplessy et al. 2007). Toward the end of stage 5.5 the seasonal contrast became less pronounced since summer insolation was gradually decreasing in high northern latitudes, while winter insolation increased (Berger 1978; Khodri et al. 2003; Berger et al. 2006). High SSTs in low latitudes in the late stage 5.5 provided intensive evaporation and moisture supply into the Arctic and surrounding land. This in turn promoted river runoff intensification and ice formation (due to freshening) in Arctic seas, leading to higher freshwater input to the North Atlantic and weakening of the deep-water convection and the thermohaline circulation (Cortijo et al. 1994). As a result, heat accumulated in low latitudes, triggering the positive feedback. E. Cortijo and co-authors (1999a) concluded that slight variations of SST and surface salinity in the North Atlantic were of minor importance before reaching the critical threshold, but after the thermohaline circulation reorganized and climate system entered the glacial inception. The second cooling step at around 115 ka was associated with the significant change in deep-water circulation (Adkins et al. 1997; Shackleton et al. 2002; Khodri et al. 2003) and transition to the AMOC glacial mode. The modeling experiments demonstrated the non-linear response of climate and THC to the small (about 0.02 Sv) freshwater input into the North Atlantic (Khodri et al. 2003).

The entire stage 5 (129–74 ka BP; Martinson et al. 1987) was punctuated by strong climatic variability. In Greenland, the glaciers advanced over two cool stadials, MIS 5.2 and 5.4, and retreated during three warm interstadials, including Eemian (Dansgaard et al. 1993). Similarly, the surface (wind-driven) and thermohaline circulation in the North Atlantic are known to be variable during stage 5 (Barash 1988; Barash and Yushina 1999a, b; Barash et al. 2002; Cortijo et al. 1994; Oppo et al. 2001). Recent studies demonstrated that variations in the thermohaline circulation lasted 1,000–3,000 years (Oppo et al. 2001), i.e., their duration was comparable to the Dansgaard–Oeschger cycles discussed below.

3.3 Last Glaciation

It is generally accepted that the last glaciation (the interval from MIS 5.4 to MIS 2) culminated at the Last Glacial Maximum (LGM), 21.5–18 cal. ka BP according to the data obtained by the GLAMAP 2000 project (Sarnthein et al. 2000a, 2003a; van Kreveld et al. 2000; Weinelt et al. 2003) or 22.5–19 cal. ka BP according to the EPILOG working group (Mix et al. 2001). Average annual SST in the North Atlantic between 30 and 50°N dropped by 6–8°C below the present values (Barash 1988; Barash and Yushina 1999b; Chapman et al. 2000). The SST decrease exceeded 7°C in the Canary upwelling area (Barash et al. 1980; Pflaumann et al. 2003). According to U^k_{37} and *Mg/Ca*-based estimates the decrease in surface water temperature in the tropics reached 1–4°C on average (Nűrnberg and Müller 2000). The thermal equator with minimal seasonal variability remained within 2–6°N in the Atlantic (Pflaumann et al. 2003).

Members of the GLAMAP 2000 (the Glacial Atlantic Ocean Mapping) project developed the technique to reconstruct the seasonal sea ice margin in the Atlantic using the foraminiferal-based SSTs estimated from SIMMAX transfer functions. They used foraminiferal assemblages from the core tops within the area of modern sea ice distribution and from the sediments of the Little Ice Age (Sarnthein et al. 2000a, 2003a; Pflaumann et al. 2003) as the reference database. The results of the project showed that the sea ice margin is an important boundary condition when modeling variations of the climate system and that during the LGM sea ice retreated into the Arctic and Fram Strait during the summer season. Hence, most part of the Norwegian–Greenland Basin was ice free (Sarnthein et al. 1995, 2003a; Hebbeln et al. 1998). During winter, sea ice expanded southward as far as 50°N, covering almost the entire Norwegian–Greenland Basin and the Labrador Sea (Fig. 3.3). Thus, the sea ice distribution points to the pronounced seasonal contrast during the LGM (de Vernal and Hillaire-Marcel 2000; Sarnthein et al. 1995, 2003a; Pflaumann et al. 2003).

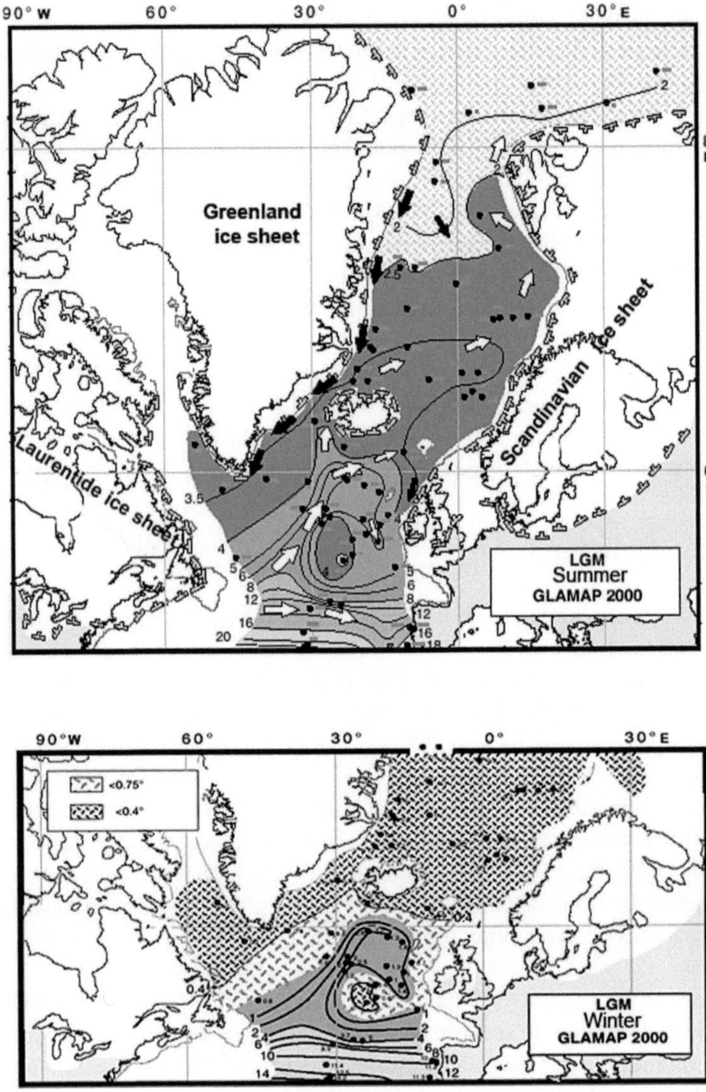

Fig. 3.3 Maximum extent of sea ice (hatched) and SIMMAX SST in the North Atlantic during LGM summer (**a**) and winter (**b**) (after Sarnthein et al. 2003a; Pflaumann et al. 2003). Sea ice limit deduced from 2.5°C isotherm in summer and from 0.4 to 0.75°C isotherms in winter. *Solid* and *open arrows* mark assumed cold and warm surface water currents. Large numbers label isotherms (°C). Southward decrease in intensity of gray color corresponds to increase in sea surface temperature. Printed with permission from AGU

The growth of the ice sheets, changes in wind stress, and advance of sea ice into the North Atlantic caused the dramatic change of the AMOC (Color Plate 3.1 and 3.2). The data and modeling results do not support the formation of

NADW in the Norwegian–Greenland Basin and cold intermediate water in the Labrador Sea at modern rates (Sarnthein et al. 1994b; Vidal et al. 1997; Lynch-Stieglitz et al. 2007). Multiple water-mass tracers including oxygen and carbon isotopes, *Cd/Ca* ratio, and radiocarbon ascertain a sharp vertical density gradient and reduced ventilation of the deep sea (Lynch-Stieglitz et al. 2007; Paul and Mulitza 2008). The Norwegian Current was not equivalent to its modern analogue due to a southward shift of the main flow of the Gulf Stream (Barash et al. 1974; Barash 1988; Duplessy et al. 1991b) and changes in the North Atlantic system of gyres. However, surface Atlantic waters penetrated the Norwegian–Greenland Basin along the Faeroe-Shetland Channel (Hebbeln et al. 1994; Dokken and Hald 1996; Rasmussen et al. 1996, 1997; Pflaumann et al. 2003) and through the Denmark Strait (Sarnthein et al. 1995, 2003a). Species diversity and total abundance of benthic foraminifers significantly decreased on the Vøring Plateau, indicating the reduction in surface water productivity (Ivanova 1998).

Northeastern warm water transport by the Gulf Stream and North Atlantic Current occurred on the northwestern fringe of the mid-latitude anti-cyclonic gyre. A cold boundary current with a temperature of 2–4°C on the eastern fringe of this gyre transported icebergs and melt water southward along the western coast of Great Britain and Europe. The Subpolar Front migrated to the south, especially in the eastern Atlantic, where it shifted from Iceland toward the coast of Portugal, to the latitude of 40–42°N (CLIMAP 1981; Barash 1988; Duplessy et al. 1991b; Zahn 1994; Pflaumann et al. 2003). Gradients in oxygen isotope composition of surface waters estimated by J.C. Duplessy and co-authors (1991b) indicate limited NADW formation at mid-latitudes, i.e., much farther to the south than at present. Convection down to 2 km occurred at 50°N, which is evidenced by high carbon isotope values of benthic foraminifers (Duplessy et al. 1988, 1991b; Labeyrie et al. 1992; Oppo and Lehman 1993; Vidal et al. 1997; Elliot et al. 2002; Color Plate 3.2). Location of convective cells during the LGM is still under discussion (Duplessy et al. 2002). Reconstruction by M. Sarnthein and co-authors (2003a), shown on Fig. 3.3, supports the idea that the ice-free submeridional channel between 50 and 60°N could promote the NADW formation on relatively shallow depths to the south of Iceland during winter convection. One of the mechanisms of deep and intermediate water formation was the brine release during sea ice formation, in particular in the Norwegian Sea (Rasmussen et al. 1996; Vidal et al. 1998; Dokken and Jansen 1999).

The maximum IRD input to the Iceland area in the North Atlantic was recorded during the LGM (Jansen et al. 2000). A number of reconstructions (Hebbeln et al. 1994; Dokken and Hald 1996) demonstrate seasonal ice-free conditions in most parts of the Norwegian and Greenland seas during the early and late stadials of the glaciation (MIS 4 and 2), when the glacier expanded on the Spitsbergen (Mangerud et al. 1998). Advection of the North Atlantic waters into the Norwegian–Greenland Basin (at least between 31.4–26.5 and 23–17.4 cal. ka BP) provided moisture for the Barents Sea ice sheet growth during the last glacial (Hebbeln et al. 1994; see Section 4.5.1).

A number of carbon isotope and geochemical reconstructions carried out during the 1980–1990s showed that the Atlantic was enriched with nutrients at depths greater than 2 km, and depleted at intermediate depths (Duplessy et al. 1988, Sarnthein et al. 1994b; Bickert and Mackensen 2004; Curry and Oppo 2005; Marchitto and Broeker 2006). This finding is supported by the increase of *Cd* content in benthic foraminifera during the LGM (Boyle and Keigwin 1987). Cadmium content is well correlated with dissolved phosphorus, which is lower in the modern deep Atlantic as compared to the Pacific (see Section 1). Assuming that this correlation was maintained during glacials, E. Boyle and L. Keigwin (1987) concluded that NADW formation was reduced during the LGM. Weakening of NADW turnover was accompanied by the simultaneous increase in the production rate of less dense, low-nutrient, and ^{14}C-rich Glacial North Atlantic Intermediate Water (GNAIW also called the upper NADW) that invaded the area above 2-km depth (Fig. 3.4; Broecker et al. 1985; Boyle and Keigwin 1987; Duplessy et al. 1988, 2002; Sarnthein et al. 1994b; Bender et al. 1994; Lynch-Stieglitz et al. 1999, 2007; Williamovski and Zahn 2000; Carry and Oppo 2005). High-nutrient and ^{14}C-poor water is likely to have widespread below 2-km depth in the North Atlantic. This water partly originated from the same area and partly from the Southern Ocean (Lynch-Stieglitz et al. 2007). The scenario of the two major water masses in the North Atlantic above and below 2-km depth is coherent with the existing ^{14}C measurements in benthic foraminifera (Keigwin 2004).

Significant decrease of the Gulf Stream water flow through the Florida Strait at intermediate depths, compensating outflow of the North Atlantic water to the south, was recorded in sediment cores from the Bahamas area and along the Florida coast (Lynch-Stieglitz et al. 1999). In general, AMOC decreased by about 30% during the LGM (Lynch-Stieglitz et al. 2007 and references therein).

Reduction in NADW formation is indirectly confirmed by the drop in the concentration of biogenic opal in the sediments from the northwestern Pacific (ODP Site 882), which most likely points to the weakening of upwelling of silicate-rich deep waters (Haug et al. 1999). The other indirect evidence is the rise in foraminiferal-based SST in the northwestern Pacific reconstructed from SIMMAX (Kiefer et al. 2001). Deep-water temperature could have been fluctuating in accordance with the changing intensity of NADW inflow into the Southern Hemisphere (Fichefet et al. 1994; Winguth et al. 1999; Duplessy et al. 2002). Oxygen isotope data on benthic foraminifera show that the deep World Ocean water (below 2.5 km) had negative temperature as a result of cooling by an average of 3°C compared to the interglacial values. The most part of the deep and bottom waters were formed in the Southern Ocean, partly due to open-ocean convection. Oxygen isotope measurements in the interstitial water showed that deep waters in the North Atlantic were 2°C cooler than in the North Pacific (Adkins et al. 2005). AABW differed from NADW in low oxygen content and lighter carbon isotope composition (Duplessy et al. 1988, 2002; Broecker 2000).

Strengthening of the deep- and bottom-water formation in the Antarctic is indicated by low carbon isotope values and cadmium content in benthic foraminifera from the South Atlantic sediment cores (Mackensen and Bickert 1999). Changes in oxygen isotope composition and salinity of pore water in bottom sediments showed

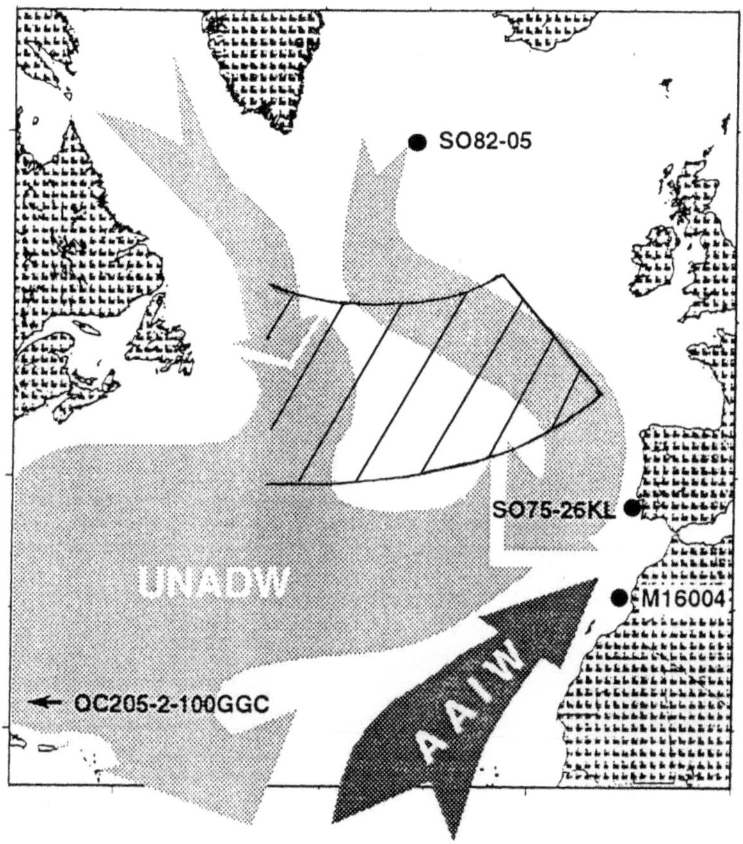

Fig. 3.4 Upper ocean circulation in the glacial North Atlantic and the location of the Heinrich–Ruddiman IRD belt (modified from Williamovski and Zahn 2000). A strong flow of the upper NADW (=UNADW) from the northern North Atlantic to the south and across the basin to the west is inferred from paleocean modeling (Seidov et al. 1996). A northward penetration of AAIW to the mid-latitude northeast Atlantic is inferred from benthic $\delta^{13}C$ and Cd/Ca data, which indicate a mid-depth hydrographic front between the Moroccan and Portuguese margins. Positions of sediment cores with $\delta^{13}C$ records studied are indicated. UNADW = Upper North Atlantic Deep Water, AAIW = Antarctic Intermediate Water

that salinity of AABW was twice higher than the modern one, and higher than that of NADW by 0.4‰ (Adkins et al. 2002). Temperature of both water masses was close to the freezing point. Increased salinity of AABW is most likely related to more intensive ice formation in the Southern Ocean (Adkins et al. 2005). Sea level drop by 120–140 m in the World Ocean during the LGM (Fairbanks 1989; Labeyrie et al. 1987; Lambeck et al. 2000; Yokoyama et al. 2000) resulted in a salinity rise of 3.16–3.6% (i.e., ~1‰), and increased oxygen isotope composition of mean oceanic water by 0.95–1.08‰ (Duplessy et al. 2002). The average global cooling was about 3–5°C (Broecker 2000; Jansen et al. 2007).

In the beginning of the LGM, the northward shift of Southern Ocean fronts compared to their present and interglacial locations (e.g., Becquey and Gersonde 2003; Gersonde et al. 2003) caused the significant reduction of the Antarctic Circumpolar Current and warm water route from the Indian Ocean into the Atlantic via the Agulhas leakage (Berger and Wefer 1996; Peeters et al. 2004). Nevertheless, the return branch of the THC is assumed to have been sustained (Color Plate 3.1) and then remarkably intensified by the end of the LGM (Peeters et al. 2004).

Modeling results demonstrated that the "glacial" ocean is characterized by the stable mode of the thermohaline circulation with the deep-water formation to the south of Iceland. Northward heat transport in the Atlantic at 20°N during the LGM was similar to the present one (i.e., 1.0×10^{15} W; Ganopolski et al. 1998; Hewitt et al. 2001). Some modeling experiments showed weakening (by 30–65%) and shallowing of the deep thermohaline circulation (Ganopolski et al. 1998; Seidov and Haupt 1997; Winguth et al. 1999; Schulz et al. 2001; Fig. 3.4), but several models show no changes or even a strengthening of the AMOC during the LGM (Weber et al. 2007) as well as strengthening of the convection in the northern and southeastern parts of the Pacific (Seidov and Haupt 1999). Freshwater input to the surface layer in the North Atlantic was highly variable compared to today. The outflow from the Arctic was very weak due to the decrease in river runoff. At mid-latitudes freshwater inflow was strongly increasing due to the shift of cyclones track to the south and sea ice melting (negative NAO index).

The conveyor could respond to the meltwater discharge into the ocean surface layer by gradual retreat to the south of the area of NADW formation and shallowing of their sinking depths. As a consequence of such "flexible reaction," the thermohaline circulation in the glacial ocean was constant and stable provided there were no large freshwater injections. The maximum freshwater discharge area located at 40–60°N served as a barrier for Atlantic water penetration to the north and caused the shift of the convection to the south. However, the "shutdown" of the conveyor, i.e., the cessation of convection in the North Atlantic during the glacial, is considered to be very unlikely (Lynch-Stieglitz et al. 2007). Moreover, the colder was the climate, the more stable was the "glacial mode" of the conveyor and more freshwater was required for the transition to the "warm interglacial mode" (Ganopolski and Rahmstorf 2001; Bard 2002; Fig. 3.5). Weakening of the thermohaline circulation affected the Antarctic and the Pacific Oceans to a lesser degree (Manabe and Stouffer 1995, 1997; Seidov and Haupt 1997, 1999).

Another more stable mode of the thermohaline circulation is typical for interstadials. For example, in the middle of MIS 3, the lightening of carbon isotope values of benthic foraminifera in the northeastern Atlantic cores points to the advection of waters of Antarctic origin up to 40°N. The 1‰ isoline of carbon isotope composition of bottom water which today outlines the distribution of NADW (Color Plate 3.2) was shifted to the north and did not reach 30ºN. High, close to modern values of $\delta^{13}C$ were particularly typical for the western part of the region as the well-oxygenated deep waters were formed in the northwestern Atlantic, while in the northeastern Atlantic NADW merged with local deep waters and AABW (Vidal et al. 1997).

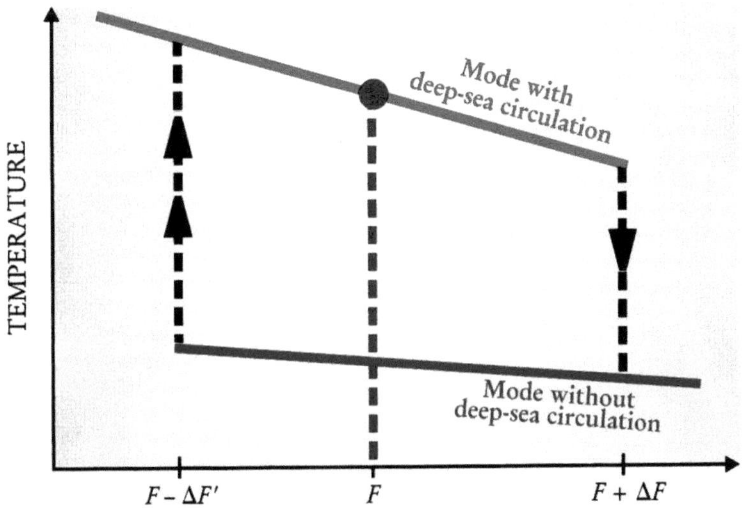

Fig. 3.5 Schematic stability diagram (hysteresis loop) for the AMOC with *solid lines* indicating stable equilibrium states and *dotted lines* unstables states (modified from Ganopolski and Rahmstorf 2001; Bard 2002). Shutdown and resumption of NADW formation are shown by *single* and *double arrows*, respectively. NADW formation and SST decrease after the additional meltwater (F) input to the North Atlantic, the critical freshwater value (F + ΔF') causes the AMOC collapse. NADW formation restarts after the strong reduction of freshwater input (F – ΔF'). AMOC = Atlantic Meridional Overturning Circulation; NADW = North Atlantic Deep Water; SST = sea surface temperature

3.4 Dansgaard–Oeschger Cycles

Relatively short-term, millennial timescale, climatic oscillations were first identified by W. Dansgaard in Camp Century ice core from Greenland (Dansgaard et al. 1971), and later in other cores from that area (Dansgaard et al. 1982, 1993; GISP 1993). During the last glaciation (MIS 2–4), 21 Dansgaard–Oeschger cycles (DO cycles) were distinguished, and in a wider time interval of 20–105 ka BP – 24 cycles were identified (Dansgaard et al. 1993; Grootes et al. 1993; Grootes and Stuiver 1997). In the 1990s these cycles were also found in the North Atlantic deep-sea cores. In particular, it was established that the total abundance of planktic foraminifers varies in phase with SST changes, and these oscillations are well correlated with DO cycles in the Greenland ice core GRIP (Bond et al. 1993; Fig. 3.6). G. Bond and co-authors assumed that the periodical collapse of the North American ice sheet occurred throughout the glaciation and that during the initial stage it was accompanied by the arrival of high number of icebergs into the North Atlantic. Later, when the glacier was retreating, the breakout of icebergs and discharge of meltwater stopped, and warm subtropical waters penetrated further to the north.

Variations in different THC proxies in the North Atlantic sediments between 11 and 75 ka BP were characterized by the dominant cyclicity of about 1,500 years,

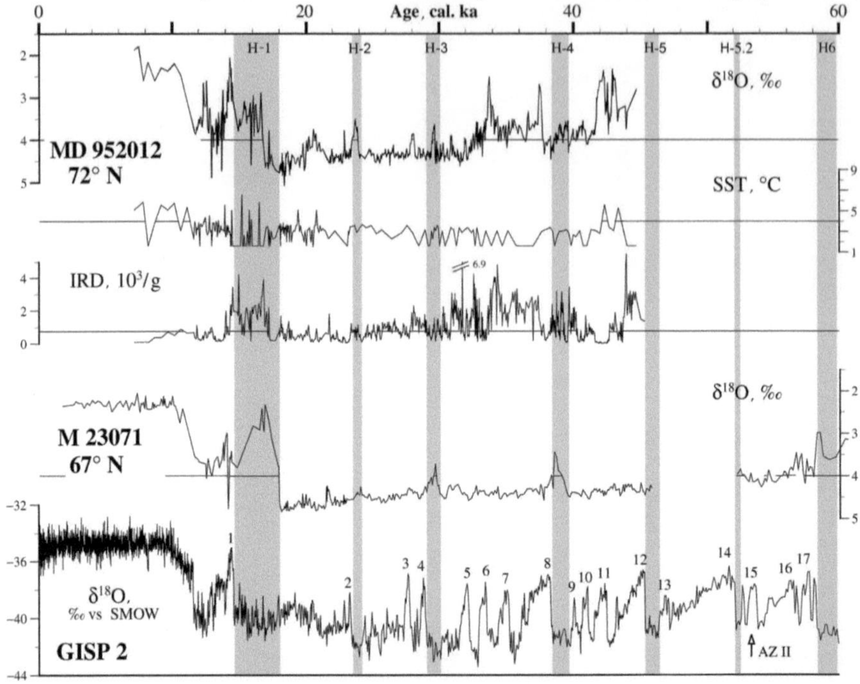

Fig. 3.6 Correlation of proxy time series in sediments cores from the northern North Atlantic, off the Barents Sea shelf, with GISP2 $\delta^{18}O$ record of temperature changes on the Greenland summit (after Sarnthein et al. 2000b). $\delta^{18}O$ minima of planktic species *N. pachyderma* (sin) reflect low-salinity meltwater spikes, where SST estimates based on planktic foraminifera census counts are low. Numbers above the bottom graph are Dansgaard–Oeschger events, *shaded bars* mark Heinrich events. SST = sea surface temperature; IRD = ice-rafted debris; AZ 2 = second ash layer

which corresponds to DO cycles in Greenland (Grootes and Stuiver 1997; Schulz et al. 1999; Labeyrie 2000; Figs. 3.6 and 3.7). During interstadials, SST and salinity at 60°N were increasing on average by 8°C, and up to 36.5‰, respectively, owing to the strong Irminger Current and vigorous thermohaline circulation. During stadials, temperature dropped by 2–4°C in phase with salinity decrease by 1–2‰. This was caused by large-scale iceberg meltwater discharge into the East Greenland Current, which "turned off" the convection in the North Atlantic (van Kreveld et al. 2000). However, in the eastern part of the Norwegian Sea, SST fluctuations reconstructed from the study of diatom and planktic foraminiferal assemblages did not exceed 1–2°C, and the Norwegian Current was weak at least during interstadials 1–4 (Jansen et al. 2001; Weinelt et al. 2003). According to the GLAMAP 2000 project data, heat transport in the surface layer over the Faeroe–Iceland sill, between 53 and 72°N, was reduced by 50–60% during interstadials compared to modern conditions and almost by 100% during stadials of DO cycles. However, during summer season, sea ice rarely penetrated further south than 70°N (Weinelt et al. 2003).

Mineral composition of the IRD and geochemical data provide evidence that terrigenous material originated from different sources around the North Atlantic during

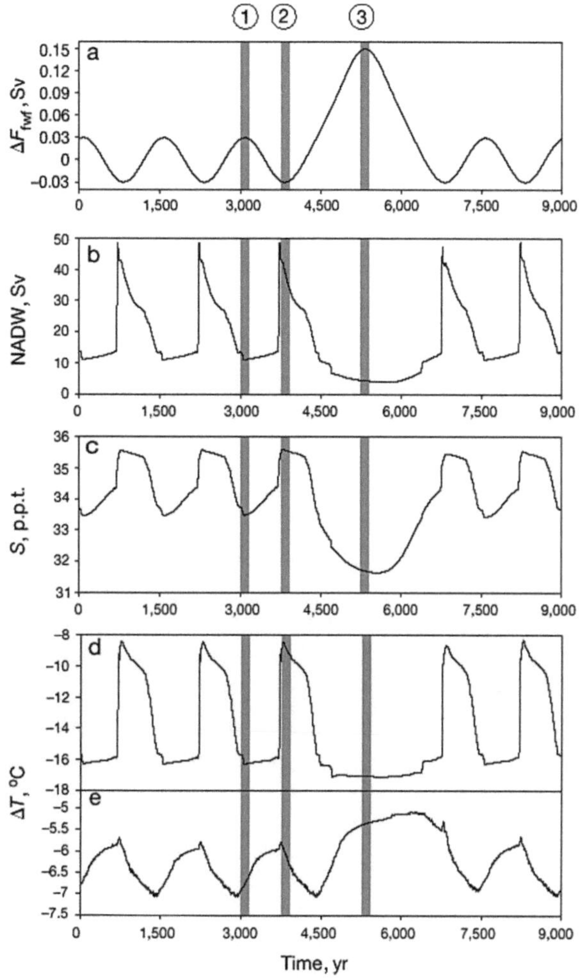

Fig. 3.7 Simulated DO and Heinrich events (after Ganopolski and Rahmsdorf 2001 with Nature permission). (**a**) Forcing, (**b**) Atlantic overturning, (**c**) Atlantic salinity (S) at 60°N, (**d**) air temperature in the northern North Atlantic sector (60–70°N), and (**e**), temperature over Antarctica (temperature values are given as the difference from the present-day climate, ΔT). The vertical bars denote the times for which the simulations were made: 1 and 2 – stadial and interstadial of DO cycle, 3 – Heinrich event

different climate cycles. It could be Greenland, Iceland, Spitsbergen, and possibly other ice sheets (van Kreveld et al. 2000). Magnetic susceptibility, genesis, and concentration of magnetic minerals are sensitive indicators of DO cycles in the North Atlantic sediments. Changes in these parameters demonstrate direct correlation with air temperature in Greenland (Rasmussen et al. 1996; Dokken and Jansen 1999; Kissel et al. 1999).

Besides, the depletion in oxygen isotope (on average by 0.4‰) and carbon isotope composition of carbonate tests of both planktic and benthic foraminifers was detected in several North Atlantic cores. This signal is considered to be a proxy of the surface water freshening in bottom environments explained by different mechanisms, such as brines release during sea ice formation in the Norwegian–Greenland Basin (Jansen and Veum 1990; Vidal et al. 1998; Dokken and Jansen 1999; Bauch and Bauch 2001; Waelbroeck et al. 2006), and freshwater advection to the deep ocean (Lohman and Gerdes 1998). M. Raymo and co-authors (1998) argued that small amounts of freshwater were merged with subducting waters to the north of the Site 983 (60.5°N, 23.5°W) and caused the weakening of the deep-water convection in the Norwegian–Greenland seas. The maximum depth of water sinking decreased, accordingly the boundary between NADW and AABW uplifted, implying that NADW was replaced by more nutrient and less ^{13}C -rich AABW. High variability of carbon isotope ratio at Site 983 allowed assuming that the boundary between the two water masses was close to the depth of this site during the glacials (Raymo et al. 1998).

Thus, the above-mentioned studies demonstrated that the Dansgaard–Oeschger cycles were asymmetrical and started with abrupt warming (interstadial or DO event), when the air temperature increased by 10–16°C (e.g., Landais et al. 2005) in Greenland and 5–8°C in North Atlantic (Sarnthein et al. 2000a; Labeyrie 2000; Shackleton 2001) within a few decades. Every interstadial was followed by gradual cooling (stadial), which usually terminated with an abrupt short-term collapse of the thermohaline circulation as a result of meltwater discharge into the ocean and water and air temperature decrease to stadial values of the previous cycle (Figs. 3.7 and 3.8; Sarnthein et al. 2000a; Labeyrie 2000; Shackleton 2001). By now, these cycles have been recorded in the most areas of the Northern Hemisphere (Bond et al. 1997; see Chapter 7). On the contrary, in the Antarctic temperature increased during the Northern Hemisphere stadials and dropped during interstadials, but amplitude

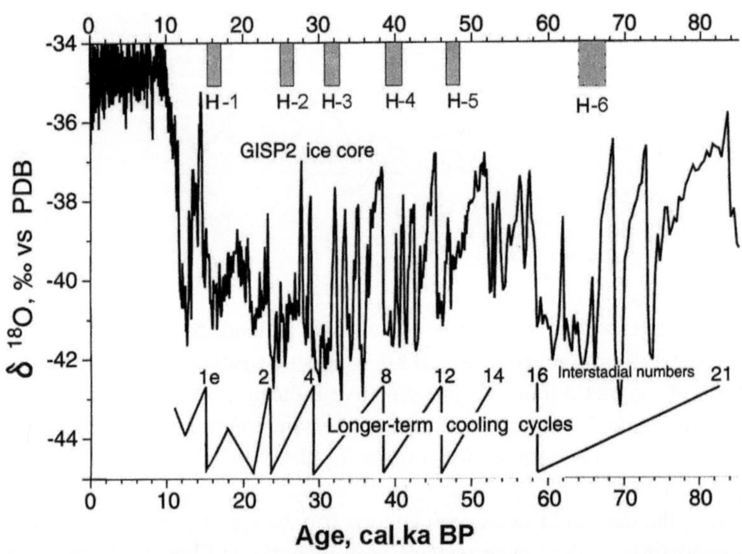

Fig. 3.8 Correlation of Heinrich events H-1–H-6 (shaded bars), interstadials of DO events (maxima in the lower curve), and Bond cycles (lower curve) with GISP2 $\delta^{18}O$ record of temperature changes on Greenland summit (after Grootes, and Stuiver 1997; Vidal and Arz 2004)

of fluctuation was low (Stocker 1998, 2002; Ganopolski and Rahmstorf 2001). Therefore, slow warming phases in the Antarctic seem to precede fast warmings in Greenland by 1–2.5 ka within the last 55 ka (Blunier et al. 1998, 1999; Blunier and Brook 2001; Brook et al. 2005). This effect of "bipolar-seesaw" is discussed in Section 7.1.

Although the time interval between the two warming phases of the two subsequent DO cycles is about 1,500 years, some cycles (8, 12, 14, 19, 20) were almost twice longer (Schulz et al. 1999; Fig. 3.6). The cause and mechanism that trigger this cyclicity are still poorly understood. According to modeling results, characteristics of DO cycles did not depend on the amplitude of external forcing, which was only a triggering mechanism, while the system changed further according to its internal dynamics (Ganopolski and Rahmstorf 2001). Since warming phases represented unstable modes within glacial stages, upon their termination the convection in the Norwegian–Greenland Basin stopped. The AMOC returned to the stable cold mode and remained in this state until the next melt-water discharge (Ganopolski and Rahmstorf 2001; Fig. 3.7).

Several hypotheses have been proposed to explain the origin of suborbital DO cycles, including variations of solar activity, differences in combination of major orbital cycles, changes in the ocean–atmosphere interaction, internal dynamic of ice sheets, and self-oscillations of the thermohaline circulation. W. Broecker and co-authors (1985) assumed that DO interstadials were intervals of renewal of the NADW formation in the Norwegian–Greenland Basin. Many authors assign a major role in these cycles to the alternation of the thermohaline circulation modes, since the delay of the warming in Greenland compared to Antarctica (more than 1,000 years) corresponds only to ocean overturning time, while the atmosphere has a faster response (Blunier et al. 1998; Broecker et al. 1990; Oppo and Lehman 1995; Rasmussen et al. 1996). Abrupt change of the climate system (sometimes within less than 10 years) by half of the amplitude of the glacial–interglacial change in temperature and other parameters is thought to be linked to the weakening or cessation of NADW formation caused by massive ice surges from the Northern Hemisphere ice sheets, while the re-establishment of the convection was triggered by subsequent salinity rise (Alley 1998). Data on the northwestern Atlantic show that brines release was typical of the termination of several DO cycles (van Kreveld et al. 2000), while in the Faeroe Islands area the intensive ice formation started much earlier, in the beginning of cold stadials (Dokken and Jansen 1999).

Variations in solar insolation are unlikely to have triggered DO cycles, since insolation seems to be relatively stable during MIS 3 when these cycles were most pronounced (Labeyrie and Elliot 1999). Several authors argued that the meltwater discharges that initiated DO cycles, especially during melting of Greenland icebergs, were caused by the internal instability of ice sheets. The latter, in turn, does not depend on constant low temperature and atmospheric precipitation, but is controlled by the relief, which allows the ice to slide and then to break off (van Kreveld et al. 2000; Sarnthein et al. 2000a). The oscillations of the Fennoscandian and Laurentide ice sheets were found to be comparable in duration with DO cycles

(Fronval et al. 1995; Rasmussen et al. 1996, 1997; Labeyrie et al. 1999; Labeyrie and Elliot 1999).There is a point of view that high amplitude cycles occurred when the ice sheets volume had reached the critical value, which was equivalent to a global sea level drop of 45 m. Slow cooling and fast warming within cycles are similar to slow growth and fast melting of ice sheets during the ~100-ka glacial cycles. For this reason M. Schulz and co-authors (1999) assumed that fast response of relatively small glaciers and of the periphery of bigger ones to environmental changes played an important role in the intensification of DO cycles. During a slow ice sheets growth their advance to the south and the erosion of the underlying periglacial deposits destabilized their margins and changed their thermal balance. During the fast deglaciation sea level rose quickly that must have strongly destabilized the marine-based ice sheets (Schulz et al. 1999), such as the Barents Sea sheet (see Section 4.5).

Nevertheless, even if the internal instability of the ice sheets might have caused and intensified DO cycles, later this signal was transmitted on large distances by the global conveyor and the atmosphere. Measurements of greenhouse gases concentration made on gas bubbles extracted from polar ice demonstrated dramatic variability in CH_4 and N_2O content at the transitions between DO stadials and interstadials occurring within 20–40 years (e.g., Blunier et al. 1998; Flückiger et al. 2004). However, these gases might have amplified but not triggered the DO cycles. This assumption is supported by a relatively negligible variability of CO_2concentrations within the DO cycles (Indermühle et al. 2000).

Experiments performed with climate models of intermediate complexity (Ganopolski and Rahmstorf 2001, 2002; Schmittner et al. 2002; Wang and Mysak 2006) demonstrated that small variation of freshwater flux into the North Atlantic could cause significant variations of the AMOC with the large temperature shifts over the northern North Atlantic comparable in magnitude and temporal dynamics to those recorded in the Greenland ice cores. So far, it has not been possible to corroborate these results using the state-of-the art climate models due to their prohibiting computational cost.

According to the new sea level reconstructions and modeling experiment by H. Arz and co-authors (2007), the strongest events of meltwater discharge corresponded to the DO interstadials; nevertheless, they did not considerably affect the AMOC as the icebergs from the southern parts of the Northern ice sheets melted at mid-latitudes. However, a smaller freshwater input to the area of the deep-water formation caused by the melting of icebergs, released from northern parts of the ice sheets during the DO stadials, strongly weakened the AMOC. These results are in line with the previous reconstructions by L. Vidal and co-authors (1997) but contradict with the data of E. Rohling and co-authors (2005) showing the 20–30-m sea level rise only during the H-events. If the age model by H. Arz and co-authors (2007) is correct and the suggestion about the stronger meltwater influx during interstadials is true, it means that its extent is more significant for changes in the AMOC and climate than its volume. This modeling experiment also demonstrated that change of the THC mode in turn considerably affect the mass balance of the surrounding ice sheets which can be an important feedback for the millennial scale variability.

3.5 Heinrich Events

DO cycles are grouped in long-lasting Bond cycles (Bond et al. 1993), which are separated by Heinrich events (Heinrich 1988, Figs. 3.8 and 3.9). The latter were first described from the North Atlantic sediments, where they were identified by a pronounced increase in ice-rafted debris content (>150 μm) within so-called IRD belt between latitudes of 40 and 55ºN (Fig. 3.4). This belt was named after H. Heinrich, as well as the events themselves, although some researchers name it after W. Ruddiman. The thickness of IRD layers in the sediments reaches several meters in the Labrador Sea and become thinner eastward, down to 1–2 cm at 10ºW (Grousset et al. 1993, 2001). From the very beginning, Heinrich events, named H-1, H-2, etc. (Figs. 3.6 and 3.8), were associated with massive ice surges from the Laurentide ice sheet covering North America and producing multiple iceberg "flotillas" (Bond et al. 1993; MacAyeal 1993). Heinrich events are documented by the depletion of the oxygen isotope composition of planktic and benthic foraminifera, the carbon isotope composition of benthic foraminifers, an abrupt increase in *Neogloboquadrina pachyderma* (sin.) content (reaching 100% of the planktic foraminiferal assemblage), as well as by the change in sediment color reflectance (Figs. 3.9 and 3.10; Bond et al. 1992, 1993; Bond and Lotti 1995; Vidal et al. 1997, 1998; Cortijo et al. 1997, 1999a; Grousset et al. 2001; Kandiano and Bauch 2003). Each event started with abrupt cooling and ended with fast warming in Greenland and North Atlantic, which made up about half the amplitude of the warming within the last glacial–interglacial cycle (Sachs 2005).

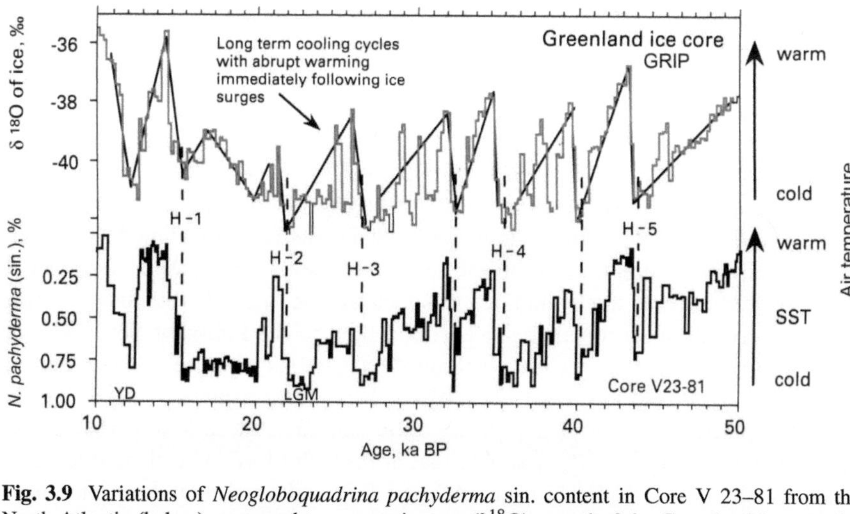

Fig. 3.9 Variations of *Neogloboquadrina pachyderma* sin. content in Core V 23–81 from the North Atlantic (below) compared to oxygen isotope ($\delta^{18}O$) record of the Greenland ice core for the last 50 ka (above). Heinrich events and Bond cycles with slow cooling and rapid warming just after intense iceberg calving are shown (modified after Bond et al. 1993)

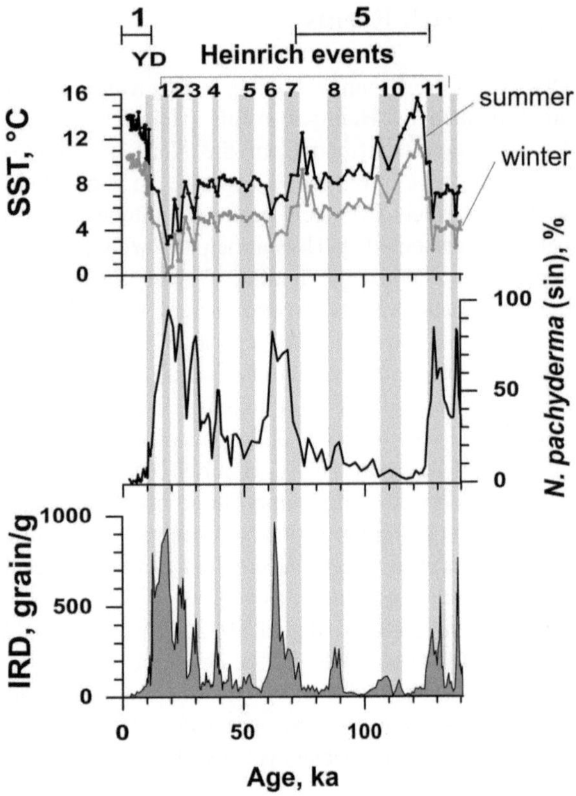

Fig. 3.10 Variations in sea surface temperature (SST) reconstructed by two methods (MAT and RAM), in content of polar planktic foraminiferal species *Neogloboquadrina pachyderma* sin., and in ice-rafted debris counts (IRD > 250 μm) during the last 140 ka studied in Core M23414 (53° 32′ N., 20° 17′ W, water depth 2,200 m) from the North Atlantic (modified after Kandiano and Bauch 2003). Younger Dryas (YD) and Heinrich events 1–11 are marked by shaded bars. The cooling events are characterized by high *N. pachyderma* sin. and IRD content

The forcing mechanisms for these events are still being debated (see review in Hemming 2004). W. Broecker and co-authors (1990) were the first to propose that Heinrich events can be caused by strong variations in salinity ("salt oscillator hypothesis"). Later the catastrophic collapse of ice sheets was suggested to be a possible trigger as it could strongly affect the convection in the North Atlantic (Broecker 1992). D. MacAyeal (1993) proposed a "binge-purge" hypothesis involving an internal instability of the ice sheet and geothermal heating as a driving force of their purging. Results of the simulations with a realistic ice sheet model by Calov et al. (2002) confirmed the validity of this mechanism. When the ice sheet above the Hudson Strait was thickening, ice was melting due to geothermal heating from below promoting iceberg calving. Icebergs drifted to the Labrador Sea, and then eastward over the ocean where the entrained terrigenous material was melting out (Bond et al. 1993; MacAyeal 1993). When the process exhausted itself, the ice sheet thickened again and a new Bond cycle started (Figs. 3.8 and 3.9). S. van Kreveld and co-authors (2000) argued that at first "flotillas" of Icelandic icebergs reached latitudes of 40–60° N in the Atlantic (Fig. 3.4), and then during the most intense Heinrich events they could have been supplemented with icebergs from the Laurentide ice sheet. The lowering of the altitude of the ice sheet caused warming in the North Atlantic as the strong cold winds blowing from it relaxed. Reduction in meltwater inflow intensified the thermohaline circulation and advection of heat accumulated in the tropics to high latitudes. Two to three km of ice accumulated on average within 5–10 ka, which roughly corresponds to the periodicity of Heinrich events (Paillard and Labeyrie 1994; Labeyrie and Elliot 1999; Broecker and Hemming 2001; Hemming 2004).

Another mechanism proposed by several authors as a trigger for Heinrich events is the sea level rise that might have affected the ice sheets (McIntyre and Molfino 1996). This mechanism has been recently investigated by J. Flückiger and co-authors (2006) in relation to variations in THC. According to their model, a collapse of the AMOC during H-events caused significant regional sea level rise in the North Atlantic and increased heat uptake of the global ocean. The latter in turn promoted further sea level rise that finally destabilized Northern Hemisphere ice shelves and ice sheets thus triggering ice surges. Ice calving and subsurface ocean warming provided positive feedbacks including further sea level rise and iceberg discharge. Finally, the large surge from the Laurentide ice sheet broke down the feedback loops. As a result, ice sheets reached a new equilibrium, the amount of ice calving and meltwater discharge decreased, and the AMOC resumed.

The relationship between Heinrich events and DO cycles is not yet well known. Iceberg calving during Heinrich events occurred after long cooling and the southward migration of the Subpolar Front and terminated with fast warming and return to the interstadial mode of the thermohaline circulation. During Heinrich events, SST in the North Atlantic between 40 and 60°N dropped by 2–8°C, and salinity by 1–3‰ (Cortijo et al. 1999a; Matul et al. 2002). To the south, beyond the Heinrich–Ruddiman IRD belt limits, at 21°N (ODP Site 658, Zhao et al. 1995), temperature decreased by 3–4°C during most severe cooling events. Such drastic cooling and correlation of IRD layers in numerous sediment cores, document the simultaneity of Heinrich events (within the age uncertainty) in the entire North Atlantic area and point to their strong influence even on distant basins. Heinrich events were followed by a fundamental reorganization of the climate system due to short-term collapse of NADW formation and heat flux to the northern North Atlantic. Ventilation of deep water significantly deteriorated (e.g., Vidal et al. 1997; McManus et al. 2004).

Six Heinrich events have been recognized during stages MIS 4–2 of the last glacial interval; they correspond to ~16.8, 22, 30, 38, 45, and 60 cal. ka BP (Bond et al. 1992; Zhao et al. 1995; Cortijo et al. 1997, 1999a; Labeyrie and Elliot 1999; Labeyrie et al. 1999; MacManus et al. 1999; Chapman et al. 2000; Broecker and Hemming 2001; Hemming 2004). The duration of every H-event is estimated between few hundreds and up to 1.2–2.0 thousand years; during this interval icebergs could break off from different ice sheets on the Atlantic periphery at different times (Vidal et al. 1997; Sarnthein et al. 2000a; Grousset et al. 2001; Figs. 3.11 and 3.12). Data on the South Atlantic and Antarctic show that cold Heinrich events in the North

Atlantic coincided with pronounced warming events in the southern regions ("bipolar seesaw" discussed in Sections 3.4 and 7.2). Outside the North Atlantic, geographic distribution of Heinrich events was different from that of DO cycles. For

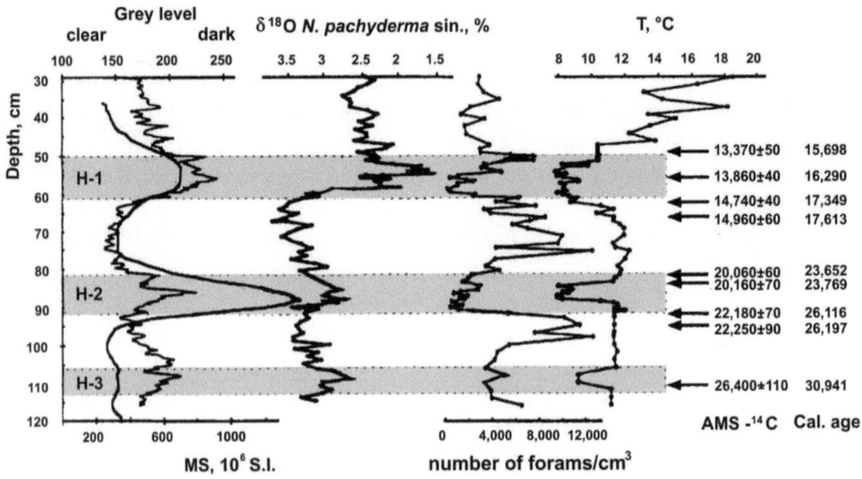

Fig. 3.11 Correlation of magnetic susceptibility and gray level of bulk sediment in core SU90-09, along with planktic oxygen isotope record on *N. pachyderma* sin., planktic foraminiferal abundance, and summer sea surface temperatures (SSTs) calculated using the modem analogue technique between 30 and 120 cm (modified from Grousset et al. 2001). Location of Heinrich layers H-1, H-2, and H-3 is shown by *stippled areas*. AMS-^{14}C ages are reported along with calendar ages

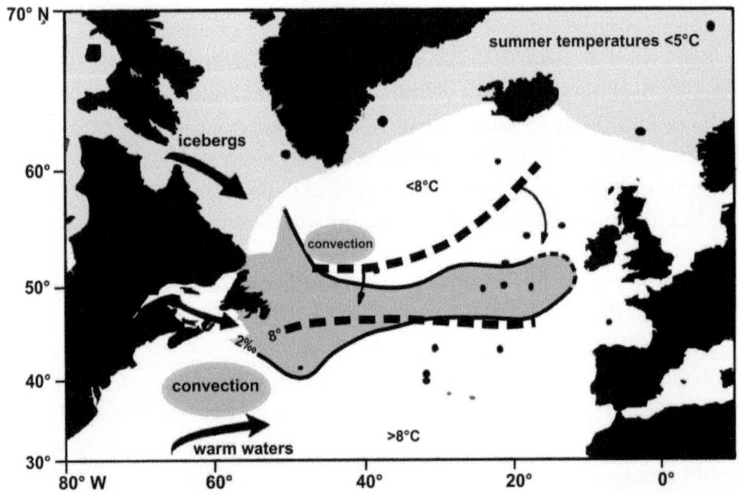

Fig. 3.12 Paleoceanographic situation during the Heinrich event H-4, about 35 ka BP, in the North Atlantic. Assumed convection areas according to oxygen isotope data are shown. Directions of iceberg drift and warm waters inflow are marked by *solid arrows*. The isoline of 2‰ bounds the zone of maximum δ^{18}O values in planktic foraminiferal tests. The isotherm of 8°C is shown for the time slice of 37 ka BP, before the iceberg calving, and for iceberg melting time, 35 ka BP (after Cortijo et al. 1997, 1999a; Vidal et al. 1997)

instance, on the Brazilian continental margin only Heinrich events were registered, while in Greenland only DO cycles were documented (Arz et al. 1998; Broecker and Hemming 2001; Hemming 2004).

Instability of the conveyor is of crucial importance for Heinrich events. According to paleoceanographic data, NADW formation decreased and even completely ceased, while that of AABW and intermediate water intensified (Maslin et al. 1995; Zahn et al. 1997; Elliot et al. 2002). During the strongest Heinrich events AABW reached 62°N (Elliot et al. 2002). Sea surface salinity in the North Atlantic dropped so much that water could not sink even down to 2 km, which is supported by modeling experiments (Paillard and Labeyrie 1994). Deep-water convection occurred outside the basin with freshened surface water. The location of convective cells was different over different Heinrich events and during the LGM (Vidal et al. 1997, 1998; Elliot et al. 1998; Broecker and Hemming 2001). This most likely evidence the arrival of melting icebergs from different ice sheets on the periphery of the North Atlantic (Bard et al. 2000; Sarnthein et al. 2000a).

The AMS-^{14}C-dated oxygen isotope records, in particular from the Fram Strait, suggest that H-1 event was caused by the collapse of the Barents Sea ice sheet and drastic meltwater discharge into the North Atlantic at 14.7–13.3 ka BP (about 17.3–15.7 cal. ka BP) (Jones and Keigwin 1988; Lehman 1991, Weinelt et al. 1991; Sarnthein et al. 1995, 2000a; Broecker and Hemming 2001). Low carbon isotope values of benthic foraminifers point to the shutdown or sharp reduction of deep convection in the North Atlantic during this time and penetration of waters of southern origin not only to the east but also to the western part of the basin. A relatively small convective cell where water could sink down to 2,500 m could only exist to the west of the Rockall Plateau between 52 and 60°N (Sarnthein et al. 1994; Vidal et al. 1997). According to M. Sarnthein and co-authors (1995), old oxygen depleted intermediate and deep waters filled the North Atlantic.

Relatively high carbon isotope values of benthic foraminifera (+0.9‰) indicate that deep convection did not stop completely during the H-4 event, about 38 cal. ka BP, but was localized to the north and south of the Heinrich IRD belt in the West Atlantic. In the eastern Atlantic, deep waters of southern origin penetrated further north than in the western Atlantic, as it happened during the LGM (Vidal et al. 1997; Cortijo et al. 1999; Fig. 3.12). Planktic foraminiferal assemblages document the dominance of Arctic waters in the surface layer of the northeastern Atlantic, while the Subpolar Front was located to the south of the Faeroe Islands (Cayre et al. 1999b; Rasmussen et al. 1996, 1997). This is in accordance with the dinocyst assemblages distribution pointing to almost year-round and extensive sea ice cover and reduced surface water salinity during Heinrich events (de Vernal and Hillaire-Marcel 2000). After the termination of Heinrich events δ^{13}C returned to previous high values, indicating resumption of NADW formation. Thus, the isotope and microfossil data show that during Heinrich events northward penetration of southern source waters was controlled by surface water parameters and thus by the intensity of the thermohaline circulation in the North Atlantic (Vidal et al. 1997). Warmer conditions were reconstructed for the northeastern Atlantic during H-3 and colder for H-2 (Cayre et al. 1999b; Rasmussen et al. 1996, 1997).

The surface water freshening signal was transferred to the deep water most likely by brines release during sea ice formation (Vidal et al. 1998). During H-3 and H-4 events, this process impacted more the intermediate and deep-water formation in the southern Norwegian Sea and northern periphery of the Atlantic, than convection in the open ocean, whereas during H-2 event brines rejection affected the other parts of the North Atlantic (Hebbeln et al. 1994; Vidal et al. 1998). During Heinrich events, favorable conditions for ice formation probably existed along almost the entire periphery of the Norwegian–Greenland Basin, including the Greenland–Iceland shelf. Ice was quickly carried southward from the Subarctic Atlantic by surface currents. Ice evacuation along with the strengthening of easterly winds associated with the ice sheets promoted the intensification of ice formation (Midttun 1985; Vidal et al. 1998).

The modeling studies by D. Seidov and co-authors (1996) and D. Seidov and B. Haupt (1997) showed a collapse of the thermohaline circulation as a result of freshwater discharge into the North Atlantic that lasted from 100 to 1,000 years. These results are consistent with extensive paleoceanographic reconstructions of M. Sarnthein and co-authors (1995). Other modeling experiments (e.g., Weaver 1995) also captured strong freshwater signal in the northern North Atlantic. It is not surprising that such abrupt changes in temperature and thermohaline circulation were reflected in the oceanic heat transport. Although on the ocean surface this effect was restricted to the North Atlantic, deep branch of the conveyor was affected as far as the eastern coast of Australia (Seidov and Haupt 1997). The model demonstrated a return flow of the ACC deep branch in the Indian and Atlantic sections of the Southern Ocean. In the North Atlantic, these waters reached 50ºN, where they upwelled during H-1 event, which agrees with the above-mentioned reconstructions. D. Seidov and B. Haupt (1997) showed that deep waters of the North Atlantic represented a mixture of AABW from the Weddell Sea and waters sinking in the eastern Indian Ocean, explaining the "old age" of the North Atlantic waters.

Thus, the above-mentioned and more recent data and modeling results provide good evidence of the global-scale changes in the THC and in particular in its deep water branch during H-events (see also Chapter 7). The global conveyor belt mode during Heinrich events differed from the stable glacial mode perhaps even more drastically than the modern one does (Seidov and Haupt 1997, 1999). However, according to these authors, intensification of convection in the northern and southeastern Pacific seems to be typical for both Heinrich and glacial modes of the circulation. D. Lund and A. Mix (1998) found the intervals of stronger ventilation in the northeastern Pacific roughly coincident with Heinrich events. Other modeling experiments (Ganopolski and Rahmstorf 2001) showed that massive meltwater release (about 0.15 Sv) during numerous iceberg discharges into the North Atlantic resulted in the temporal "shutdown" of the conveyor and warming in the Antarctic. However, it did not cause further cooling in Greenland, since it was not affected by warm branch of the thermohaline circulation even before Heinrich events, and temperature on its surface was already low. Detailed accelerator mass spectrometry dating (Fig. 3.11; Vidal et al. 1997; Grousset et al. 2001; Chapman et al. 2000) confirmed that Heinrich events had started after coolings in Greenland and therefore could not have caused them. Coolings associated with Heinrich events were especially well pronounced in the subtropical Atlantic, in particular off the Iberian margin, where DO cycles were weak (Bard et al. 2000). Since "shutdown" mode was unstable, AMOC was spontaneously returning to stable "cold" mode after the cessation of meltwater discharge, like it happened during DO cycles (Ganopolski and Rahmstorf 2001). J. Adkins and co-authors (2005) argued that the transition from the stable mode with stratified water column represented by strongly freshened water on the surface, shallow and relatively freshened Atlantic water mass below and heavy saline AABW on the bottom, to active meridional circulation could happen only due to geothermal heat influence. The possibility of the significant contribution of the thermobaric component to the instability of the glacial mode of the thermohaline circulation has currently been studied by modelers.

3.6 Deglaciation and the Holocene

Deglaciation is a complex process, generally controlled by the interaction between solar insolation, global thermohaline circulation, internal dynamics of ice sheets, permafrost, and greenhouse gases in the atmosphere (Blunier et al. 1998; van Geel et al. 1999; Jansen et al. 2001). It was suggested that the Southern Ocean warming and corresponding sea ice retreat preceding the Greenland warming have strengthened the AMOC (Knorr and Lohmann 2003). In particular, the AMOC resumption was associated with the southward shift of the Southern Ocean fronts and strengthening of the Agulhas leakage since the end of the LGM (Peeters et al. 2004). The resulting increase in the northward oceanic heat transport contributed to the reduction of the ice sheets and to the meltwater input into the North Atlantic. The modeling experiments demonstrated that the freshwater discharge could weaken but not shutdown the NADW formation due to the stabilizing effect of the different convection sites (Knorr and Lohmann 2004).

During the deglaciation or Termination I, important climatic events such as Bølling and Allerød warmings and coolings of the Older and especially of the Younger Dryas were registered in North Atlantic bottom sediments (Fig. 3.13). During Termination II that is during the transition from the penultimate glaciation to the Eemian, variations in the location of convective cells in the North Atlantic and Norwegian–Greenland Basin were recorded and correlated with SST changes. NADW turnover varied during both terminations (Lototskaya and Ganssen 1999; Oppo et al. 2001). According to carbon isotope data, convection in the Norwegian Sea did not reach water depths greater than 2 km during warm intervals of Terminaton I as a result of the freshening of surface waters and diminishing density (Bauch et al. 2000; Oppo et al. 2001). Penetration of warm Atlantic water into the Norwegian–Greenland Basin and further into the Arctic between 18 and 14.5 cal. ka BP (Dokken et al. 2001) contributed to ice sheets melting, which is confirmed by our Barents Sea data (Ivanova and Murdmaa 2001; Ivanova et al. 2002; Murdmaa et al. 2006; Section 4.5).

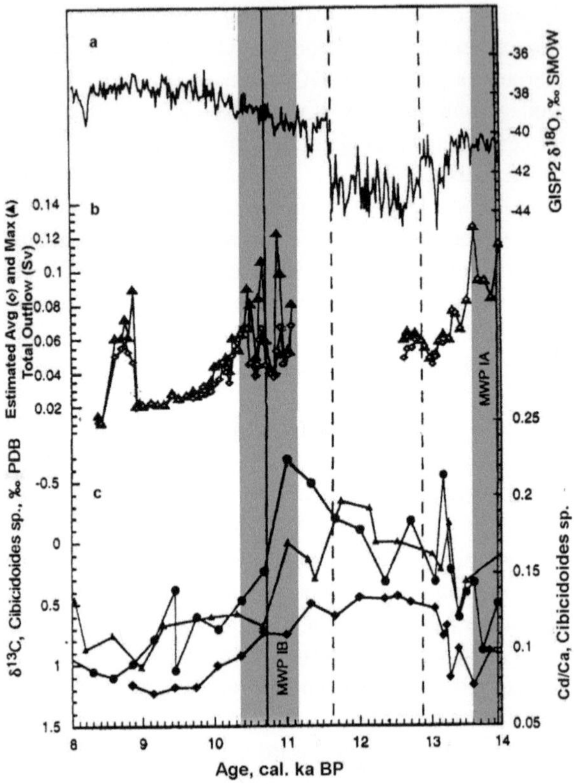

Fig. 3.13 Oxygen isotope record of GISP2 ice core (**a**) compared with (**b**) estimated average (*diamonds*) and maximum total outflow (*triangles*) from the southern and eastern Laurentide ice sheet (modern Great Lakes area), and with (**c**) the carbon isotopes values and Cd/Ca ratios measured in the tests of benthic foraminifera from the deep North Atlantic (modified after T. Moore et al., 2000). (**c**) *Filled triangles* represent core $\delta^{13}C$ values in core KN31-GPC5 (33°41'N; 57°37'W; w.d. 4,583 m), *circles* represent core ENI20-GGCI (33°40'N, 57°37'W; w.d. 4,450 m) $\delta^{13}C$ values, *diamonds* represent core ENI20-GGCI Cd/Ca values. Two vertical *dashed lines* define the limits of Younger Dryas in the GISP2 ice core. The vertical *grey* bands define the time intervals of meltwater pulses (MWP 1A and 1B).

SST and salinity reconstructions from the area west of Scotland (Austin and Kroon 2001) showed that northward heat and salt transfer from the North Atlantic within the last 15 ka reached its maximum during the Bølling onset. Benthic carbon isotopes, Cd/Ca ratios in benthic foraminifers, Nd isotopes and $^{231}Pa/^{230}Th$ ratios of deep-sea sediments from different locations in the Atlantic also indicate the intensification of NADW formation and rapid resume of the AMOC at ~14.6 cal ka BP (e.g., Boyle and Keigwin 1987; Piotrowski et al. 2004; McManus et al. 2004). In the North Atlantic, the entire water column was filled by radiocarbon-enriched water like today (Robinson et al. 2005). The rapid resumption of the oceanic overturning was associated with the abrupt onset of the Bølling-Allerød warming that extended throughout much of the Northern Hemisphere (Clark et al. 2002). However, according to several studies (Veum et al. 1992; Koç et al. 1993; Bauch et al. 2001) the thermohaline circulation was unstable with periodical advection of Atlantic waters to the Norwegian–Greenland Basin either in the surface or in the subsurface layer. W. Berger and E. Jansen (1995) proposed a hypothesis of estuarine circulation in the Norwegian Sea with meltwater outflow into the western part of the basin and Atlantic water inflow on the east due to the Coriolis Effect.

The Younger Dryas (YD, 13–11.5 cal. ka BP according to Bard et al. 2000 or 12.9–11.6 cal. ka BP according to IPCC report 2007) was the first relatively short-term cold event during the deglaciation recorded in multiple terrestrial and marine sections. It has been studied by paleoclimatologists and paleoceanographers for many years, since it separates two phases of deglaciation or Termination I (Duplessy et al. 1981b; Jansen and Veum 1990). For instance, a large set of palinological data on northern Eurasia summarized by V.A. Klimanov (1990) showed considerable cooling (on average by 5ºC in July and 10ºC in January) and decrease in precipitation (on average by 100–300 mm compared to today). In the North Atlantic, the Subpolar Front had modern-like location and was noticeably shifted to the south only during the YD (Ruddiman and McIntyre 1981; Koç et al. 1993). However, up to now data on this event have been very controversial. In order to explain it, several hypotheses were proposed including fast freshwater discharge into the North Atlantic from the North American glacial lake Agassiz (Rooth 1982) or meltwater flux from the Laurentide ice sheet into the St. Laurence River via the Lake Champlain (Broecker et al. 1985).

The freshwater draining into the North Atlantic caused an abrupt partial reduction in the AMOC at the start of the YD. Several paleoceanographic proxies including $\delta^{13}C$ values, *Cd/Ca* ratios, *Pa/Th* ratios, *Nd* isotope water-mass tracer and sediment grain size suggest that the AMOC slowdown during the YD was similar to the LGM state and unlike the near collapse at H-1 (Boyle and Keigwin 1987; Sarnthein et al. 1994b; Manighetti and McCave 1995; McManus et al. 2004; Keigwin 2004; Piotrowski et al. 2005). A concomitant shoaling of NADW is inferred from the replacement of the radiocarbon-enriched NADW by the radiocarbon-depleted AABW below ~2,500 m (Keigwin 2004).

The so-called 8.2 cal. ka cooling event was also suggested to result from the accumulation and then fast drainage of meltwater from ice-dammed lakes at the southern margin of the collapsing Laurentide ice sheet (Barber et al. 1999). Subsequently, several authors (e.g., Broecker and Hemming 2001) assumed that the collapse of convection in the North Atlantic during the Younger Dryas and 8.2 cal. ka event might be triggered by the sudden freshening of the surface waters. It implies that the mechanism of these abrupt coolings could be analogous to that of Heinrich events. 8.2 cal. ka cooling event was revealed first in the Greenland and lake Agassiz ice cores (Alley et al. 1997; Koerner 2001), and also inferred from the appearance of coldwater ostracods in sediments of the Ammersee Lake in southern Germany (von Grafenstein et al. 1998). During this event the temperature in the central Greenland was by 4–8°C lower than today (Alley et al. 1997; Mayewski et al. 2004). In the Norwegian Sea, weakening of the convection and Atlantic water penetration were identified from diatom assemblages, reduction in oxygen isotope composition of benthic foraminifera before and after the Younger Dryas and during the 8.2 cal. ka cooling (Fig. 3.14; Koç et al. 1993; Bauch et al. 2001; Risebrobakken et al. 2003). Perturbation of the thermohaline circulation led to the warming at low latitudes (Rühlemann et al. 1999; Labeyrie 2000).

The 8.2 cal ka event was identified by different proxies at several location, for example, by increased quartz-to-plagioclase ratio on the Reykjanes Ridge (Moros

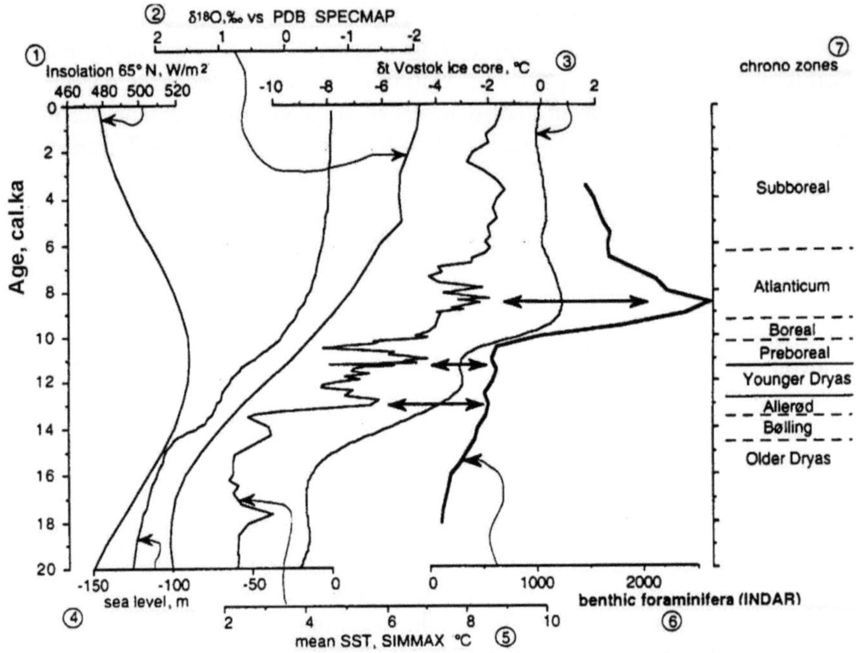

Fig. 3.14 Correlation of global and regional paleoclimatic and paleoceanographic parameters during the last 20 ka (after Nees 1997, with permission from Grzybowski Foundation): (1) insolation at 65°N; (2) SPECMAP record; (3) temperature record at the *Vostok* station, Antarctica; (4) sea level; (5) sea surface temperature over the Rockall Plateau; (6) accumulation rates of benthic foraminiferal tests in the Norwegian Sea; and (7) chronostratigraphic zones

et al. 2004), by SST drop reconstructed in the annually laminated lake sediments from Estonia (Veski et al. 2004), and by changes in benthic assemblages and depletion of benthic oxygen isotope ratio in the Arctic fjord Van Mijenfjorden on Svalbard (Hald and Korsun 2008).

However, some authors found that 8.2 cal ka event is muted (e.g., Hall et al. 2004) or not highly significant (Andersen et al. 2004; Bendle and Rosell-Melé 2007) in sediment records. The discrepancy can be associated with the decoupling between surface and subsurface conditions and corresponding proxies (Moros et al. 2004; Jansen et al. 2008). Nevertheless, some authors (e.g., Rohling and Pälike 2005) believe that 8.2 cal ka event has a compounded nature and hence could not cause the significant perturbation of AMOC.

For many years Holocene climate was considered to be more stable than the glacial one. Nonetheless, G. Bond and co-authors (1997) revealed fluctuations in abundance and species diversity of planktic foraminifera, IRD content in sediments, oxygen and carbon isotope composition of foraminiferal tests within the Holocene with a cyclicity of about 1,500 years, resembling DO cycles, but characterized by lower amplitude. The temperature drop probably did not exceed 2°C. Cool waters with drifting ice and IRD arrived from the north and flowed southward down to the latitude of Great Britain. Highest concentration of coarse grain-size fractions was recorded at 1,400, 2,800, 4,200, 5,900, 8,200, 9,400, 10,300, and 11,100 years ago. The Subpolar Front shifted southward by about 5° latitude compared to its present location. In the tropical Atlantic, off the west coast of Africa, planktic foraminiferal assemblages reveal several ~1,000-year long cooling intervals within the Holocene, which are synchronous with SST variations in the Subpolar North Atlantic (de Menocal et al. 2000).

The finding of this cyclicity in the Holocene (see also Mayewski et al. 2004) led to reconsideration of DO cycles forcing mechanisms, and allowed to assume that they could not be caused by the instability of ice sheets, but were linked to some particular glacial processes in the climate system, probably intensified by the North Atlantic thermohaline circulation (Bond et al. 1997). M. Raymo and co-authors (1998) showed that such periodicity existed also in the Early Pleistocene. Analogous Holocene cyclicity in current's velocity variations close to Iceland was revealed from the mean size of sortable silt (Bianhi and McCave 1999), moreover the currents were found to slowdown, for instance during the Little Ice Age. These data reflect variations in the thermohaline circulation during the Holocene.

The heat transfer during the Holocene (1.0×10^{15} W) did not differ from the glacial times (Ganopolski and Rahmstorf 2001). According to modeling results, even weak climatic fluctuations could cause high amplitude short-term warmings because of non-linearity of the AMOC response. However, when the system was in the warm interglacial mode with convection to the north of Iceland, it became insensitive to weak influence of 1,500-year cycles. Although these cycles are present in some of the Holocene time series, they were not reinforced by the instability of the thermohaline circulation and remained weak. Modern "interglacial" type of circulation is unstable only on the periphery. The transition to this mode is relatively easy to trigger, which agrees well with alternation of cold and warm episodes in DO cycles (Ganopolski and Rahmstorf 2001). Nonetheless, not all researchers agree that periodicity of AMOC variations during the Holocene was constant and comparable to DO cycles. In some Holocene sections from the North Atlantic studied in detail, a persistent cyclicity was not distinguished at all (e.g., Risebrobakken et al. 2003).

The warmest conditions in the surface layer of the Norwegian–Greenland Basin were attributed to the time interval 10–6 cal. ka BP, with an exception of the above-mentioned cold event around 8.2 cal. ka BP (Koç et al. 1993; Bauch et al. 2001; Klitgaard-Kristensen et al. 2001; Andersen et al. 2004; Moros et al. 2004; Jansen et al. 2008). On the boundary between the Norwegian and Barents seas, at 75°N, winter and summer SST reached maximum values at ~10.5–8 cal. ka BP and at about 2 cal. ka BP (Sarnthein et al. 2003b, Hald et al. 2007; Ivanova et al. 2008a; see also Section 4.5.3 and Color Plate 4.8). The duration of the climate optimum decreased from south to north, i.e. from 3,200 years at 63°N down to 1,000 years at 80°N (Koç et al. 1993; Duplessy et al. 2001). According to several authors (Kerwin et al. 1999 and references therein), summer SSTs were higher than the modern ones by 2–4°C in the Norwegian Sea and by 0.5–3°C in the Labrador and Greenland Seas, and Hudson and Baffin Bays at about 6 ka BP. Some authors consider this optimum to be associated with the Holocene maximum of Atlantic water inflow (Koç et al. 1993), while others assume that only after 7 cal. ka BP modern conditions characterized by enhanced water exchange with the Arctic, lower SSTs, and deep convection started to develop (Bauch et al. 2001; Lubinski et al. 2001). This led to the strengthening of the frontal zones and meridional gradient between water masses at about 5 cal. ka BP. According to H. Bauch and co-authors (2001), the climatic optimum preceding the onset of strong convection indicates that the postglacial warming was not directly controlled by the thermohaline circulation, but was caused by the Early Holocene insolation maximum at high latitudes. Yet, our data on the Barents Sea (see Section Section 4.5.3; Duplessy et al. 2001), as well as data of J.C. Duplessy and co-authors (1992) on the North Atlantic show the link between the optimum, forced by preceding summer insolation maximum, and the intensification of the global thermohaline circulation. Warming started in lower latitudes, and subsequent cooling in higher latitudes, that is in the Arctic. In the Nordic Seas, the optimum started earlier and was shorter in the areas near the modern sea ice limit than to the south, where it was most clearly expressed between 8 and 6 cal. ka BP. (Jansen et al. 2001, 2008).

In the Early–Middle Holocene (8–6 cal. ka BP), predominance of intervals with positive NAO index was detected along with the rise of SSTs in the Norwegian Sea and close to Iceland, reflecting the increase of meridional heat flux to high latitudes. The intervals of positive NAO index are associated with the shift of the Subpolar Front in the North Atlantic to the east and increase in content of boreal planktic foraminifer species *Turborotalia quenqueloba* in the Norwegian Sea. The eastward Atlantic water flow along the Norway narrowed but intensified at the same time (Jansen et al. 2001). However, temporal relationship between the thermohaline circulation and the NAO needs more profound investigation. This linkage is obviously complicated, since positive NAO index implies not only increased heat flux to high latitudes but also increased atmospheric precipitation, in particular in snow, which promotes the growth of glaciers (see Section 4.5.4).

3.7 Summary

Pronounced millennial-scale variability of the AMOC during the last ice age is recorded in paleoceanographic proxies over the North Atlantic. The glacial–interglacial and abrupt sea surface temperature and salinity changes associated with H-events, DO cycles and the YD controlled deep convection in the Nordic Seas and Labrador Sea.

The thermohaline circulation was more vigorous and SSTs were higher than modern values during the Eemian and the Holocene optimum. During rather stable interglacial mode of the AMOC, a major part of NADW formed in the Nordic Seas. Therefore, the AMOC was the most important factor controlling the global circulation and to a high extent the global climate.

During the glacials, the AMOC was periodically turned out from its stable mode and perhaps temporally collapsed at least during some of the H-events. Location of convection sites and depths of surface water subduction varied during the LGM,

Heinrich events (especially H-1 and H-4), and possibly during some of the DO stadials, Younger Dryas, and the cooling event at 8.2 cal. ka BP. NADW turnover reduced dramatically, while the formation of intermediate water and AABW intensified. Due to the southward sea ice propagation brines rejection became one of the major mechanisms of intermediate and deep-water formation in the North Atlantic. Inflow of the Atlantic waters to the Arctic, Antarctic, and other oceans sharply decreased leading to the re-arrangement of the global thermohaline circulation.

The Southern Ocean warming and associated sea ice retreat, southward shift in the oceanic fronts and strengthening of the ACC, and Agulhas leakage might have forced the AMOC resumption at the beginning of terminations. The re-arrangement of the THC was accompanied by the increase in the northward heat transport that promoted to the northern ice sheets melting.

Chapter 4
Influence of the Global Thermohaline Circulation on Paleoceanographic Events in the Eurasian Arctic Seas

Abstract At present, the Eurasian Arctic seas exhibit strong AMOC influence and, in turn, affect the NADW formation due to variability of freshwater outflow to the North Atlantic. The heat and salt exchange between the Atlantic and Arctic varied widely on glacial–interglacial, millennial, and shorter time scales. The postglacial history of the western Eurasian seas comprises three major phases. The proximal glaciomarine sediments (marine diamicton) of the early deglaciation (ca. 19–15 cal. ka BP.) are generally barren. However, recent data allow us to admit relatively persistent, although temporarily variable, Atlantic water penetration into the Arctic seas already since 17–16 cal. ka BP. The distal glaciomarine sediments of the late deglaciation (ca. 15–11.4 cal. ka BP.) contain diverse and abundant benthic microfossils only below the main Atlantic water pathways, in the southwestern Barents Sea, on the Svalbard shelf, and on the Laptev Sea slope. In the Barents Sea, paleoenvironments changed dramatically through the YD–Preboreal transition, with the development of high biological productivity and normal marine sedimentation since ~11.4 cal. ka BP. The transition was more gradual on the eastern Kara Sea shelf submerging during the Early Holocene transgression. The Holocene time series in the Arctic seas document several millennial and centennial-scale events related to the variability of the Atlantic water input and NAO, including the Holocene thermal maximum, cooling at 8.2 cal. ka, Medieval Warm Period, Little Ice Age, and recent warming.

Keywords Western Eurasian seas · Late Weichselian · Ice sheet · Deglaciation · River runoff · Brines · Atlantic water input · Foraminiferal assemblages · YD–Preboreal transition · Medieval Warm Period · Little Ice Age

4.1 Introduction

The understanding of the Late Quaternary paleoceanographic events in the Arctic Basin was significantly improved and supplemented in recent years as the influence of variations in the North Atlantic thermohaline circulation has been taken into account (Fig. 4.1; Hald et al. 1999; Knies and Stein 1998; Knies et al. 1999;

E.V. Ivanova, *The Global Thermohaline Paleocirculation*,
DOI 10.1007/978-90-481-2415-2_4,

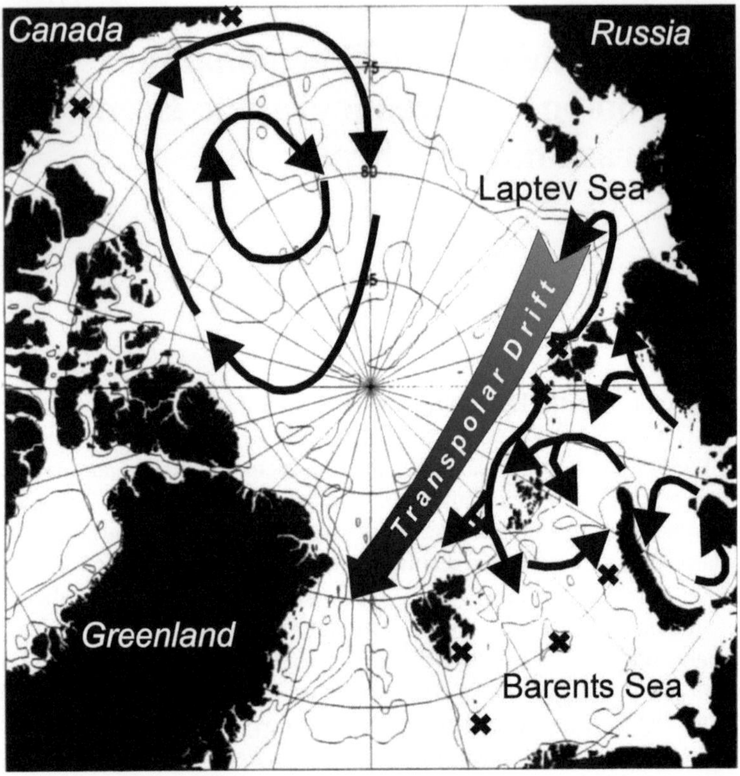

Fig. 4.1 Generalized surface water circulation in the Arctic Ocean. The *wide arrow* shows the main Arctic water outflow into the North Atlantic by the Transpolar Drift. *Crosses* indicating cascading sites in the Barents Sea are plotted after Ivanov and Shapiro (2002)

Lubinski et al. 1996, 2001; Polyak et al. 1997, 2002b; Polyak and Solheim 1994; Duplessy et al. 2001, 2005; Ivanova et al. 2002). Intervals of increased Atlantic water inflow along the continental margin of the Arctic Ocean were revealed (Knies et al. 1999; Boucsein et al. 2000). New materials obtained by Russian expeditions over the last few years, including those conducted by the Shirshov Institute of Oceanology RAS, allowed more detail reconstructing the linkage between paleoceanographic events in the Arctic seas and the global conveyor. The author took part in the 11th and 14th cruises of the Research Vessel (R.V.) *Akademik Sergey Vavilov* (1997, 1998), the 35th Cruise of the R.V. *Boris Petrov* (2000), and in the Russian–Norwegian expedition (Cruise 63 of the R.V. *Professor Shtokman*, 2004). More than 60 sediment cores have been collected during these cruises, together with high-frequency seismic records and hydrological data from the Barents and Kara seas (Ivanova 2004; Murdmaa et al. 2006; Stein and Stepanets 2001). This chapter concerns the paleoceanographic reconstructions for the last 10–17 ky, mainly based on the study of reference core sections from the above-listed expeditions (Color Plate 4.1, Fig. 4.2).

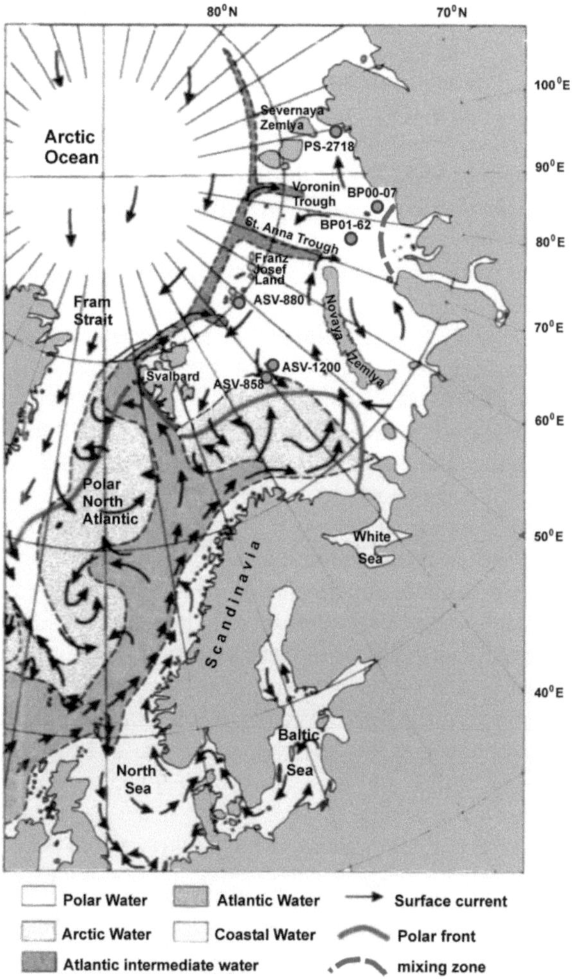

Fig. 4.2 Surface water circulation in Barents and Kara seas (after Mosby 1938) and location of the reference sediment cores studied by the author

4.2 An Overview of the Modern Arctic Hydrology

The upper part of the water column of the Arctic Ocean comprises Polar surface waters, which are separated from the underlying Atlantic-derived waters by the distinct halocline at a depth of 50–200 m (Aagaard et al. 1981). The halocline is formed by freshwaters, flowing into the surface layer from the river runoff and sea ice melting. The upper part of the surface water above the halocline has a salinity of 33.1 psu, temperature close to freezing point (–1.8°C), and a distinct peak in nutrient content. The lower part of the freshened layer is characterized by a salinity of about 34 psu, temperature close to freezing point, and a distinct minimum in nutrient content. According to the data available, the upper halocline water of the Arctic Ocean comes from the Chukchi Sea, while the lower halocline water arrives from the Barents and Kara seas (Jones et al. 1991). As a result, the surface layer accumulates a great amount of fresh waters, which eventually are transported by the Transpolar Drift into the East Greenland Current via the Fram Strait (Fig. 4.1). These waters are involved into mixing in the center of convective gyre of the Greenland and Labrador seas, and in such way they influence the unstable surface water salinity budget in this region. That is why variations in the freshwater input can significantly affect the intensity of deep-water convection (Aagaard et al. 1985; Aagaard and Carmack 1989), as well as the formation of deep water, and consequently influence the deep-water circulation in the World Ocean (Schlosser et al. 1994).

Salinity of the Arctic surface water depends on the river water inflow into vast Arctic shelves, where their amount exceeds the total annual discharge. Among Arctic shelf seas, the Kara Sea receives the greatest river runoff volume. Unlike other shelves, it is characterized by a continuous northward submeridional gradient of surface water salinity (Brezgunov et al. 1982; Burenkov and Vasilkov 1994; Stephantsev and Shmelkov 2001). In the Barents Sea, the freshening effect of river runoff is considerably less pronounced. In particular, this is supported by the heavier oxygen isotope values of the planktic foraminifera *Neogloboquadrina pachyderma* sin. in the water samples and surface sediments from the Arctic Ocean, northward of the Barents Sea continental slope as compared to the adjacent areas of the Arctic (Bauch et al. 1995; Spielhagen and Erlenkeuser 1994). Moreover, when moving with the Transpolar Drift to the west, into the Fram Strait, freshened waters from the Arctic seas mix with meltwater of sea ice which they carry out (Schlosser et al. 1994; Ivanov 2001). Atlantic waters penetrate through the Faeroe–Shetland and Faeroe–Iceland passages into the Norwegian Sea and then are divided into two branches, flowing further into the Barents Sea in the surface layer and around Spitsbergen, along the continental slope of Eurasia in the intermediate layer (Fig. 4.2, Color Plate 4.2). The core of Atlantic waters in the western Arctic lies at depth of 200–250 m and is characterized by temperature of about 2.5°C and salinity of 34.82–34.98 psu (Timofeev 1960; Ivanov 2001).

4.3 Modern Hydrology and Sedimentation Environments in the Barents and Kara Seas

In the Barents Sea, modern hydrological environment is defined by the interaction of Arctic and Atlantic waters, supplemented by the formation of cold saline waters (brines) during sea ice freezing on shoals and their accumulation in depressions owing to the cascading (Midttun 1985; Pfirman et al. 1994; Byshev et al. 2001a, b; Ivanov 2001; Ivanov and Shapiro 2002, 2005, Color Plate 4.1, Figs. 4.1 and 4.2). Atlantic waters flow into the Barents Sea from the north at intermediate depth through the Franz Victoria and St. Anna Troughs, and from the west through the Bear Island Trough as a warm surface current, representing the continuation of the North Cape Current (Novitskiy 1961; Loeng 1991; Pfirman et al. 1994) (Color Plate 4.1). According to the distribution of seasonal surface water temperatures (Matishov et al. 1998), these currents represent the main carriers of heat from the Atlantic. About 2 Sv of Atlantic waters inflow the Barents Sea that is from 1/3 to 1/2 of their total amount brought to the Arctic Ocean (Rudels et al. 1994).

The vertical structure of the water column in the northern Barents Sea troughs (Color Plate 4.2) is characterized by the succession of three different water masses. The surface water layer down to depth of <100 m is occupied by cold Arctic water with reduced salinity, coming into the trough from the northeast. Temperature and salinity values, measured in September in the surface layer at Station ASV-880, were −1.2°C and 32.5 psu (Color Plate 4.2). The surface water mass is subject to mixing during winter convection and is stratified into warm surface layer and cold subsurface layer in summer. The intermediate layer is composed of warmer Atlantic-derived waters with higher salinity (0.5–1.0°C, 34.8 psu) inflowing from the northwest. The bottom water (depths exceeding 200 m) represents the cold-water mass with increased salinity (<0°C, 34.9 psu) generated as a result of brines formation during sea ice freezing, their downslope descending from shallow depths, and mixing with the Atlantic water by the way. Above the trough flanks, the intermediate Atlantic water borders with Arctic waters demonstrating pronounced lateral temperature gradients at this boundary. Atlantic water flow from the north does not penetrate into the Central Deep owing to the sill that separates the Franz-Victoria Trough from southern depressions (Pfirman et al. 1994). In situ CTD records obtained in September 1998 and in July 2004 at Stations ASV-1200 and PSh-5147, respectively, support this suggestion (Color Plate 4.2).

Since the branch of the North Cape Current enters the Barents Sea from the west and crosses it in a northeastern direction, the central and eastern parts of the sea are receiving already strongly modified Atlantic surface waters (Pfirman et al. 1994; Byshev et al. 2001a, b; Ivanov 2001). Average winter multiannual pack ice margin runs close to Stations ASV-858 and ASV-1200, while summer margin is located slightly to the north from Station ASV-880. Average summer surface water temperature at Station ASV-880 is close to 0°C, and at stations ASV-858 and ASV-1200 it is about +4°C (Matishov et al. 1998). The latter two stations, most likely, are located in the zone of seasonal migrations of the Polar Front.

Part of the Atlantic waters flows from the northwest into the Kara Sea through the St. Anna Trough, to the east of Franz Josef Land (Tantsyura 1959; Pfirman et al. 1994; Ivanov 2001). However, due to intensive heat release during mixing, the heat storage of these waters in the Severnaya Zemlya area comprises only 40% of its initial value (Treshnikov and Baranov 1972; Nikiforov and Shpaykher 1980). Surface water circulation in the Kara Sea is cyclonic. Arctic waters move from the north to the south along the coast of Novaya Zemlya Archipelago, and then turn to the east, mix with Ob' and Yenisei river waters and inflow into the Arctic Ocean and into the Laptev Sea (Burenkov and Vasilkov 1994). Extremely strong river discharge, about 1,100 km^3(Treshnikov 1985), results in the strong freshening of surface and even bottom waters in the southern part of the sea. Vertical structure of the water column is characterized by distinct summer picnocline, separating relatively warm surface river waters (>0°C, <25 psu) from cold and saline (<0°C, >25 psu) bottom waters (Pivovarov et al. 2003). Bottom water temperature usually does not exceed

Fig. 4.3 Bottom-water salinity in the southeastern Kara Sea (after Simstich 2001) and location of sediment cores studied by the author

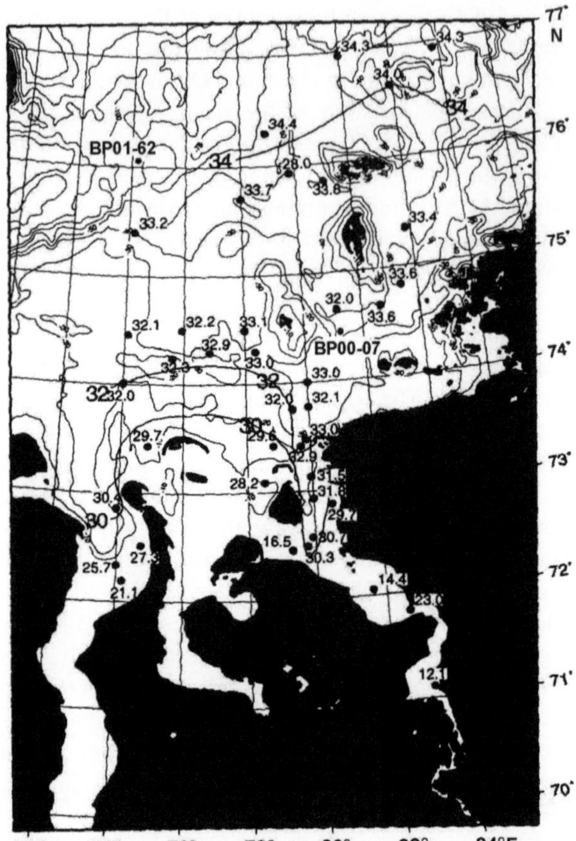

0°C except for shallow-water areas close to estuaries (Pavlov et al. 1996). Salinity of bottom waters in the Kara Sea ranges from 14.4 psu in the Yenisei estuary in September and up to 34.3 psu in the northern part of the sea (Stephantsev and Shmelkov 2001; Fig. 4.3). These salinity values result from the strongly modified Atlantic water influence, reaching the southern part of the sea along the submerged paleoriver deltas (Dmitrenko et al. 2001; Harms et al. 2003). Heavy saline waters sink during ice formation in polynias in winter also contributing to the salinity of bottom waters (Pavlov and Pfirman 1995; Bauch et al. 2003).

Temperature and salinity of bottom waters demonstrate strong seasonal and synoptical variability, especially in the Yenisei estuary where the halocline reaches sea bottom in the beginning of autumn (Dobrovol'skiy and Zalogin 1982; Stephantsev and Shmelkov 2001). Hydrological and geochemical investigations show that the vast area of marine and riverine waters mixing is located at the latitude of 74°N (Burenkov and Vasilkov 1994; Burenkov et al. 1995; Lisitzin et al. 1995, 2000; Stephantsev and Shmelkov 2001; Sukhoruk et al. 2001; Churun and Ivanov 1998).

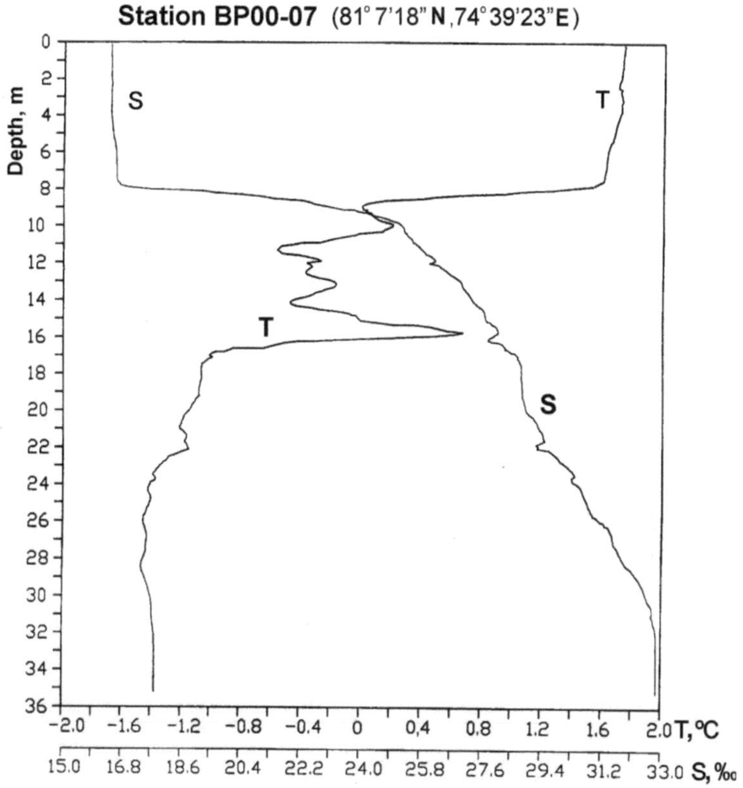

Fig. 4.4 Vertical temperature and salinity records for September at Station BP00-07 (74°39.23'N, 81°07.18'E) measured during Cruise 35 of the R.V. *Boris Petrov* (from Stephantsev and Shmelkov 2001, courtesy by L. Stephantsev)

Core BP00-07, studied by the author, was retrieved near this area. Salinity and temperature measured at this site in September 2000 are shown in Fig. 4.4 (Stephantsev and Shmelkov 2001).

Bioproductivity of the Barents Sea is to a high degree controlled by the seasonality of the development of phyto- and zooplankton at the end of summer – beginning of autumn, during day light hours, and maximum retreat of pack ice into the Arctic Ocean. This fact was already established by the pioneer hydrobiological investigations in the Barents Sea (Zenkevich 1963). During the 11th Cruise of the R.V. *Akademik Sergey Vavilov,* the phytoplankton bloom was observed in early September at the sea ice margin, close to the Station ASV-880 (Vedernikov and Gagarin 1998). The effect of the sea ice marginal-melting zone on phytoplankton development has been thoroughly studied from different areas of the Arctic and North Atlantic (e.g., Carstens et al. 1997). Presence of Atlantic waters below the pycnocline amplifies the phytoplankton bloom not only at the sea surface but also in the subsurface waters (Matishov et al. 1999). The chlorophyll content

0.20–0.15 mgC m^{-2}day^{-1}measured in the surface layer near stations ASV-880 and ASV-860 and characteristic for mesotrophic areas is due to the presence of Atlantic waters (Vedernikov and Gagarin 1998). The highest productivity is observed in the southwestern part of the sea, especially in near-shore waters off the Kola Peninsula, and within a small area to the south of the Franz Josef Land (Vinogradov et al. 2000; Kopelevich et al. 2002), where core ASV-880 is retrieved. During spring, particularly in April, productivity in the ice-free waters in the central and eastern parts of the sea significantly increases (up to 200–500 mgC m^{-2} $^{-}$day^{-1}and higher).

Regional differences in hydrological structure of the water column within the Barents Sea are defined by the different taxonomic composition of foraminiferal assemblages from the surface sediment layer and from the Holocene deposits as a whole (Color Plate 4.2). Routes of the subsurface and intermediate Atlantic waters inflow can be traced by the presence of cold-water planktic species *N. pachyderma* sinistral, which is mainly restricted to this water mass (Polyak and Solheim 1994; Lubinski et al. 2001). The so-called Atlantic benthic foraminiferal species: *Cassidulina teretis, Pullenia bulloides, Trifarina angulosa,* and some others also indicate the Atlantic water inflow (Fig. 4.5, Korsun et al. 1994; Polyak and

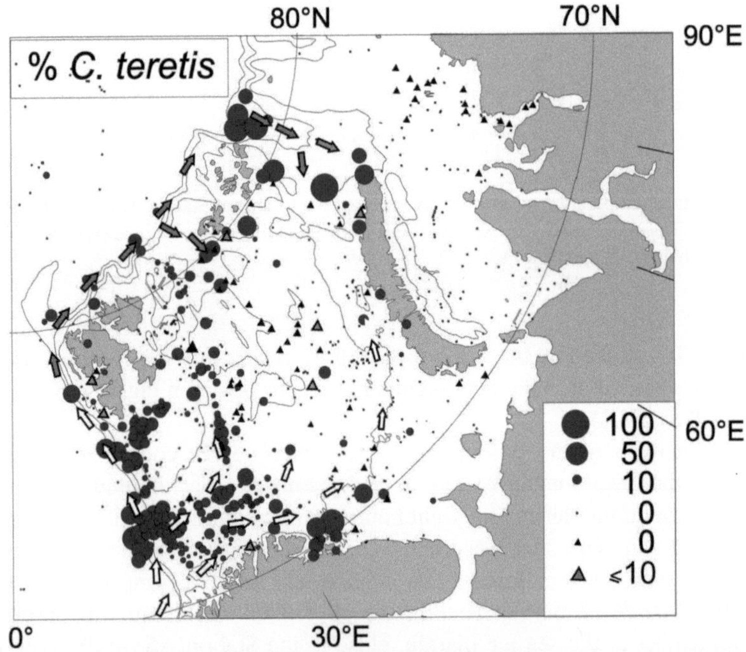

Fig. 4.5 Distribution of the "atlantic" foraminiferal species *Cassidulina teretis* in the core tops from the Barents Sea (after Lubinski et al. 2001 with additions). The data from Lubinski et al. (2001) are marked by *circles*, the data by E. Ivanova and E. Ovsepyan (unpublished) are shown by *triangles*

Solheim 1994; Hald and Steinsund 1996; Pogodina and Tarasov 2002). Both groups of species are rarely found in the eastern part of the sea, where Atlantic waters practically do not reach. Foraminiferal assemblages of the southwestern Barents Sea, where Atlantic waters are distributed throughout the water column, are distinguished by the predominance of boreal planktic species *Turborotalita quinqueloba* and *N. pachyderma* dextral over the polar species. They are also characterized by the low percentage of typical Arctic shelf benthic species (*Elphidium excavatum* forma *clavatum, Cassidulina reniforme* and others), as it is shown by our data obtained during the Cruise 63 of R.V. *Professor Shtokman.*

The study of suspended matter distribution in the water column of the Barents Sea (Aibulatov et al. 1999; Lisitzin 2001) revealed surface and bottom layers of increased turbidity (nepheloid layers). The leading role of the bottom nepheloid layer in the lateral transport of fine-grained sediments from shoals into depressions was established. The terrigenous material is supplied into the Barents Sea both from nearby (surrounding archipelagoes) and distant (Eurasian rivers runoff, ice rafting in the Arctic Ocean, remote aeolian dust) sources. A lot of fine-grained (pelitic) material is carried into the sea by glaciers meltwater as "glacial milk." The supply of "glacial milk" by glaciers of the Franz Josef Land likely plays an important role in the sedimentary material budget of the Franz Victoria Trough. Grain-size fractions coarser than 0.05 mm in depressions of the open sea mainly represent a result of sea ice and iceberg rafting (Murdmaa et al. 2001, 2006).

Sediments in the Kara Sea are predominantly fine grained (Müller and Stein 1999) with an organic matter content between 0.5 and 2%, which increases in the Ob' and Yenisei estuaries (Vinogradov et al. 1994; Vetrov and Romankevich 2003). However, a major part of the organic matter cannot be consumed by benthic organisms, in particular by foraminifers, since rivers discharge a large amount of old organic matter (Boucsein et al. 2000). Oxygen content in the near-bottom water layer and surface sediments is higher in the northern part of the sea (Galimov et al. 1996). Sedimentation processes and bioproductivity in the southern part of the sea is mainly controlled by Ob' and Yenisei Rivers runoff. These rivers provide a supply of about 22×10^6tons of suspended matter and 8×10^6tons of dissolved and suspended organic matter (Gordeev et al. 1996). A major proportion of dissolved and suspended matter accumulates in front of estuaries, and does not reach further north than the mixing zone of marine and riverine waters (Fig. 4.6; Lisitzin et al. 1995, 2000). High content of nutrients and labile organic matter in the river waters provide high bioproductivity in the southern part of the sea, reaching 300 mgC $m^{-2}day^{-1}$. In the open Kara Sea, to the north of the mixing zone, bioproductivity decreases down to <50 mgC/m^2/day (Vedernikov et al. 1993). However, excess suspension content in the river runoff may cause bioproductivity decrease in estuaries and a shift of the maximum productivity zone to the north (Vedernikov et al. 1993). In the northern part of the sea, increased primary production is linked to the sea ice margin in spring and summer. The same is true for the Barents Sea and for the northern North Atlantic (Carstens et al. 1997; Matishov 1995).

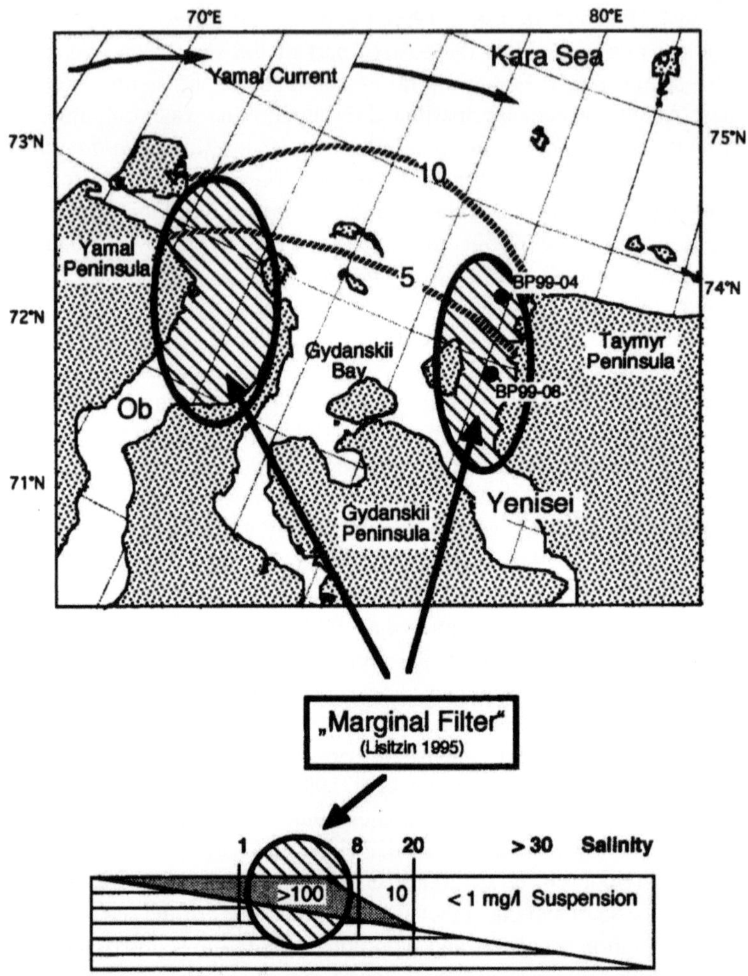

Fig. 4.6 Scheme of the marginal filter off the Ob and Yenisei river estuaries, where 90–95% of the suspended matter discharged by the rivers settles (after Lisitzin 1995; Stein 2001). Isohalines 5‰, 10‰ for surface water and location of coring stations by the R.V. *Boris Petrov* are shown in the map above

4.4 Stratigraphy and Correlation of Postglacial Sediments

Stratigraphic units which we distinguished together with I.O. Murdmaa in postglacial sediments from the Barents Sea (Ivanova 1999, 2002a, 2003, 2004; Murdmaa and Ivanova 1999; Ivanova and Murdmaa 2001; Ivanova et al. 2002; Murdmaa et al. 2006) briefly characterized below are supplemented with micropaleontological data from several sections studied in detail (Fig. 4.7; Color Figs. 4.1 and 4.3; Tables 4.1

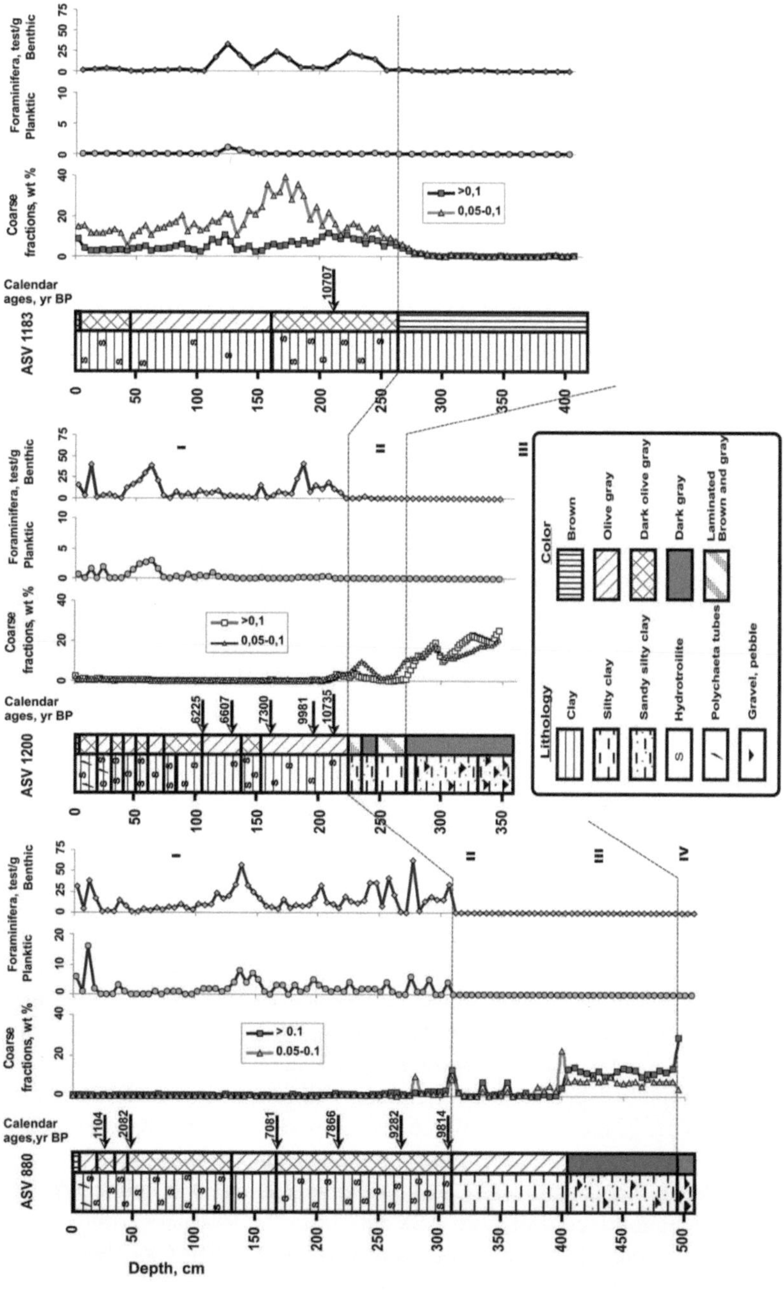

Fig. 4.7 Correlation of stratigraphic units in the sediments cores retrieved along the longitudinal transect in the Barents Sea (for location see Color Plate 4.1). Proxy records: content of coarse grain size fractions, wt %; planktic and benthic foraminifera (tests/g of dry sediment). Stratigraphic units: (I) Holocene; (II) Late deglaciation phase; (III) Early deglaciation phase; (IV) Last (Late Weichselian) glaciation (modified from Ivanova et al. 2002).

Table 4.1 Location, water depth, core recovery, and thickness of units of sediment cores from the Barents and Kara seas used in this chapter

	Coordinates				Thickness of units, m		
Core	N	E	Water depth, m	Core length, m	I	II	III
ASV-880	79°55.5′	47°08.2′	388	5.08	3.10	0.94	0.90
ASV-1200	75°54.4′	41°00.5′	308	3.60	2.25	0.47	0.88
ASV-858	75°50.6′	39°54.8′	312	3.56	2.07	0.49	1.00
ASV-1183	71°28.3′	40°48.3′	330	4.17	2.63	1.54	–
ASV-987	76°12.3’	62°29.2’	170	5.95	5.95	–	–
ASV-1310	75°06.1′	53°21.7′	228	4.20	3.89	0.31	–
PSh-5147	76°4.5′	56°8.32′	314	2.53	0.42	2.11	–
PSh-5157	78°55.8	41°53.3	461	2.20	2.20	–	–
PSh-5159R	71°22.77′	22°49,50′	414	500	137	363	–
PS-2718-6	77°31.5°	97°04.4°	153	7.80	7.80	–	–
BP00-07/5	74 °65.8′	81 °14.1′	42.8	6.32	6.32	–	–
BP01-62/6	76 °20.1	74 °20.3	119.8	5.80	5.80	–	–

Note: (–) unit is not recovered.

and 4.2). Three core sections from the Kara Sea studied by the author recover only Younger Dryas and Holocene sediments, as confirmed by radiocarbon datings (Table 4.3). Data on the interstitial water salinity determined by N.V. Pimenov and G.A. Pavlova are taken from Murdmaa et al. (1998a). Stable isotope (delta ^{18}O, ^{13}C) measurements and AMS-^{14}C datings are performed in the *Laboratoire des Sciences du Climat et de l'Environnement*, CNRS-CEA (France), Leibniz Laboratory for Age Determination and Isotope Research at Kiel University (Germany), Poznań Radiocarbon Laboratory (Poland), laboratories of the Woods Hole Oceanographic Institution, Arizona University, and Geochron (USA) (Levitan et al. 2000; Duplessy et al. 2001; Ivanova et al. 2002; Simstich et al. 2004; Polyak et al. 2004; Ivanova et al. 2008b; Ivanova and Risebrobakken, unpublished; Simsthich and Spielhagen, unpublished). AMS ^{14}C dates for cores PS-2718 (Levitan et al. 2000), BP01-62/6 (Simsthich and Spielhagen, unpublished), ASV-1310, and ASV-858 (Ivanova and Murdmaa, upublished) were corrected by myself for a reservoir effect of 440 years (Mangerud and Gulliksen 1975) and converted to calendar years BP applying the program CALIB 5.0.1 (Stuiver and Reimer 1993) and using the marine calibration dataset (Hughen et al. 2004). In each cores, age time scales (models) were obtained by linear interpolation between the dated tie points and assuming a modern age for the sediment surface. The calendar calibrations for several other cores were performed earlier (Duplessy et al. 2001, 2005; Ivanova et al. 2002; Simstich et al. 2004; Tables 4.2 and 4.3) using the CALIB 4.3 program (Stuiver and Reimer 1993). For the Last Millennium record of Core ASV-987 from the Russian Gavan' Fjord (Polyak et al. 2004, Table 4.4) calendar ages were also calculated by CALIB 4.3 program with the local reservoir correction $\Delta R = 85$ years previously established by S. Forman and L. Polyak (1997). Calendar ages for the recently dated cores PSh-5151 from Isfjorden, Svalbard, and PSh-5159R from the southwestern Barents Sea (were

Table 4.2 AMS radiocarbon datings and calendar year calibrations in the Barents Sea cores

Lab Code	Depth in core (cm)	^{14}C age (yr. BP)	Error 1 sigma (yr.)	Mean cal. age (yr. BP)	Material dated
ASV-880 (Duplessy et al. 2001)*					
Gif 99457	25	1550	60	1104	*Yoldiella lucida*
Gif 99385	46	2440	60	2082	*Yoldiella lenticula*
Gif 99386	166	6560	80	7081	*Yoldiella fraterna*
Gif 99458	216	7420	80	7866	*Yoldiella*spp.
Gif 99387	266	8830	90	9282	*Yoldiella fraterna*
Gif 99388	305	9150	90	9814	*Yoldiella fraterna*
ASV-1200 (Ivanova et al. 2002)*					
Gif 99802	110	5810	70	6225	*Yoldiella* sp.
Gif 99803	120	6180	60	6607	mixed benthic forams
Gif 99804	165	6770	70	7300	*Yoldiella fraterna*
Gif 99805	195	9220	80	9981	*Yoldiella lucida*
Gif 99806	205	9450	80	10082	*Yoldiella* sp.
Gif 99807	210	9550	80	10331	*Yoldiella intermedia*
ASV-1183 (Ivanova et al. 2002)*					
Gif 100083	218	9820	100	10707	*Yoldia* sp.
ASV-858 (Ivanova and Murdmaa, unpublished)**					
Poz-20412	215	9380	50	10045	Bivalves
PSh-5151 (Ivanova et al. 2008b, Ivanova and Risebrobakken unpublished)****					
Poz-20251	22 – 24	830	35	370	*Yoldiella lenticula*
Poz-16593	92 – 94	1805	35	1320	*Yoldiella*sp.
ASV-1310 (Ivanova and Murdmaa, unpublished)**					
Poz-20413	143	8670	50	9280	Bivalves
Poz-20414	241	9220	50	9970	Bivalves
Poz-20415	329	9890	50	10765	Bivalves
PSh-5159R (Ivanova et al. 2008b, Ivanova and Risebrobakken unpublished)***					
Poz-20399	12 – 14	635	30	220	*Lenticulina* sp.
Poz-19991	117	10010	50	10905	*Thyasira dunbari* +*Yoldiella* aff. *intermedia*
Poz-16594	240 – 242	12150	70	13526	Bulk benthic foraminifera
Poz-19992	332 – 334	13550	70	15437	Bulk benthic foraminifera

Notes: Mollusk species were identified by E.M. Krylova and I.O. Murdmaa. Radiocarbon ages were converted to calendar ages with the program (*) – CALIB 4.3, (**) – CALIB 5.0,1, (***) – CALIB 5.0.2 with $\Delta R = 105 \pm 24$, (****) – CALIB 5.0.2 with $\Delta R = 71 \pm 21$. Lab. code: Gif – CFR now LSCE CNRS-CEA, Gif-sur-Yvette (France), Poz – Poznań Radiocarbon Laboratory (Poland). For core location see Table 4.1, Fig. 4.2 and Color Plate 4.1, for details and references see text (Section 4.4).

calculated by B. Risebrobakken (Table 4.2) applying CALIB 5.0.2 program with $\Delta R = 105 \pm 24$ and 71 ± 21, respectively according to Mangerud and co-authors (2006).

Bottom moraine deposits (glacial till) of the last (Late Weichselian) glaciation of the North Hemisphere are only recovered in the core catcher sample of Core

Table 4.3 AMS radiocarbon datings and calendar year calibrations in the Kara Sea cores

Lab Code	Depth in core (cm)	^{14}C age (yr. BP)	Error 1 sigma (yr.)	Mean cal. age (yr. BP)	Material dated
BP00-07/5 (Simstich et al. 2004*)					
KIA 16376	35	1115	25	640	*Portlandia arctica*
KIA 16377	155	3130	30	2840	*Portlandia arctica*
KIA 16378	245	3965	30	3890	*Portlandia arctica*
KIA 16379	365	5300	30	5610	*Portlandia arctica*
KIA 16380	495	6510	35	6950	*Portlandia arctica*
KIA 16381	625	7605	45	8000	*Portlandia arctica*
BP01-62/6 (Simstich and Spielhagen, unpublished**)					
KIA 18169	20	1225	20	720	*Portlandia arctica*
KIA 18170	120	4405	25	4510	*Portlandia arctica*
KIA 18171	210	7020	30	7480	*Leionucula bellotii*
KIA 18172	340	9065	40	9710	*Leionucula bellotii*
KIA 18173	540	10250	40	11200	*Hiatella arctica*
PS 2718/6 (Levitan et al. 2000**)					
GifA 98227	120	3510	60	3350	*Bivalves*
GifA 98228	250	5680	70	6040	*Bivalves*
GifA 98240	400	7340	80	7470	*Bivalves*
GifA 98234	750	10450	90	11540	Bivalves

Notes: Radiocarbon ages were converted to calendar ages with the program (*) – CALIB 4.3, (**) – CALIB 5.0.1. Lab. code: Gif – CFR now LSCE CNRS-CEA, Gif-sur-Yvette (France), KIA – Leibniz Laboratory at Kiel University (Germany). For core location see Table 4.1 and Fig. 4.2, for details and references see text (Section 4.3). R. Spielhagen and J. Simstich are kindly acknowledged for providing the unpublished AMS-^{14}C dates from core BP01-62/6.

Table 4.4 AMS radiocarbon datings and calendar year calibrations in core ASV-987 from Russian Gavan' Fjord, Barents Sea(Polyak et al. 2004)

Lab code	Depth in core (cm)	Material dated	^{14}C age BP	Calibrated age AD (1σ)	Median probability age AD
AA34305	41–44	*Thyasira*sp., *Yoldiella*sp.	post-bomb		1950
AA35039	175–180	*Nuculana tenuis*	795±40	1538–1651	1592
AA39567	304–306	*Thyasira*sp., shell detritus	919±40	1440–1527	1488
AA35038	399–401	*Macoma*sp.	990±45	1394–1485	1434
AA39568	474–476	*Macoma*sp.	1068±40	1333–1420	1377
GX-24917	559–561	*Macoma*sp.	1260±40	1184–1291	1229
AA39569	584–586	shell detritus	1301±40	1133–1263	1190

Notes: Radiocarbon ages were converted to calendar ages by the program CALIB 4.3 with ΔR = 85 after (Forman and Polyak 1997). Lab. code AA – Arizona AMS Facility; GX – Geochron Laboratories. For core location see Table 4.1 and Fig. 4.18.

ASV-880 (Fig. 4.7; Color Plate 4.3) from the Franz Victoria Trough. Here, they consist of a very dense (water content 17%) dark gray unsorted pebble loam with black shale fragments up to 12 cm in diameter and with a low-salinity interstitial water (13‰). We found in the sediment several tests of the polar planktic foraminiferal species *N. pachyderma* sin. possibly contaminated from above. According to existing ideas about timing of the last glaciation (e.g., Martinson et al. 1987), we assume that the age of the till top may be about 18–17 cal. ka BP.

Dark gray sandy mud with pebble and gravel admixture overlies the glacial till in Core ASV-880 (Unit III in Fig. 4.7; Color Plate 4.3) and occurs as a basal layer in many other sections. We interpret this diamicton unit as a proximal glaciomarine facies of the initial deglaciation phase. The relationship between grain-size fractions suggests a superposition of two independent sedimentation mechanisms: iceberg rafting of coarse debris (IRD) and settling of an extremely fine-grained suspended material likely delivered by glacier meltwater ("glacier milk") under very calm, mainly sub-ice hydrodynamic conditions. The interstitial water salinity ranges within 28–32 psu that is much higher than that of the underlying till. Rather rare microfossils are represented by both extant and reworked foraminifera of various preservation states. The reworked Mesozoic–Cenozoic benthic foraminiferal tests range from "fresh" to recrystallized ones (as determined by L.F. Kopaevich and L.V. Polyak). Very rare *N. pachyderma* sin., bivalve shells, and polar pteropoda species *Limacina helicina* are found. Recent benthic foraminifera are represented by *Elphidium* spp., *C. reniforme, Islandiella norcrossi, Haynesina orbiculare, Cibicides lobatulus, Nonion labradoricum,* and some other species. This unit is not dated because of the scarcity of carbonate microfossils. So far as it lies between the till of last glaciation and dated deposits of the late deglaciation phase, we assume that its age is approximately 18–15 cal. ka BP.

The Unit III is overlain by Unit II corresponding to the late deglaciation phase from Bølling to Younger Dryas. It consists of visually homogeneous greenish gray clay with several thin sandy interbeds in the Franz Victoria Trough (ASV-880) and laminated silty clay with a low sand content in many other sections, especially in the Persey Trough (Core ASV-1200, Fig. 4.7). The salinity of interstitial water varies within 31–34‰. Benthic foraminifera in the Unit II are represented by the same species, as in the Unit III. Besides, rare rather well-preserved calcareous benthic and planktic foraminiferal tests of species characteristic for the North Atlantic and Norwegian Sea are also found in the Franz Victoria Trough (Core ASV-880) including: *P. bulloides, Cibicides wuellerstorfi, C. teretis, Melonis barleeanus, N. pachyderma* dex., and *T. quinqueloba.* Interbeds with increased foraminifera content occur in Unit II of cores ASV-1183, ASV-1200, and PS-5147 from the Central Deep and Persey Trough. *C. reniforme* and *E. clavatum* predominate in these interbeds. Reworked cretaceous radiolarians and foraminifera (identified by L.V. Polyak) are found in some samples, for example, in Core PS-5147. Unit II is dated by AMS-^{14}C in several sections from the Barents Sea (Polyak and Solheim 1994; Polyak et al. 1997; Hald et al. 1999; Lubinski et al. 1996; 2001). Its age estimates from 15–14 to 11.4 cal. ka. In Core 1183, the lighter values of $\delta^{18}O$ and the most negative values of $\delta^{13}C$ are measured in the upper part of the unit (Fig. 4.8).

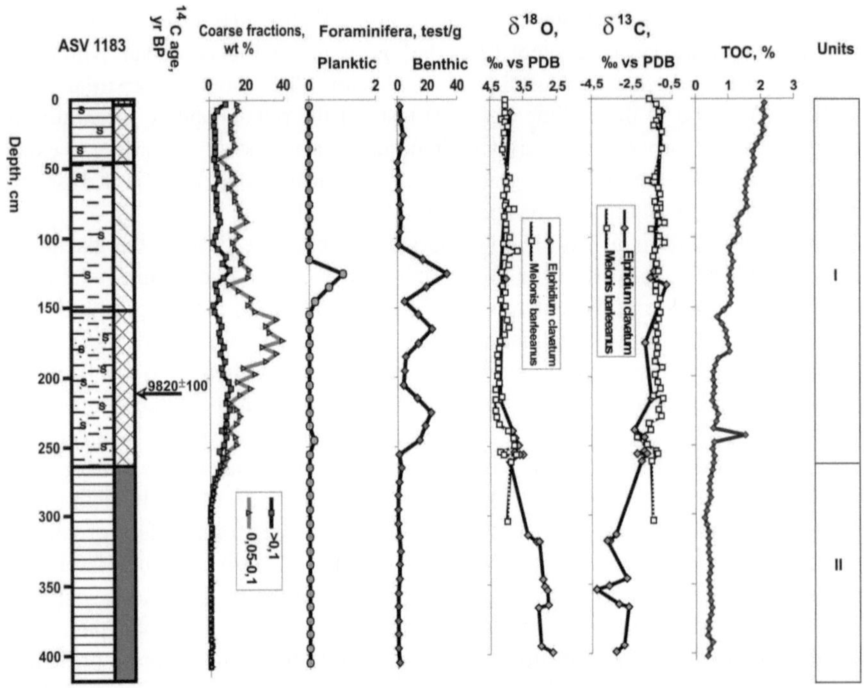

Fig. 4.8 Core ASV-1183 from the southern Barents Sea (for location see Color Plate 4.1, for lithological symbols see Fig. 4.7). Proxy records: content of coarse grain-size fractions, wt.%; planktic and benthic foraminifera (tests/g of dry sediment), variations of $\delta^{18}O$ and $\delta^{13}C$ on benthic species *E. clavatum* and *M. barleeanus*, and TOC content. Stratigraphic units: (I) Holocene; (II) Late deglaciation phase (modified from Murdmaa et al. 2006). For legend see Fig. 4.7

In the Ingøydjupet Depression, southwest Barents Sea, the AMS-^{14}C-dated sediments of Unit II (Table 4.2) are represented by silty clay with thin fine-grained sand laminae and lenses. According to our AMS-^{14}C dates (Table 4.2; Fig. 4.9), prior to Bølling foraminiferal assemblages are strongly dominated by the typical Arctic species *Elphidium clavatum* and *C. reniforme* whereas the Bølling–Allerød and Younger Dryas intervals contain rather diverse fauna with very abundant "atlantic species" (Fig. 4.9; Color Plate 4.7; Chistyakova et al., in press). The abundance of arctic species increases by the end of the Younger Dryas.

The Holocene Unit I (Figs. 4.7, 4.8; Color Plate 4.3) is composed of olive gray clay with iron sulfide (hydrotroilite) and commonly abundant *Polychaeta* tubes. The lower part of the unit from the Central Deep (Core ASV 1183, Fig. 4.8) is enriched in very fine sand. The interstitial water salinity varies within the limits measured in present bottom waters (32–34‰). Unit I contains abundant and diverse benthic foraminiferal assemblages, as well as rare pteropoda *Limacina helecina*, bivalve, gastropoda, ostracoda shells, and sponge spicules. Calcareous benthic foraminifera strongly predominate over planktic foraminifera throughout the unit. They also prevail over agglutinated benthic foraminiferal tests, which are rapidly destroyed during early diagenesis (Alve and Murray 1995; Hald et al. 1999; Hald and Korsun

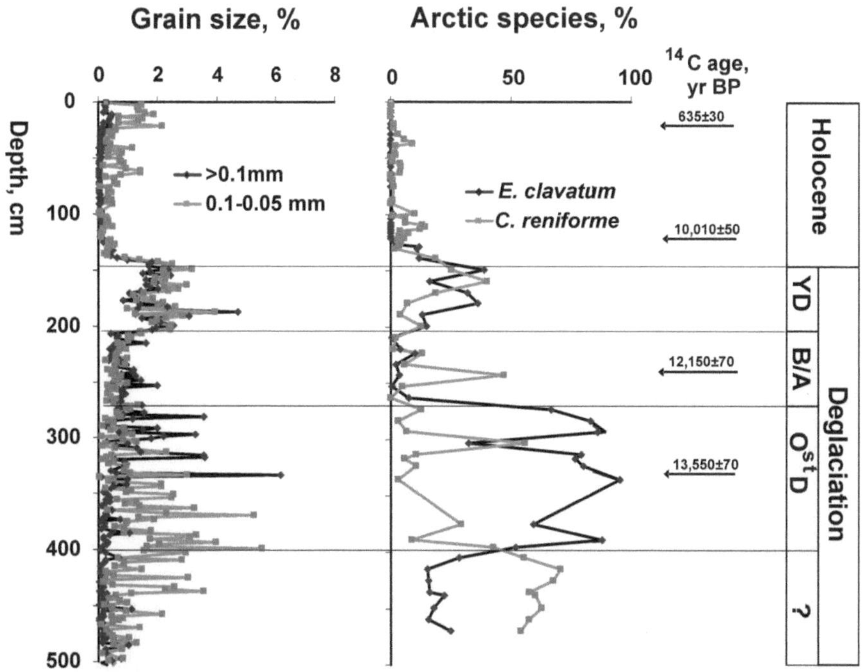

Fig. 4.9 Chronostratigraphy, distribution of two coarse fractions, and dominant arctic species versus depth in Core PSh-5159 N/R from the Ingøydjupet Depression (for location see Color Plate 4.1, for calibration of AMS-^{14}C ages see Table 4.2). Data from (Chistyakova et al., in press)

1997; Polyak and Solheim 1994; Polyak et al. 2002a). The preservation state of planktic and benthic foraminifera varies even within a single sample. Foraminiferal assemblages of the Barents Sea are known to be strongly affected by dissolution, and foraminifera are totally absent in many samples or even within entire core intervals (Polyak and Mikhailov 1996; Polyak and Solheim 1994; Steinsund and Hald 1994).

The polar species *N. pachyderma* sin. dominates among the planktic foraminifera throughout the Unit I in the central and northeastern Barents Sea, whereas only rare tests of boreal species *N. pachyderma* dex. and *T. quinqueloba* occur here. Unlike this, foraminiferal assemblages from the southeastern Barents Sea and the western Spitsbergen margin are dominated by boreal species of planktic foraminifera and "atlantic" species of benthic foraminifera, especially in Core PSh-5159 (Color Plate 4.7). Numerous radiocarbon datings confirm that the age of Unit I does not exceed 10 ^{14}C ka, that is 11.4 cal. ka (Fig. 4.7, Table 4.2; Polyak and Solheim 1994; Polyak et al. 1997; Hald et al. 1999; Lubinski et al. 1996, 2001; Ivanova et al. 2002).

The Holocene section from the Franz Victoria Trough is distinctly subdivided into three intervals according to changes in benthic foraminiferal assemblages (Appendix, Table 1, Fig. 4.10, Color Plate 4.7) dated by AMS-^{14}C (Duplessy et al. 2001). *C. reniforme* dominates in the lower interval (310–160 cm, 9.8–6.9 cal. ka BP); close percentages of *C. reniforme, Buccella* spp., and *E. clavatum* are noted in

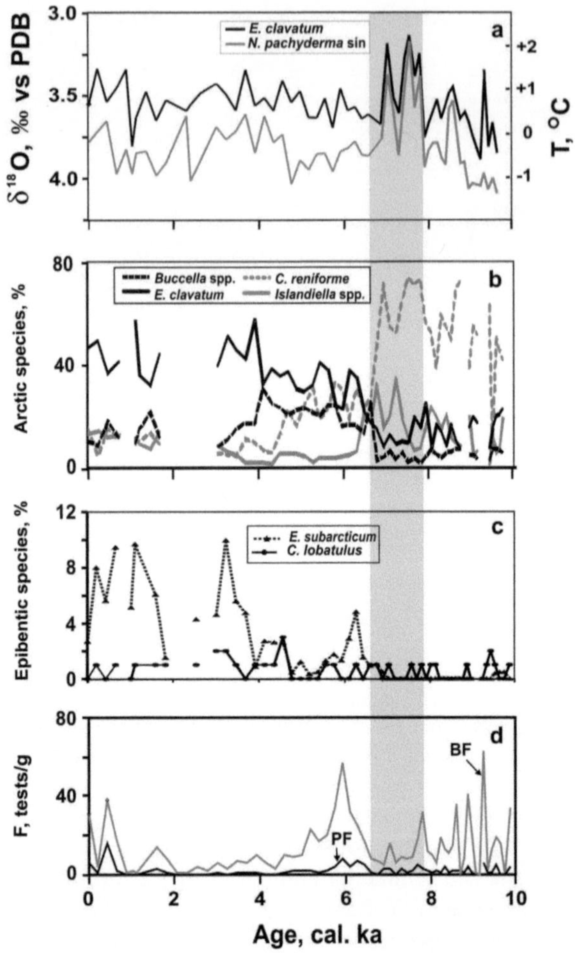

Fig. 4.10 Core ASV-880 from the Franz Victoria Trough (for location see Color Plate 4.1): (**a**) local variations in $\delta^{18}O$ for the planktic species *N. pachyderma* sin.and benthic species *E. clavatum,* with subsurface and bottom waters temperature records calculated from the isotope data (Duplessy et al. 2001); (**b**) percentage of common benthic foraminiferal species in the Holocene interval of the section; (**c**) percentage of epibenthic species; (**d**) planktic and benthic foraminiferal abundance (tests/g of dry sediment). The gaps in the species records result from foraminiferal dissolution. The Holocene thermal optimum is marked by gray band at ~7.8–6.9 cal. ka BP, when the trough was filled with Atlantic water

the middle interval whereas the upper interval is strongly dominated by *E. clavatum* well adapted to harsh environments.

A distinct maximum of the "atlantic" species *C. teretis* is fixed at the lower boundary of the Holocene in some sections from the northern Franz Victoria Trough (Lubinski et al. 1996, 2001; Murdmaa et al. 2005), and in the upper part of the Unit I in Core PSh-5155 from the southwestern Franz Victoria Trough (Akhrimenko and Ivanova 2005). However, a low foraminiferal content in most cores from the northern Barents Sea, for example, in Core PSh-5157 from the southern part of the trough, does not allow us to subdivide the Holocene unit into subunits according to changes of dominating species.

The same three intervals are distinguished in the Holocene section from the Persey Trough (cores ASV-858 and ASV-1200), and rather well dated in Core ASV-1200 (Ivanova et al. 2002; Table 4.2). *E. clavatum* and *C. reniforme* dominate in the Lower Holocene sediments from the Core ASV-1200 (220–135 cm, 10.7–6.9 cal.

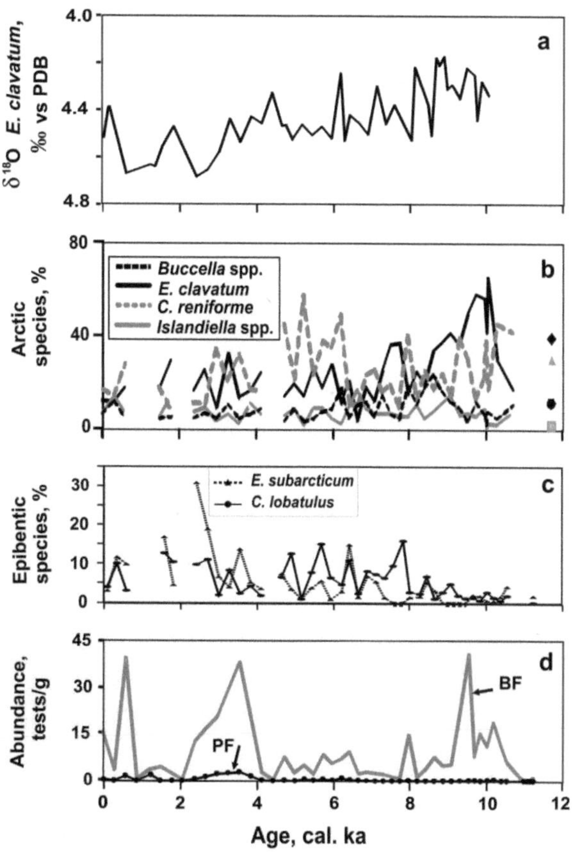

Fig. 4.11 Core ASV-1200 from the Persey Trough (for location see Color Plate 4.1): (**a**) local variations in $\delta^{18}O$ for the benthic species *E. clavatum* (Duplessy et al. 2005); (**b**) percentage of common benthic foraminiferal species in the Holocene interval of the section; (**c**) percentage of epibenthic species; (**d**) planktic and benthic foraminiferal abundance (tests/g of dry sediment). The gaps in the species records result from foraminiferal dissolution

ka BP; Figs. 4.11 and 4.12; Appendix, Tables 2 and 3). Foraminiferal assemblages from the middle part of Unit I (135–75 cm, 6.9–4.1 cal. ka BP) are dominated by *C. reniforme* (in Core ASV-1200), or contain close percentages of *E. clavatum, C. reniforme, I. norcrossi/helenae, Buccella* spp.*, M. barleeanus, C. lobatulus, N. labradoricum,* and *E. subarcticum* (in Core ASV-858). Several intervals of strong foraminiferal dissolution and a slight domination of *E. clavatum* и *C. reniforme* over other species are noted in the upper part of the unit in both cores (75–0 cm, 4.1–0 cal. ka BP). The Holocene unit is less than 50-cm thick in Core 5147 from the western Persey Trough. *E. clavatum, C. reniforme, I. Norcrossi,* and *Buccella* spp. represent the most abundant species in foraminiferal assemblages at this site, whereas "atlantic" species are scarce.

Even a more intense dissolution of foraminiferal tests is noted in Core ASV-1183 from the southern margin of the Central Deep. Because of this, we could not subdivide the Holocene section into intervals. Moreover, only one radiocarbon date is obtained for this core (Table 4.2, Ivanova et al. 2002) owing to a very low content of both foraminiferal tests and mollusk shells. The foraminiferal assemblages found

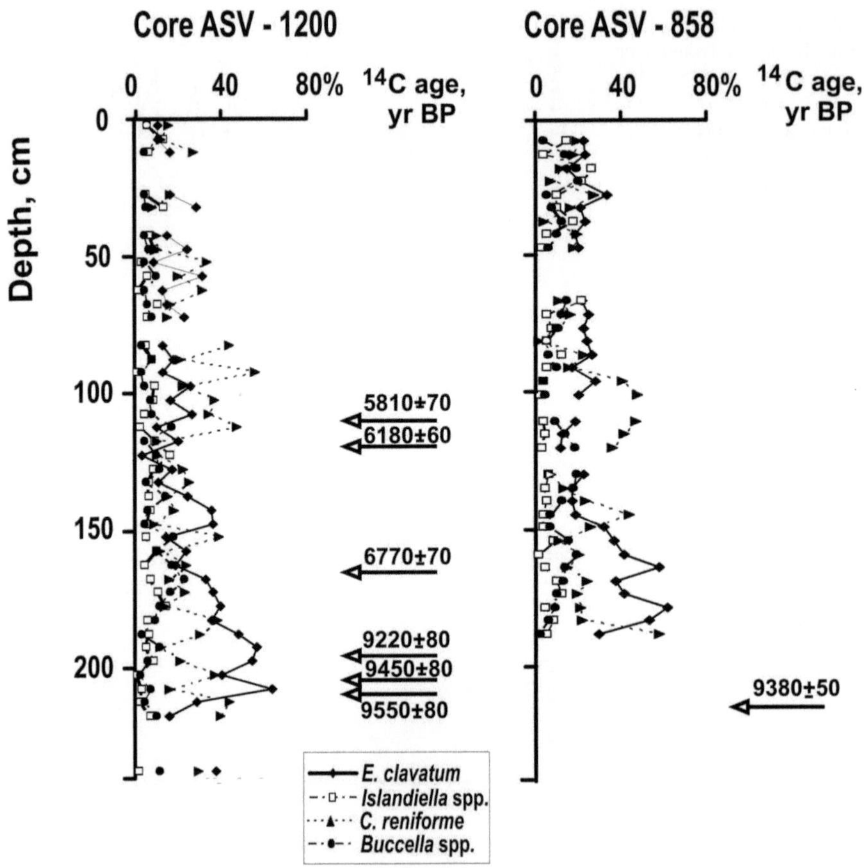

Fig. 4.12 Relationship between the abundance (%) of common benthic foraminiferal species in Holocene sediments from the Persey Trough (cores ASV-1200 and ASV-858). The gaps in the species records result from foraminiferal dissolution

in this core are typical for the southern Barents Sea. They are dominated by *C. reniforme* and *M. barleeanus.* Other arctic species like *E. clavatum, C. lobatulus, I. norcrossi, E. incertum, E. subarcticum,* and *I. helenae* are rather abundant, and agglutinated species *Adercotryma glomerata, Reophax* spp., and *Prototrochammina karika* persist to occur. Tests of planktic foraminifera are very scarce.

The Holocene assemblages of the southwestern part of the Barents Sea strongly differ from those at other locations by a pronounced dominance of the "atlantic" species throughout the section in Core PSh-5159 (Color Plate 4.7; Ivanova et al. 2007a; Chistyakova et al. in press). The preservation of foraminiferal tests and percentage of different species vary significantly.

The author studied three Holocene sections from the Kara Sea (Tables 4.1 and 4.3, Fig. 4.2) where typical species of the Arctic shelf, *E. clavatum* and *C. reniforme,* prevail (Fig. 4.13). Two of the cores, BP00-07/5 and BP01-62/6, were obtained within the framework of the Russian–German SIRRO Project, during cruises 35 and

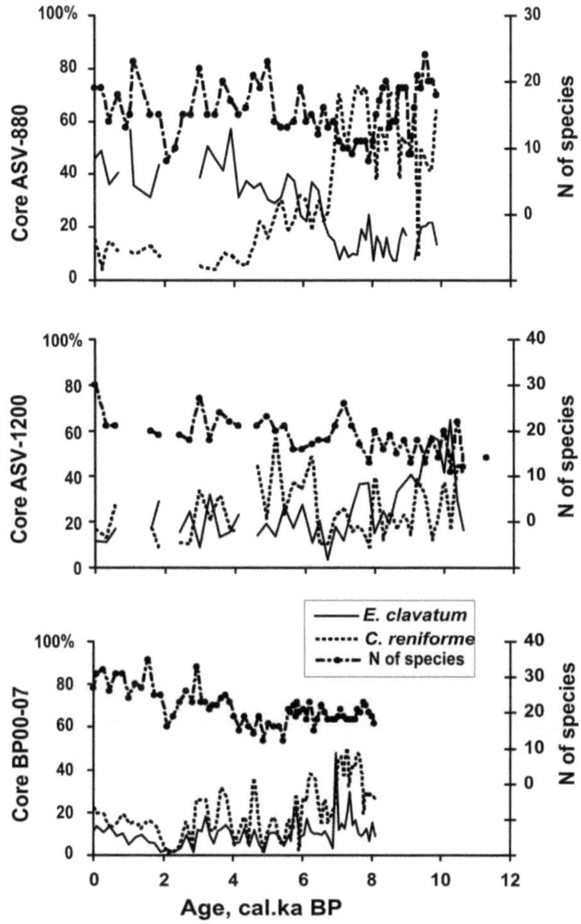

Fig. 4.13 Variations in faunal portion of two dominant species and in number of benthic foraminiferal species in Holocene sediments from the Barents and Kara seas. The gaps in the species records result from foraminiferal dissolution

36 of the R.V. *Boris Petrov*, respectively. The most complete section is recovered by Core BP00-07/5 retrieved northward of the Yenisei River estuary, near the present position of the hydrological front and the river water/seawater mixing zone, at the water depth 42.8 m. Core BP01-62/6 was taken northward of the mixing zone, and it reflects variations in paleoceanographic parameters of the seawater itself. Core PS 2718/6 was retrieved earlier by the German R.V. *Polarstern* near the Vilkitsky Strait, almost at the same latitude.

Core BP00-07/5 recovered a lithologically uniform section of bioturbated olive gray mud (Stein and Levitan 2001). Benthic foraminifera are represented by 55 calcareous and 20 agglutinated species, commonly from 15 to 30 species in one sample (Fig. 4.13, Appendix, Table 4). Preservation of foraminiferal tests varies depending on the dissolution intensity. Abundance of foraminifera ranges from 50 to 2430 tests per 1 g of dry sediment, with minimum values between 55 and 355 cm below sea floor (i.e. 1.0–5.5 cal. ka BP). It should be noted that the time interval of low abundance corresponds to that of rather high values in cores from the Barents Sea (Figs. 4.7 and 4.14).

The lower and upper intervals of the section are characterized by a high foraminiferal abundance. Agglutinated foraminiferal tests are often destroyed and they rarely occur in the lower part of the core, below 335 cm. However, in some layers from the upper part, agglutinated foraminifera contribute up to 60–83% to the total foraminiferal abundance. Dominant agglutinated species are *P. karika, Recurvoides turbinatus, Reophax* spp., whereas *Elphidium asklundi/incertum, E. clavatum, C. reniforme,* and *H. orbiculare* most commonly occur among the calcareous species (Appendix, Table 4).

In order to reveal variations in paleoceanographic parameters during the Holocene, first of all in salinity, common calcareous species are subdivided into two groups: (1) species preferring the normal marine salinity and (2) species well adapted to decreased salinity. The Group 1 includes *C. reniforme, I. norcrossi, I. helenae,* and N. *labradoricum* ; and the Group 2 comprises *E. bartletti, E. asklundi, E. incertum, E. groenlandicum,* and *H. orbiculare.* Unfortunately, the ecology of most agglutinated species is still poorly known (Polyak et al. 2002a; Alve and Murray 1995), thus it seems hardly possible to attribute them to any of the above groups.

The subdivision of calcareous species into groups is based both on the author's studies of their distribution in surface sediments of the Kara Sea (Ivanova 2001) and on published data (Khusid 1996; Khusid and Korsun 1996; Korsun 1998, 1999; Polyak et al. 2002a). Common species *E. clavatum* and *Buccella* spp. are not included into any of the groups owing to contradictions in the data on their ecology. Because of this, the author's grouping differs somewhat from that suggested by L. Polyak and co-authors (2002a) for the Kara Sea. *Pyrgo williamsoni* as well as some species of genera *Quinqueloculina* and *Miliolinella* likely may be attributed to the Group 1, but this needs additional studies of their ecological preferences.

The relationship between two groups in Core BP00-07/5 is shown in Fig. 4.14. Inervals of the dominance of Group 1,630–500 cm (about 8.1–7.0 cal. ka BP) and 90–0 cm (1.6–0 cal. ka BP) roughly correspond to the layers with high foraminiferal abundance. The upper interval contains all species of this group, whereas *C. reniforme*, a typical species of Arctic shelves, strongly dominates in the lower layer. High amount of *Quinqueloculina* spp. and *Miliolinella* spp. are also counted in the lower layer. The middle part of the section (500–90 cm, 7.0–1.6 cal. ka BP) is generally characterized by prevailing of Group 2 and roughly corresponds to the interval of low total foraminiferal abundance. Therefore, the section is subdivided into three subunits according to both dominance of one ecological group and difference in foraminiferal abundance. The pattern found was confirmed by plotting the abundance of several common species (tests per 1 g of dry sediment) versus age (Fig. 4.14). The diagram clearly shows high values of *E. clavatum* and *C. reniforme* in the lower and upper parts of the section whereas along with those, high values of the representatives of Group 2: *H. orbiculare, E. bartletti* and *E. asklundi* occur in its middle part, thus supporting the above suggestions.

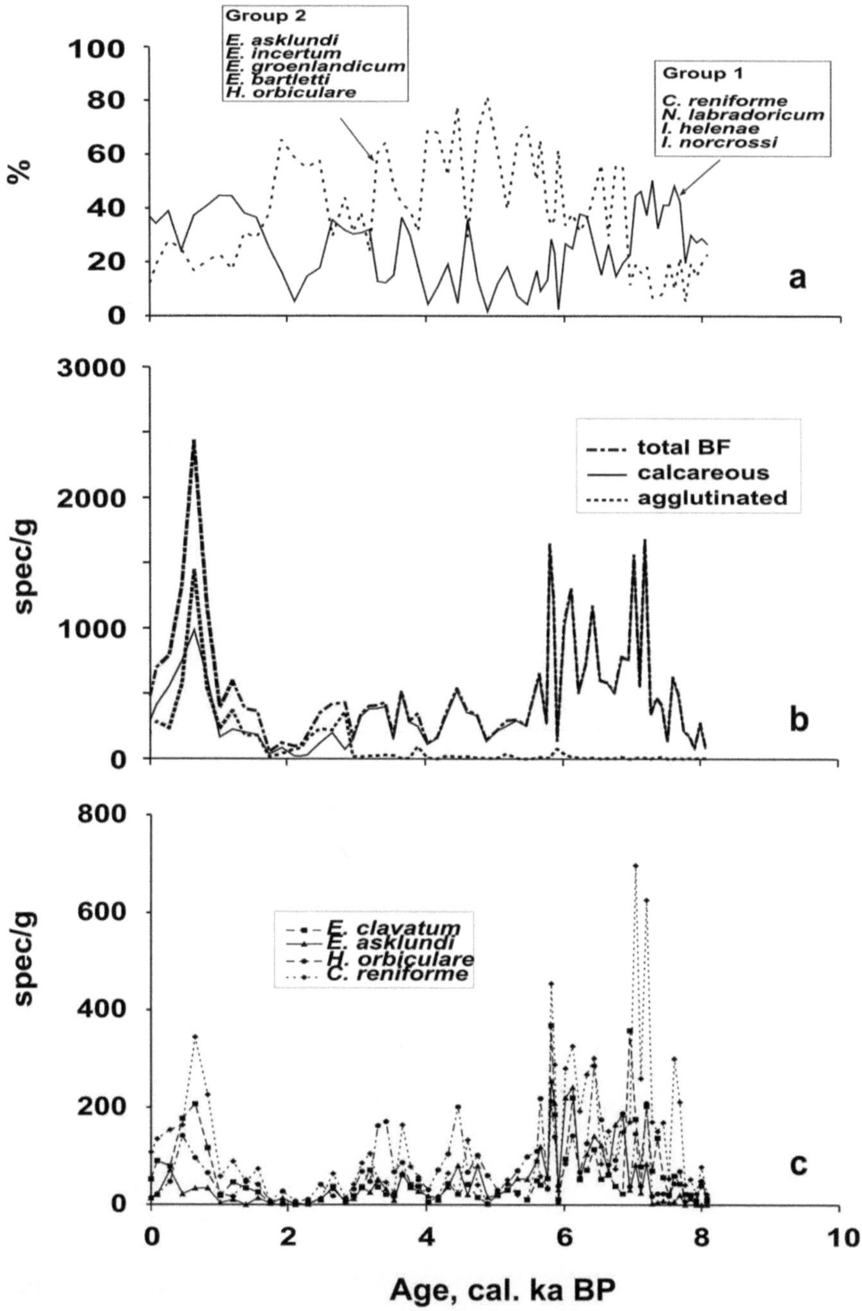

Fig. 4.14 Benthic foraminiferal records in Core BP00-07/5 from the southeastern Barents Sea: (**a**) relationship between two groups of benthic foraminiferal species: (1) species preferring normal marine salinity (*C. reniforme, I. norcrossi, I. helenae, N. labradoricum*); (2) species tolerant to decreased salinity (*E. bartletti. E. asklundi, E. incertum, E. groenlandicum, H. orbiculare*); (**b**) abundance of calcareous and agglutinated foraminifera (tests/g of dry sediment); and (**c**) abundance of indicative species (tests/g of dry sediment)

Core BP01-62/6 penetrated the Holocene section of bioturbated olive gray mud (Stein and Stepanets 2001), and recovered the Younger Dryas according to radiocarbon dates (Table 4.3). In this core, agglutinated foraminifera dominate only in the upper 40 cm (that is during the last 800 years), where they are represented mainly by *Reophax* spp. (Appendix, Table 5). Group 1 prevails throughout the section, except for the level 480 cm, and an increased content of Group 2 only occurs in the lower part of the core (580–580 cm, 10.7–12.0 cal. ka BP, Fig. 4.15). Such a relationship between the two groups reflects the location of Core BP01-62 within the area of normal seawater distribution, northward of the mixing zone characterized by Core BP00-07/5.

Unlike two cores described above, Core PS 2718/6 contains foraminiferal assemblages with dissolution features only in its lower part. The upper part of the core was lost during coring, and the interval 120–340 cm is almost barren, practically does not contain foraminifera, although there have been found some mollusk shells used for the AMS-^{14}C dating (Table 4.3, Levitan et al. 2000). Paleontologically barren layers occur within the interval 340–780 cm. Preservation of foraminiferal tests becomes better only in the interval 670–780 cm. Therefore, foraminiferal

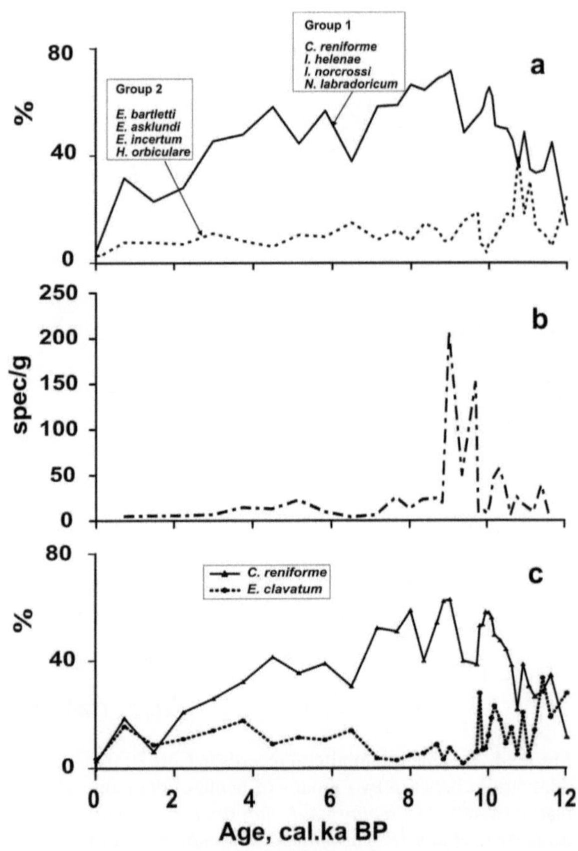

Fig. 4.15 Benthic foraminiferal records in Core BP01-62/6 from the eastern Kara Sea: (**a**) relationship between two groups of calcareous benthic foraminifera (see Fig. 4.14); (**b**) total abundance of benthic foraminifera (tests/g of dry sediment); and (**c**) percentage of *E. clavatum* and *C. reniforme*

assemblages were studied only in the interval 7–12 cal. ka BP that includes the Younger Dryas/Holocene boundary (Fig. 4.16). *E. clavatum* and *C. reniforme* typical for the Arctic shelves are the most abundant species in the interval 340–780 cm (Appendix, Table 6), the replacement of *E. clavatum* dominance by *C. reniforme* one approximately corresponds to the lower boundary of the Holocene. Figure 4.16 demonstrates the relationship between above-described foraminiferal groups 1 and 2 in this core. As in Core BP01-62/6 taken at the same latitude, the Group 1 prevails

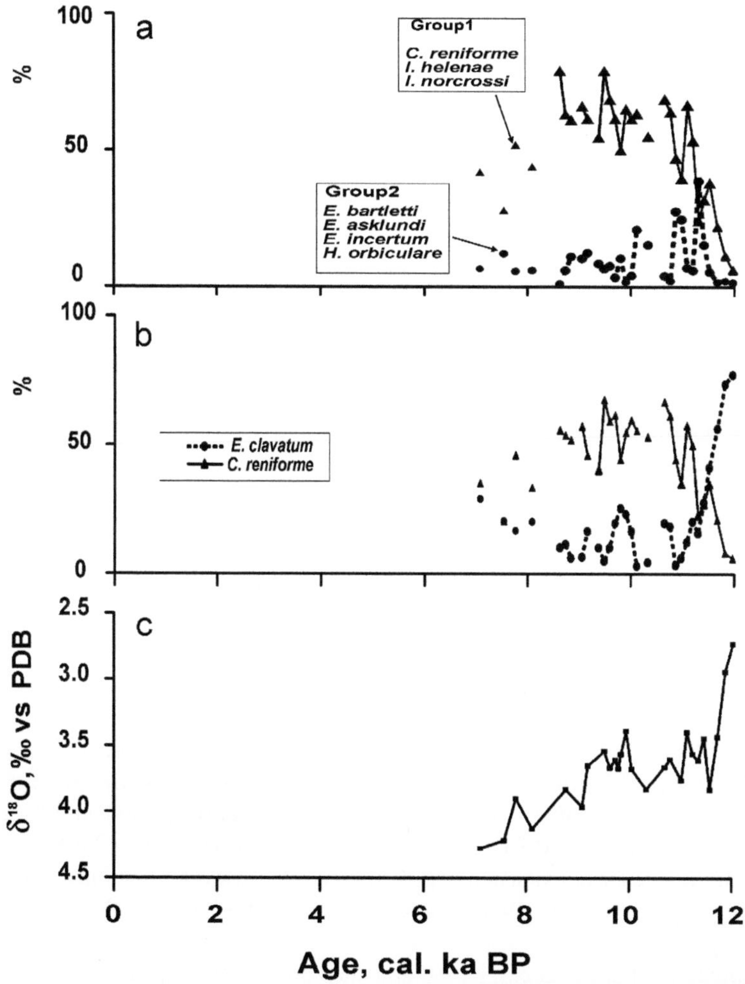

Fig. 4.16 Benthic foraminiferal and oxygen isotope records in Core PS 2718/6 from the eastern Kara Sea: (**a**) relationship between two groups of benthic foraminifera (see Fig. 4.14); (**b**) percentage of *E. clavatum* and *C. reniforme*; (**c**) $\delta^{18}O$ *E. clavatum* (unpublished data by E.V. Ivanova and J.-C. Duplessy). The upper part of core PS2718/6 was lost during the coring and the middle part is barren

throughout the entire time interval 7–12 cal. ka BP, except for the level 11.3 cal. ka BP (Color Plate 4.4). The Group 2 content is somewhat increased only in the lower part of the section, within the time interval 10.1–11.4 cal. ka BP.

4.5 The Influence of the Atlantic Water Input on the Paleoenvironments in the Barents and Kara Seas

4.5.1 The End of the Last Glaciation

The pebble loam recovered at the bottom of Core ASV-880 and interpreted as a basal moraine deposit (glacial till) is evidence of the grounded ice sheet in the Franz Victoria Trough. The low interstitial water salinity suggests absence of seawater at this site during the till deposition. The single finding of *N. pachyderma* sin. may possibly be explained by a later penetration of seawater through cracks into the moraine deposit (Ivanova and Murdmaa 2001). The presence of moraine deposits in Core ASV-880, along with our geophysical and geological data from the Sedov Trough (Pavlidis et al. 2001), supports the idea about united grounded ice sheet of the last glaciation. Some Russian authors assume that it extended over Spitsbergen, Franz Josef Land, Novaya Zemlya archipelagoes, and adjacent sea bottom, but not over the southern Barents Sea (e.g., Matishov 1984; Pavlidis 1992; Samoylovich et al. 1993; Tarasov et al. 1999, 2000). However, a majority of authors believe that the ice sheet covered the entire Barents Sea with surrounding archipelagoes and the eastern part of the Kara Sea (e.g., Lambeck 1995; Elverhøi et al. 1998; Mangerud et al. 1999, 2002, 2004; Svendsen et al. 1999, 2004; Thiede and Mangerud 1999), possibly except for the so-called Pechora Sea (Polyak et al. 2000a, Gataullin et al. 2001; Bauch et al. 2002) (Color Plate 4.5).

According to many authors, the Barents–Kara ice sheet united with the Scandinavian ice sheet was up to 2–3-km thick (Lambeck 1995, Forman et al. 1995; Landvik et al. 1998; Elverhøi et al. 1998). However, the modeled ice thickness over the shelf did not exceed 2 km (Svendsen et al. 2004). Zeeberg with co-authors (2001) suggest that the ice sheet, less than 1-km thick, existed on the Barents Sea shelf in 30–18 cal. ka BP. Warm Atlantic water inflow to the Norwegian Sea and further to the north up to the northern Spitsbergen, 31–26.5 cal. ka BP, promoted the ice sheet growth (Hebbein et al. 1994). Atlantic water likely penetrated to the Kara Sea as well, but the ice sheet extension was restricted here to western and possibly northern areas (Landvik et al. 1998; Mangerud et al. 1999, 2002, 2004; Polyak et al. 2000a, 2002c; Stein et al. 2002; Color Fig. 4.6). The ice sheet edge reached the shelf break at about 21.5 cal. ka BP and stayed there up to 18 cal. ka BP (Vorren and Laberg 1996). It isolated the Barents Sea from the Arctic Ocean in the north. Therefore, regardless the subsurface Atlantic water penetration into the Arctic since the LGM (Vogt et al. 2001; Nørgaard-Pedersen et al. 2003; Wollenburg et al. 2007; Rasmussen et al. 2007), it could not enter into Barents and Kara seas through Franz Victoria and St. Anna troughs.

Emerging of a considerable part of the shelf (Fig. 4.5) assumes a significant progradation of river deltas to the north (Stein et al. 2002; Fig. 4.6). The maximum ice sheet extension scenario results in ice damming of the Ob-Yenisei river discharge and formation of an ice lake in the central Kara Sea (Polyak et al. 2000b). Climate aridization is fixed on the surrounding land during LGM (Velichko et al. 1997; Serebryanny et al. 1998; Velichko 1999; Forman et al. 1999; Polyak et al. 2000a).

4.5.2 Deglaciation

The early phase of deglaciation in the eastern Arctic and on the continental margin of Spitsbergen started at 18.6 and 17.4, respectively, as inferred from a rapid lightening of planktic oxygen isotope values and a contemporaneous minimum of the carbon isotope composition (Jones and Keigwin 1988; Hebbeln et al. 1994, Stein et al. 1994; Knies et al. 1999).

The Barents Sea ice sheet retreated from the Franz Victoria and St. Anna troughs likely 18–15 cal. ka BP (Polyak et al. 1997; Hald et al. 1999; Lubinski et al. 2001), thus opening the marine passages. This event is marked in our Core ASV-880 by the transition from the basal moraine deposit (glacial till) to proximal glaciomarine facies of Unit III, which conformably overlaps different bottom topographic features (Color Plate 4.3). We suggest the glaciomarine origin of Unit III based on a "fresh" appearance of planktic and benthic foraminiferal tests and pteropod shells belonging to extant species, granulometry of fine size fractions strongly dominated by the finest clay (<0.001 mm), and salinity of interstitial water close to considerably freshened seawater (Murdmaa et al. 2006).

Lubinski with co-authors (2001) argue that the ice sheet decay started first in the southern Barents Sea, then developed in the St. Anna Trough and after in the Franz Victoria Trough. Most researchers believe that the deglaciation commenced with iceberg calving in deep troughs about 18 cal. ka BP, and a major part of the sea was already ice free at 14 cal. ka BP. They correlate the beginning of deglaciation with a lightening oxygen isotope signal in the Fram Strait at 17.4 cal. ka BP (Knies et al. 2000). Troughs and shelf depressions released from the ice sheet to 15 cal. ka BP. Glaciers only retained on archipelagoes and shoals (Vorren and Laberg 1996; Elverhøi et al. 1998; Landvik et al. 1998; Knies et al. 1999; Color Plate 4.5).

The proximal glaciomarine sediments accumulated in conditions of intense fine-grained suspended material delivery with glacier meltwater (so-called glacier milk), and iceberg rafting of coarse debris (Murdmaa and Ivanova 1999; Ivanova et al. 2002; Murdmaa et al. 2006). The meltwater flows and icebergs might derive Mesozoic and Cenozoic microfossils into glaciomarine sediments, along with terrestrial material of glacier erosion. Indeed, reworked microfossils are found in studied cores.

We relate the decay of ice sheet and iceberg calving to the global warming and the World Ocean level rise. At that time, the northern Barents Sea was perhaps permanently covered with drifting sea ice, as evidenced by domination of the fine clay fraction in the grain-size spectra of Unit III sediments (Murdmaa and Ivanova

1999; Ivanova and Murdmaa 2001; Pavlidis et al. 2001; Murdmaa et al. 2006). The time of Unit III deposition, that is the early deglaciation phase, presumably corresponds to the Heinrich event H-1 (17.3–15.7 cal. ka BP) in the North Atlantic likely related to the meltwater outflow from the Barents Sea shelf (Sarnthein et al. 1994b, 1995, 2000a). Therefore, the intense Barents Sea ice sheet melting substantially affected the global thermohaline circulation resulting in a "shutdown" of the AMOC, as discussed in Section 3.4. This assumption is supported by the new three-dimensional (3D) reflection seismic imaging in the southwesten Barents Sea. The seismic-derived seafloor geomorphology points to a more dynamic behavior of the ice sheet than was previously suggested (Andreassen et al. 2008).

Calving of about 510 km^3icebergs per year is assumed, the value comparable to that of the annual Yenisey River discharge, 560 km^3per year (Vorren and Laberg 1996). This iceberg calving event lasted only about 500 years (Jones and Keigwin 1988) and is reflected by a considerable lightening of the benthic and planktic oxygen isotope values in the sediment cores from the North Atlantic (Jones and Keigwin 1988; Sarnthein et al. 1992; Koç and Jansen 1994; Hebbeln et al. 1994), as well as from the Lomonosov and Heckel ridges, Arctic Ocean (Nørgaard-Pedersen et al. 1998). Most of the icebergs melted in warm Norwegian Sea waters within the Vøring Plateau area.

Considering the deglaciation, we have to take into account changes in the bathymetry of the sea owing to glacioisostatic uplift of the region and glacioeustatic sea level rise (Fairbanks 1989; Lambeck 1995; Kaplin and Selivanov 1999). The uplift was much faster in the southern Barents Sea. Because of this, the northern shelf was by 200-m deeper than today during the late deglaciation phase (15–11.4 cal. ka BP), whereas depths of the southern Barents Sea were close to present values. Depth of the Bear Island Trough reached 400 m, and Atlantic waters might easily penetrate into the sea. Even at the beginning of the Holocene, the northern shelf was still by 100-m deeper than today (Polyak et al. 1995; Vorren and Laberg 1996; Forman et al. 1997; Lubinski et al. 2001). This time interval includes Bølling and Allerød warmings, as well as Early Dryas and Younger Dryas coolings (see Section 3.5). Many authors just attribute the major melting of the Barents Sea ice sheet to the Bølling and Allerød warmings (e.g., Vorren and Laberg 1996). However, several studies (e.g., Sarnthein et al. 2000a; Lubinski et al. 2001; Murdmaa et al. 2006) suggest much earlier rapid disintegration of the ice sheet as was mentioned above.

The late deglaciation phase (Unit II, 15–11.4 cal. ka BP) was characterized by a distal glaciomarine environment. At that time, the glacier front retreated inland from the Novaya Zemlya coast (Forman et al. 1999; Zeeberg et al. 2001). Gataullin and co-authors (2001) argue that it happened after the Younger Dryas. It is suggested that remains of the Barents Sea ice sheet were melted on shoals and southern Novaya Zemlya islands at the beginning of Holocene (Vorren and Laberg 1996; Landvik et al. 1998). Retreat of the glacier front from the coastline resulted in almost total cessation of the iceberg rafting of coarse debris (Murdmaa et al. 2006). Settling of fine-grained suspended material derived from glacier meltwater became the main sedimentation mechanism. Bioproductivity was generally low owing to severe sea ice conditions (Murdmaa and Ivanova 1999; Ivanova et al. 2002).

The northern part of the sea was perhaps permanently covered by sea ice, both locally formed and drifted from the Arctic Ocean. Polar surface water, in spite of an intense input of subsurface Atlantic water during the Bølling/Allerød interstadial, was also noted in cores NP94-51 and JM02-440 taken off the northern and western Spitsbergen coast, respectively (Ślubowska et al. 2005, 2007). The surface waters seem to be also cold and seasonally covered by sea ice in the southwestern Barents Sea even during the Bølling/Allerød warming as follows from the extremely scarce occurrence of planktic foraminifera in Core PSh-5159R (Chistyakova et al. in press).

Paleoenvironments in the northern Franz Victoria and Persey troughs and in the Central Deep and Ingøydjupet Depression in the southern Barents Sea differed considerably. Our cores recovered relatively thin Unit II sections from Franz Victoria and Persey troughs, as well as from other local depressions of northern and central Barents Sea. They mainly consist of fine-grained laminated (and color banded) sediments with sandy laminae, which suggest pulsating sedimentation by low-density gravity flows (Murdmaa et al. 2006). At the same time, up to 20–30-m (or even 100 m) thick bodies of fine-grained sediments accumulated in deeps of the southeastern Barents Sea (Gataullin et al. 2001; Murdmaa et al. 2006), as well as in the Ingøydjupet Depression in the southwestern part of the sea according to the high-resolution seismic records obtained during the cruise 63 of the R.V. *Professor Stokman*. The latter is confirmed by our AMS-^{14}C datings in core PSh-5159R (Table 4.2).

The pulsating sedimentation likely created an unfavorable environment for both macro- and microbenthos (Murdmaa and Ivanova 1999; Ivanova and Murdmaa 2001) in the major part of the basin. This suggestion is consistent with existing data on the seasonal development of benthic foraminifera near the modern glaciers front (Korsun and Hald 2000). Rather rare occurrence of extant foraminiferal species may also be related to dissolution of their calcareous tests. Sea ice freezing brines and nepheloid flows carry dissolved oxygen and carbon dioxide from the sea surface to the bottom of troughs, thus promoting dissolution of calcareous microfossils (Steinsund and Hald 1994; Hald and Steinsund 1996).

Meanwhile, presence of "atlantic" planktic and benthic foraminiferal tests in the Unit II sediments suggests episodic Atlantic water penetration into the Barents Sea via the troughs (Murdmaa and Ivanova 1999; Ivanova and Murdmaa 2001). This suggestion is supported by our data on cores ASV-1183, PSh-5147, and ASV-1200, as well as those by other authors (Polyak and Solheim 1994; Lubinski et al. 1996, 2001; Hald et al. 1999) about thin interbeds in Unit II sediments with relatively abundant benthic foraminifera including species characteristic for Atlantic waters.

On the contrary, the diverse Bølling–Allerød benthic assemblages of the Ingøydjupet Depression are dominated by the "atlantic species," whereas the percentage of the arctic species increases through the Younger Dryas (Color Plate 4.7; Chistyakova et al. in press). High abundance of warm water benthic foraminifera indicate the persistent presence of the Atlantic water in the bottom layer of the southwestern Barents Sea whereas the scarcity of planktic foraminifera most likely points to very low SST and seasonal ice cover.

The latter assumption is in line with the data from the western and northern shelf of Svalbard (Ślubowska et al. 2005, 2007). Our planktic foraminifera-based estimate in Core PSh-5159R provides SST as low as 6°C for the YD–Preboreal transition that is ~6°C below the present value (Ivanova et al. 2007a).

However, in the major part of the Barents Sea area, more favorable conditions for production and/or preservation of benthic foraminiferal fauna appeared episodically. The basin became perhaps partially ice free for short periods during the late deglaciation phase, possibly due to polynya formation where high productivity developed. Hence, the deglaciation was characterized by unstable paleoceanographic conditions. The basin was filled with considerably freshened seawater due to glacier meltwater delivery, as well as river water inflow via White, Pechora, and Kara seas, as evidenced by decreased interstitial water salinity in Unit II sediments (Murdmaa et al. 1998b, 2006). The strong freshening is also confirmed by depleted oxygen and carbon isotope values of benthic foraminifera species *E. clavatum* from Core ASV-1183 from the southern part of the Central Deep (Fig. 4.8, Ivanova et al. 2002; Murdmaa et al. 2006). The freshwater isotope signal was likely transferred to the bottom water by sea ice freezing brines formed over nearby shoals (Bauch and Bauch 2001; Duplessy et al. 2001).

The meltwater isotope signal is also noted in the northern Barents Sea at 12.8 and 11.4 cal. ka BP (Lubinski et al. 2001), at ca. 13 and 12 cal. ka on the eastern slope of the Laptev Sea (Spielhagen et al. 2005), and as short episodes in several locations of the Arctic Ocean within the interval 18.8–13 cal. ka BP (Stein et al. 1994; Nørgaard-Pedersen et al. 1998; Poore et al. 1999). It is assumed that the freshwater signal reflects not only the Barents Sea ice sheet melting but also an increased influence of the Ob, Yenisei, and other Siberian rivers runoff due to their seaward propagation prior to the Younger Dryas and at the YD/Preboreal transition (Polyak et al. 2000b; Lubinski et al. 2001).

An intense glacier melting during the late deglaciation phase was amplified by warm Atlantic water inflow into the Barents Sea via the deep troughs. These events likely correspond to the intensified thermohaline circulation in the North Atlantic. To test such an assumption, it is necessary to date levels with abundant foraminiferal fauna within the late deglaciation unit. This was not done up-to-date, because of insufficient biogenic calcium carbonate content in the sediments. However, our new dates of benthic foraminiferal records in Core PSh-5159R (Table 4.2; Fig. 4.9; Chistyakova et al. in press) and similar data by M. Ślubowska-Woldengen and co-authors (2007) from the Svalbard shelf allow us to admit rather persistent subsurface Atlantic water input into the Barents Sea along the Norwegian and western Svalbard coasts through the late deglaciation and especially during the Bølling–Allerød warming. New micropaleontological data from the western continental slope of the Laptev Sea point to a persistent although temporarily variable inflow of the Atlantic-derived water since ~16 cal. ka (Taldenkova et al. 2008). As a whole, deglaciation of the eastern Barents Sea was completed prior to the Holocene, 10 ^{14}C ka or 11.4 cal. ka BP (Polyak et al. 1995; Forman et al. 1996; Hald et al. 1999).

Oxygen isotope composition of benthic foraminifera *E. clavatum* from Core PS 2718/6 became considerably heavier at the YD–Preboreal transition, 11.8–11.4 cal. ka BP (Fig. 4.16). This signal likely reflects a bottom water salinity increase during the gradual sea level rise. The suggestion well corresponds to an increase in abundance of Group 1 foraminifera and to the alteration of *E. clavatum* dominance by that of *C. reniforme* in cores PS 2718/6 and BP01-62/6 (Figs. 4.15 and 4.16).

4.5.3 The Long-Term Holocene Events

The paleoceanographic situation at the deglaciation to Holocene boundary, about 11.4 cal. ka BP (Bard et al. 1990), changed dramatically. The glaciomarine sedimentation with prevailing lateral sediment transport by gravity flows was replaced by marine sedimentation of hemipelagic type with pelletal settling of fine-grained suspended material. The biological productivity substantially increased, as evidenced by the appearance of planktic foraminifera and an increase in abundance and species diversity of benthic foraminifera (Fig. 4.7). Shelf depressions were inhabited by macrobenthos including polychaeta that resulted in a development of bioturbation (Ivanova and Murdmaa 2001). Paleoceanographic environments in the Kara Sea, especially in its eastern part, substantially differed from those in the Barents Sea (Ivanova 2003, 2006, 2007).

In the southwestern Barents Sea, pronounced subsurface water warming of about 6°C is estimated by foraminiferal assemblages in cores PSh-5159R and PSh-5159N (Ivanova et al. 2007a; Chistyakova et al. in press) over the YD–Preboreal transition. The content of coarse grain-size fraction drastically decreased at the same level (Fig. 4.9). The subsurface waters were very warm until ~7.5 cal. ka BP with rather negligible oscillations in consistent with the SIMMAX-derived SST record in core M-23258 about 2° to the north of our core (Sarnthein et al. 2003b; Color Plate 4.1). In Core PSh-5159N, the Early Holocene exhibits high amount of *C. teretis* (up to 60%) and planktic foraminifera (Color Plate 4.7; Chistyakova et al. in press) which are generally considered to be indicators of the Atlantic water penetration into the Barents Sea (Polyak and Solheim 1994; Lubinski et al. 1996; Hald et al. 1999).

Planktic and benthic oxygen isotope records in the Core ASV-880 from the northern Barents Sea (Duplessy et al. 2001) reflect a rather abrupt, during a few centuries, increase in temperature of subsurface (from about –0.5°C to +1.5°C) and bottom (from about –1 to +1°C) waters in the Early Holocene (Fig. 4.10) related to the global warming and Atlantic waters inflow via the Franz Victoria Trough. This conclusion is supported by the presence of *C. teretis* and planktic foraminifera in the sediments. Stratification of the water column at site ASV-880 weakened due to the warming, and the subsurface water temperature reached its present values at about 8.7 cal. ka BP (Fig. 4.10). According to Lubinski et al. (2001), both subsurface and bottom waters in the northern Franz Victoria Trough and the Southern St. Anna Trough were very cold during 10–8.3 cal. ka BP, when a considerable warming already occurred in the Norwegian Sea (see Section 3.5), and the Holocene maximum of surface water and air temperature is noted in the Spitsbergen region, North

Sea, Norwegian Sea, and the western Barents Sea margin (Salvigsen et al. 1992; Lubinski et al. 1999; Klitgaard-Kristensen et al. 2001; Sarnthein et al. 2003a; Hald et al. 2007).

Increased sand content in Lower Holocene sediments suggests ice rafting of the debris (Figs. 4.7 and 4.8). Intense glacier melting continued on the surrounding archipelagoes and meltwater discharge delivered abundant terrigenous material. Glaciers retreated from the coastal area off the northwestern Novaya Zemlya, southward of the Russian Gavan' Fjord, about 10.2 cal. ka and 9.7–9.3 cal. ka BP (Forman et al. 1999). Maximum sedimentation rates for the Holocene (~30–100 cm/ka) are fixed at the beginning of this time interval, and then they slowed down likely owing to a decrease in material supply from melting glaciers (Murdmaa and Ivanova 1999; Ivanova et al. 2002; Murdmaa et al. 2006). High sedimentation rates are noted in the St. Anna Trough 10–8 ^{14}C ka BP, or about 11–9 cal. ka BP (Hald et al. 1999).

M. Hald and co-authors (2001) postulated that the decrease in sedimentation rates about 8 ka BP possibly resulted from the sea level rise. In fact, existing data from the Franz Josef Land and Novaya Zemlya show that the sea level rise substantially exceeded the glacioeustatic uplift of these islands in the Early Holocene (Zeeberg et al. 2001). The transgression continued up to 6 cal. ka BP, when the sea level rise ceased.

The biological productivity varied widely. Maximum abundance of both planktic and benthic foraminifera is noted in the Franz Victoria Trough, whereas the species diversity is somewhat higher in the Persey Trough (Fig. 4.11, Ivanova 1999, 2002a). Judging by a considerable depletion of the oxygen isotope composition of planktic and benthic foraminifera in the Franz Victoria Trough, the Early Holocene terminated with an optimum expressed in warming of subsurface and bottom waters 7.8–6.9 ka BP (Fig. 4.10). Warm Atlantic waters filled practically the entire trough (Duplessy et al. 2001). Percentage of the "Atlantic" species *C. teretis* considerably increased in the axial zone of the trough (Lubinski et al. 1996, 2001). Edge of the drifting sea ice retreated to the north that prolonged the summer ice-free season over the trough, as indicated by a high content of *I. helenae/ norcrossi* content and low percentage of *E. clavatum* in benthic foraminiferal assemblages (Fig. 4.10).

Sedimentation rate at the site of Core ASV-880 increased (Ivanova et al. 2002a) likely due to accelerated glaciers melting on the Franz Josef Land under conditions of climate warming that resulted in abundant fine-grained material delivery with the meltwater discharge. Accumulation rates and concentration of both planktic and benthic foraminifera decreased simultaneously (Fig. 4.10), perhaps because of the sea ice edge retreat. Therefore, the Holocene temperature optimum or so-called Holocene thermal maximum (Kaufman et al. 2004; Moros et al. 2004) in the northern Barents Sea began about 2 ka later than at 60s latitudes of the North Atlantic (see Section 3.5; Koç et al. 1993; Duplessy et al. 2001, 2005) and then at the western margin of the Barents Sea (Sarnthein et al. 2003b; Hald et al. 2007). Data from the St. Anna Trough (Hald et al. 1999) support this suggestion. The timing of the temperature optimum in the northern Barents Sea coincides with that of the modern Transpolar Drift formation and acceleration of West Greenland and Labrador currents (Rahman and Vernal 1994). Data from Spitsbergen point to the air temperatures higher than today during 10–5.6 ka BP. This led to glaciers reduction on this archipelagoe and also in the Franz Josef Land (Salvigsen et al. 1992; Svendsen and Mangerud 1997; Lubinski et al. 1999).

The sea over the Persey Trough was likely ice free most part of the year. The increase in *I. helenae/norcrossi* and *Buccella* spp. content in the Early Holocene interval is also noted here indicating a prolongation of the ice-free season (Figs. 4.11 and 4.12). Relatively high content of *M. barleeanus* and *C. teretis* is an evidence of Atlantic waters which flowed into the trough from the southwest, as the passage from the north was blocked by a sill (Pfirman et al. 1994). The optimum is poorly expressed in the oxygen isotope records on planktic and benthic foraminifera for cores ASV-1183 and ASV-1157 from the southern Barents Sea (Fig. 4.8; Duplessy et al. 2005; Murdmaa et al. 2006).

J.-C. Duplessy with co-authors (2005) explained a considerable difference in benthic oxygen isotope values in these sections by a strong transformation of the Atlantic water in the central and especially southeastern Barents Sea. The Atlantic water mixing with sinking cold and saline Arctic water formed during brines rejection creates the Barents Sea bottom water. In general, the Early Holocene is characterized by minor variations in the oxygen isotope composition of bottom waters in the central and southeastern Barents Sea (Fig. 4.11; Levitan et al. 2003; Duplessy et al. 2005; Murdmaa et al. 2006).

In Core ASV-1310 from the Western Novaya Zemlya Trench, relatively low oxygen isotope ratio on *E. clavatum* (Ivanova and Duplessy, unpublished) indicates substantial freshwater input transferred to the sea floor by brines whereas rare occurrence of *C. teretis* (foraminiferal counts by T.A.Khusid) points to the subsurface Atlantic water penetration into the eastern part of the sea in the earliest Holocene. At the end of Early Holocene, relatively high amount of *N. labradoricum* in benthic assemblages may indicate an increase in bioproductivity which allow us to speculate about a nearby position of the Polar Front ~9.3–7.7 cal. ka BP. (Ivanova et al. 2008a).

In the Middle Holocene, the subsurface to bottom water layer in the southwestern Barents Sea became slightly cooler than before, as it follows from our temperature estimates and decrease in Atlantic species amount in Core PSh-5159N (Color Plate 4.7; Chistyakova et al. in press). However, the environmental conditions seem to be rather stable over about 4 ka. This roughly corresponds to the data by M. Sarnthein with co-authors (2003b) and M. Ślubowska-Woldengen with co-authors (2007).

In the Franz Victoria Trough, we found the interval after the optimum with a relatively high organic carbon, planktic, and benthic foraminiferal content in the sediments and increased percentages of *Buccella* spp. and *N. labradoricum* (Duplessy et al. 2001; Fig. 4.10), known as indicators of high productivity and fresh organic matter supply to the bottom (Khusid and Korsun 1996; Polyak and Mikhailov 1996; Polyak et al. 2002a).

The increase in productivity was likely related to a southward shift, toward the Core ASV-880 site, of the summer sea ice edge 6.5–4.5 cal. ka BP. Hydrotroilite is practically absent in sediments from this interval that possibly points to a well-ventilated bottom water (Ivanova and Murdmaa 2001). The oxygen isotope record on *E. clavatum* for Core ASV-880 reflects a stronger stratification of the water column that likely results from sinking of cold brines which transported oxygen to the trough bottom (Fig. 4.10; Duplessy et al. 2001). Subsurface waters of the Franz Victoria Trough remained rather warm, at least in summer, when *N. pachyderma* sin. proliferates (Kohfeld et al. 1996). Carbon isotope values of *N. pachyderma* sin. mark the presence of transformed Atlantic waters in the subsurface layer of the water column (Duplessy et al. 2001).

Benthic foraminiferal assemblages from the Persey Trough point to a much lower productivity compared to contemporaneous assemblages from the Franz Victoria Trough (Fig. 4.11). This is indicated by a lower content of *Buccella* spp. and higher content of *C. reniforme* in cores ASV-1200 and ASV-858 relative to those in Core ASV-880 (Figs. 4.11 and 4.12). *C. reniforme* and *M. barleeanus* dominate in sediments of the same age recovered by Core ASV-1183 in the Central Deep.

The Late Holocene conditions in the southwestern part of the sea varied considerably. Benthic foraminiferal assemblages point to bioproductivity fluctuations, while variations in the ratio of boreal and subpolar species of planktic foraminifera monitor SST oscillations up to 3°C (Chistyakova et al. in press).

In general, the Late Holocene is characterized by a noticeable dissolution of calcareous microfossils on the Barents Sea floor, especially in the southeastern part of the basin (Polyak and Mikhailov 1996) and the Western Novaya Zemlya Trench (Khromova and Ivanova 2002; Murdmaa et al. 2006). Strong dissolution might have been caused by more intense sea ice freezing with abundant brines formation than before. Sinking of dense brines with high oxygen content down from shoals to depressions in the northern and central parts of the sea resulted in formation of bottom waters aggressive to calcareous foraminiferal tests (Steinsund and Hald 1994; Hald and Steinsund 1996).

Increase in abundance of the opportunistic species *E. clavatum* in foraminiferal assemblages, especially in our northernmost Core ASV-880, probably reflects a southward advance of the summer ice edge (Hald et al. 1994; Polyak and Solheim 1994; Polyak and Mikhailov 1996; Duplessy et al. 2001). *E. clavatum* became the dominant species in benthic foraminiferal assemblages of the Franz Victoria Trough and likely of the northern margin of the Barents Sea as a whole. A support to this suggestion comes from the data on species distribution in Core NP94-51 north off Svalbard (Ślubowska et al. 2005) which resembles *E. clavatum* record in Core ASV-880 (Fig. 4.17). Variations in planktic and benthic foraminiferal abundance, as well as in percentage of individual species (e.g. *N. labradoricum*) point to short-term productivity blooms, which were possibly related to changes in Atlantic water inflow, summer sea ice edge position, and winter convection. An enhanced content of epifaunal species *C. lobatulus* and *Elphidium subarcticum* in Middle and Late Holocene assemblages from the Persey and Franz Victoria troughs (Figs. 4.10 and 4.11) seem to indicate intensified bottom currents activity, as it was inferred from previous suggestions on the species ecology (Hald and Korsun 1997; Polyak et al. 2002a).

Changes in hydrological conditions and surface water productivity of the Barents Sea during the Holocene have been reflected by variations in abundance of planktic

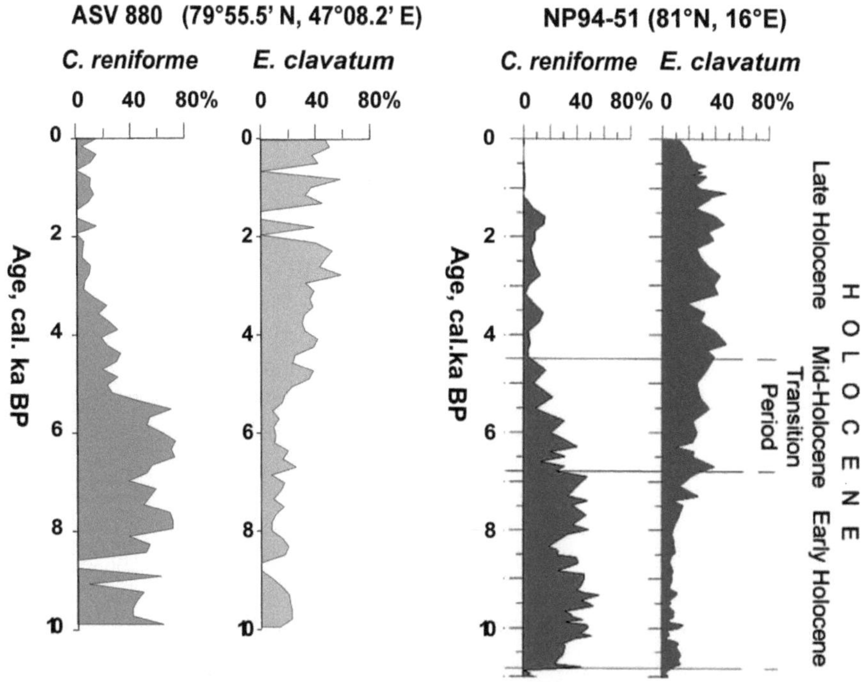

Fig. 4.17 Variations in time series of *E. clavatum* and *C. reniforme* in cores ASV-880 (data by the author) and NP 94-51 (Slubowska et al. 2005) from the northern Barents Sea

foraminifera mainly represented by the polar species *N. pachyderma* sin. (Fig. 4.10). High planktic foraminiferal abundance in the Holocene sediments of Core ASV-880 compared to other cores from the northern, central, and eastern parts of the sea, reflects its geographic position within the area of Atlantic water distribution and intense winter convection (Color Plate 4.2). According to oxygen isotope data (Spielhagen and Erlenkeuser 1994; Bauch et al. 1997), *N. pachyderma* sin. grows and reproduces mainly at about 100-m water depth. In the Franz Victoria Trough, Atlantic waters occupy that depth (Color Plate 4.2). Winter convection apparently promotes the species proliferation in the water column and further accumulation.

Atlantic waters are considerably transformed due to mixing with local Barents Sea waters near the Polar Front, where cores ASV-858 and ASV-1200 have been retrieved (Color Figs. 4.1 and 4.2). The surface water temperature is considerably higher here (Fig. 4.2) resulting in a weaker winter convection. Therefore, nutrients content, phytoplankton productivity, and hence food supply to the subsurface water, where *N. pachyderma* sin. thrives, are lower in summer (Ivanova 2002a). This suggestion is consistent with published data on *N. pachyderma* sin. distribution in the Arctic Ocean (Kohfeld et al. 1996; Bauch et al. 1997; Wollenburg and Mackensen 1998), and is supported by chlorophyll-*a* measurements performed in the Cruise 11 of the R.V. *Akademik Sergei Vavilov* (Vedernikov and Gagarin 1998). A very low planktic and benthic foraminiferal abundance in Core ASV-1183 from the Central Deep (Fig. 4.8) might be explained by a strong dissolution of their tests.

As noted above, the transition from brackish to marine environments at the YD–Preboreal boundary is ascertained in the eastern Kara Sea by an increased occurrence of the Group 1 species in cores PS 2718/6 and BP01-62/6 (Figs. 4.15 and 4.16, Color Plate 4.4). Such evidence has been also obtained from the southwestern Kara Sea (Polyak et al. 2000b). Ob and Yenisei river mouths extended seaward due to the low sea level stand (from –50 to –30 m at the beginning of Holocene, according to Fairbanks 1989), and the area of Core BP00-07 location represented an emerged shelf at that time (Stein et al. 2002; Color Plate 4.6). R. Stein with co-authors (2002) argued that the marginal filter developed on the present open shelf where sedimentation rates accelerated owing to intense terrigenous material delivery by the river runoff and coastal wave erosion. In the Laptev Sea, the maximum Lena River discharge is reconstructed for the time slice 11.7 cal. ka BP, and an intensified Atlantic water inflow is assumed at 11.2 (Polyakova et al. 2005) and 10.8–7.7 cal. ka BP (Bouscein et al. 2000). During the Holocene, Ob, and Yenisei river paleovalleys were submerged due to a gradual sea level rise (Stein et al. 2004).

Extremely high benthic foraminiferal amount (Fig. 4.14) counted in the lower part of Core BP00-07/5 from the eastern Kara Sea noticeably exceed the values obtained in sections from both the Barents Sea and the northern Kara Sea areas (Figs. 4.10, 4.11, and 4.15). At present, the core site in paleoriver channel is very shallow (38 m). This site water depth was approximately invariable since 8 cal.ka BP as a result of the compensative effects of the global sea level rise, rapid sedimentation, and isostatic uplift (Simstich et al. 2004). High foraminiferal abundance in the lower Holocene sediments likely resulted from a nearby position of the seawater and river water mixing zone, as well as the assumed marginal filter, up to 7.5 cal. ka BP (Stein et al. 2002). Intense summer sea ice melting and/or considerable seasonal contrasts at the end of early Holocene might serve as alternative reasons. It is assumed that the sea ice melting might result in local phytoplankton blooms, providing food supply for foraminifera (Matishov 1995; Polyak et al. 2002a).

Predominance of Group 1 ("marine") species, first of all *C. reniforme,* in the time interval 8.1–6.9 cal. ka BP (Fig. 4.14, Color Plate 4.4) also characterized by abundant foraminiferal tests and high sedimentation rates likely reflects not only an increase in productivity but also transformed Atlantic water inflow via the St. Anna Trough during the Early Holocene transgression (Ivanova 2002b). Submarine channels possibly served as routes for the marine water inflow to the site location with the reversal currents (Simstich et al. 2005). This suggestion is consistent with the modeling experiments (Harms et al. 2003) and the above-mentioned data from the Laptev Sea (Bouscein et al. 2000) and Barents Sea (Hald et al. 1999; Duplessy et al. 2001). Moreover, increase in the surface water salinity from 7–8 to 11–13 psu during the Early Holocene, 9.3–7.5 cal. ka BP, is reconstructed by the diatom assemblages in Core BP99-04/7 retrieved from the water depth of 32 m to the south of Core BP00-07 (Polyakova and Stein 2002).

The shift in dominance from Group 1 to Group 2 in Core BP00-07/5 about 6.9 cal. ka BP (Fig. 4.14) might result from weakening of the Atlantic water inflow to the Kara Sea via the St. Anna Trough. This suggestion well agrees with our data on the Franz Victoria Trough (Fig. 4.10, Duplessy et al. 2001). Except for the interval 2.3–1.8 cal. ka BP, the apparent minimum of foraminiferal abundance in the middle part of the section, 5.6–1.4 cal. ka BP, cannot be explained by dissolution that widely varies throughout the core. Dilution by terrigenous material as a reason of the low foraminiferal number is also excluded, because sedimentation rate decreases upward from its maximum value in the lower part of the section (Simstich et al. 2005). Hence, it seems that the rough coincidence of the intervals of low total foraminiferal abundance and predominance of Group 2 can be interpreted as an evidence of some bottom water freshening and Group 2 tolerance to restricted food resources. Relatively high abundance of some species of Group 2 in the same interval (Fig. 4.14) supports the assumption. On the other hand, published data on the foraminiferal ecology suggest that the species of Group 1 strongly depend on food supply (Hald and Steinsund 1996; Polyak et al. 2002a), perhaps except for relatively opportunistic *C. reniforme. E. clavatum,* and *C. reniforme* dominate in many samples from this part of the section (Fig. 4.14), as it is typical for the western Arctic seas in the Holocene (Color Plate 4.2, Fig. 4.13).

According to diatom analysis of Core BP99-04/7, surface water salinity decreased and the proportion of freshwater species in diatom assemblages increased in the southeastern part of the Kara Sea later, about 6 cal. ka BP (Polyakova and Stein 2002, 2004), that is only few centuries later than the increase in proportion of low salinity tolerant foraminiferal Group 1 in Core BP00-07/5 (especially keeping in mind the age uncertainty). However, maximum occurrence of marine ostracods until ~3.5 cal. ka (A.Yu. Stepanova, personal communication 2007) and relatively higher (but still low prior to 4 cal. ka) oxygen isotope values measured on ostracoda and benthic foraminifera in Core BP00-07/5 (Simstich et al. 2005) seem to contradict the foraminiferal data. The discrepancy may result from a complicated interaction of many factors near the frontal zone at 74°N: river runoff oscillations and related nutrients delivery, variations in the intensity of Atlantic water inflow via the channels, seasonal fluctuations of hydrological and hydrochemical parameters, migration of the mixing zone and marginal filter, changes in bottom water corrosivness relative to biogenic carbonate, and interspecies competition. Moreover, the foraminiferal data seem to be statistically more reliable due to much higher abundance of foraminiferal tests in this core as compared to the ostracod shells.

Relatively light oxygen isotope values of Early Holocene benthic foraminiferal tests possibly can be explained by a slightly higher temperature of transformed Atlantic water, as compared to present values near the site, and/or by the well-known effect of $\delta^{18}O$ lightening owing to rejection of brines sinking onto the bottom (e.g., D. Bauch and H. Bauch 2001). The bottom water cooling after the Early Holocene optimum led to a minor increase in benthic oxygen isotope values, not more than by 0.2–0.5‰ with a temperature fall by 1–2°C according to N. Shackeleton's equation (1974). Therefore, the oxygen isotope data on the Core BP00-07/5 (Simstich et al. 2004, 2005) possibly need to be revised considering the above-mentioned effects.

The earlier increase in percentage of benthic species adapted to low salinity (~6.9 cal. ka BP), as compared to the rise in number of freshwater diatoms (~6 cal.

ka BP, according to Polyakova and Stein 2002) might result from their suitability to restricted food resources, as it was noted above. The interval of low TOC content, ~7.5–2 cal. ka BP, detected in sediments of Core BP99-04/7 (Polyakova and Stein 2002, 2004), well corresponds to the interval of low foraminiferal abundance ~5.6–1.4 cal. ka BP in Core BP00-07/5 (Fig. 4.14), except for its lower part (7.5–5.6 cal. ka BP), where the benthic fauna likely still felt an auspicious influence of seawater.

It is noticeable that the decrease in foraminiferal abundance in Core BP00-07/5 (Fig. 4.14) correlates with the time of increase in precipitation on Taymyr Peninsula (Andreev and Klimanov 2000) and in freshwater diatoms occurrence about 6 cal. ka BP in Core BP99-04/7 (Polyakova and Stein 2002). If the salinity decrease inferred from diatom data occurred due to enhanced river runoff, it must have led to a northward migration of the mixing zone that, in turn, promoted development of foraminiferal species of Group 2 adapted to low salinity and scarce food resources. However, the mixing zone never shifted across 76°N latitude, as shown by continuous prevailing of Group 1 species in Core BP01-62/6 (Fig. 4.15, Color Plate 4.4; Ivanova 2002b, 2003a).

The return change in dominance from Group 2 to Group 1 in Core BP00-07/5 at ~2 cal. ka BP corresponds to the oxygen isotope data (Simstich et al. 2002, 2005) and suggests a bottom water salinity rise. It was accompanied by an increase in foraminiferal number ca. 1.4 cal. ka BP. Occurrence of very abundant agglutinated foraminifera in the interval of last 1.2 ka points to weak diagenetic reactions and likely to a strong dissolution of calcareous tests (Alve and Murray 1995; Steinsund and Hald 1994; Polyak et al. 2002a). This is supported by a visual evaluation of the foraminiferal tests preservation. The species related to the seasonal ice cover and high biological productivity, *N. labradoricum, I. helenae,* and *I. norcrossi,* occur in the foraminiferal assemblage. Both our data (Fig. 4.10; Duplessy et al. 2001) and the results of other studies (Ślubowska et al. 2005; Polyak et al. 2002b) confirm an intensification of subsurface Atlantic waters inflow along the Eurasian continental slope and their penetration into the Western Arctic seas during the Last Millennium, or even since ~3 cal. ka BP (Taldenkova et al. 2008).

4.5.4 The Short-Term Holocene Events

The short-term cooling about 8.2 cal. ka BP fixed in the oxygen isotope record of Core ASV-880 affected both subsurface Atlantic water and bottom water in the Franz Victoria Trough (Fig. 4.10, Duplessy et al. 2001). In the southeastern Barents Sea, the cooling is ascertained by low surface SST and salinity during the summertime and longer sea ice season reconstructed by dinoflagellate cysts (Voronina et al. 2001). This event might substantially affect the thermohaline circulation (see Section 3.5) resulting in cooling and weakening of Atlantic waters flow around the Spitsbergen before their penetration into the Arctic Ocean and more intense heat delivery at middle latitudes (Duplessy et al. 2001).

Second short-term cooling took place in the northern Barents Sea within the thermal optimum, about 7.2 cal. ka BP (Fig. 4.10, Duplessy et al. 2001). It was related to a somewhat stronger stratification of the water column that suggests a restricted sea ice freezing brines formation on the shelf. This event is still poorly documented. Hald and Aspeli (1997) consider it as termination of the climatic optimum in the southern Barents Sea. Other authors correlate the cooling event with the mountain glaciers advance in Norway (Nesje et al. 1991), with the amplitude twice less than that during the 8.2 cal. ka cooling (Johnsen et al. 1992). However, a relative warming and decrease in iceberg rafting are noted in the North Atlantic at middle latitudes (50–65°N) during the same time (Duplessy et al. 1992; Bond et al. 1997). Thus, the cooling event was likely limited to high latitudes (Duplessy et al. 2001). Short-term events are also found in the middle-late Holocene (e.g., Voronina et al. 2001) but they should be better investigated.

We ascertained even shorter events in Core ASV-987, 6-m long, raised from the Russian Gavan' Fjord, northwestern Novaya Zemlya, during Cruise 11 of the R.V. *Akademik Sergei Vavilov* (Fig. 4.18). The section covers last 800 years, from 1170 to 1997 (Table 4.4; Murdmaa et al. 2000, 2003, 2004; Ivanova et al. 2003a; Polyak et al. 2004). The modern sedimentation in the fjord is controlled by the balance of the Shokal'sky glacier discharging into the fjord, that is by the relationship between summer melting and winter snow accumulation maintaining the glacier growth. In turn, these processes depend on summer and winter air temperatures and cyclonic activity. The winter snowfall is directly related to the frequency of North Atlantic cyclons (Chizhov 1968; Serreze et al. 1993), hence to the NAO index (see Section Section 1.2.1). Winter temperature on the western Novaya Zemlya coast correlates with the Atlantic water inflow that is controlled by the atmospheric circulation, first

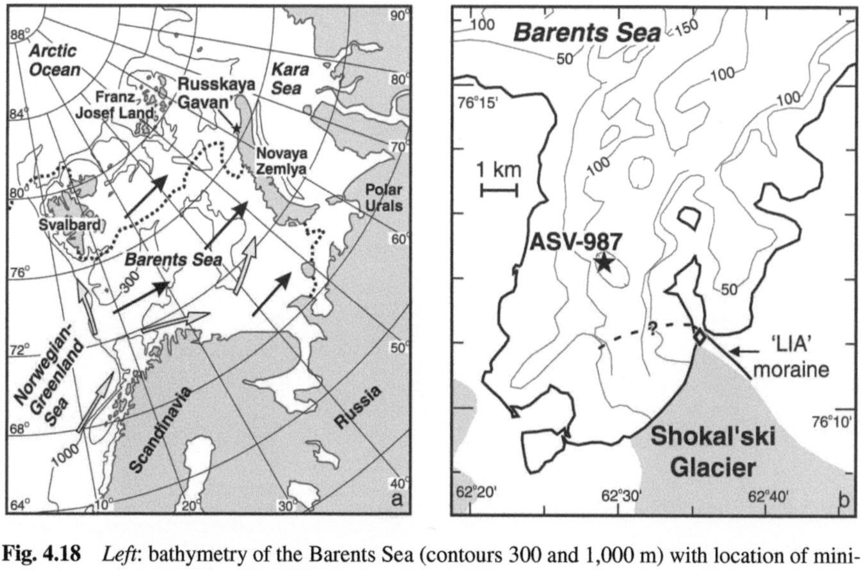

Fig. 4.18 *Left*: bathymetry of the Barents Sea (contours 300 and 1,000 m) with location of minimal winter sea ice margin (*dotted line*), direction of Atlantic water flows (*gray arrows*) and winter cyclones (*black arrows*) after (Murdmaa et al. 2004 with Elsevier permission). *Right*: index map of the Russian Gavan' Fjord with bathymetry in 50-m intervals. Diamond north of glacier shows location of ^{14}C-age constraint from the moraine after (Zeeberg 2001)

of all, by intensity of the Icelandic low and its eastward shift (Ikeda 1989; Zeeberg 2001; Zubakin 1990). Position of the sea ice margin in the Barents Sea (Zubakin 1990) and intensity of water exchange with the Russian Gavan' Fjord also depend on the Atlantic water inflow.

We distinguish three stratigraphic units in the Core ASV-987 section, based on lithology and grain-size distribution of sediments, abundance of micro- and macrobenthos, species composition of both planktic and benthic foraminiferal assemblages, oxygen isotope composition of foraminiferal tests, as well as AMS radiocarbon dates (Polyak et al. 2004; Murdmaa et al. 2004). The units correspond to the end of the Warm Medieval Period (MWP, about 1170 – 1400), the Little Ice Age (LIA, about 1400 – 1900), and the twentieth century warming (Fig. 4.19). These climatic events have been recorded earlier in many regions of the Earth, based on both paleoclimatic and historical data (e.g., Monin and Shishkov 1979; Overpeck et al. 1996; Velichko et al. 1999; Broecker 2000, 2001; Oglivie and Jonsson 2001; Klimenko et al. 2001).

Benthic foraminiferal species, *C. reniforme* and *E. clavatum,* characteristic for proximal glaciomarine environments (Polyak and Solheim 1994; Korsun et al. 1995;

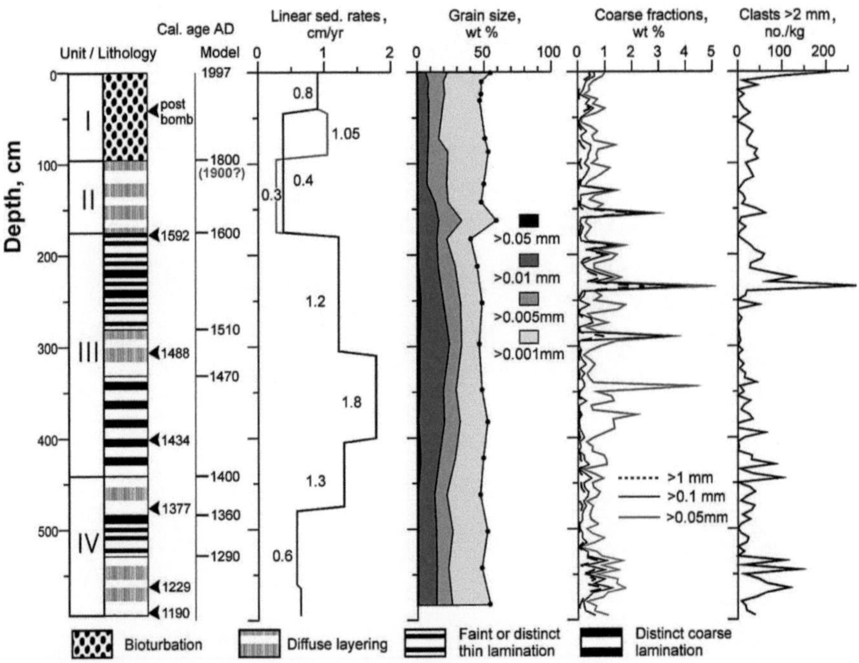

Fig. 4.19 Core ASV 987: stratigraphic subdivision, lithology, calibrated 14 C ages (determined and interpolated for unit boundaries), sedimentation rates estimated between the 14 C age determinations (two alternative models for units I and II), grain-size distribution, coarse fraction content (wt. %), and counted number of >2 mm rock clasts per 1 kg of dry sediment (after Murdmaa et al. 2004 with Elsevier permission)

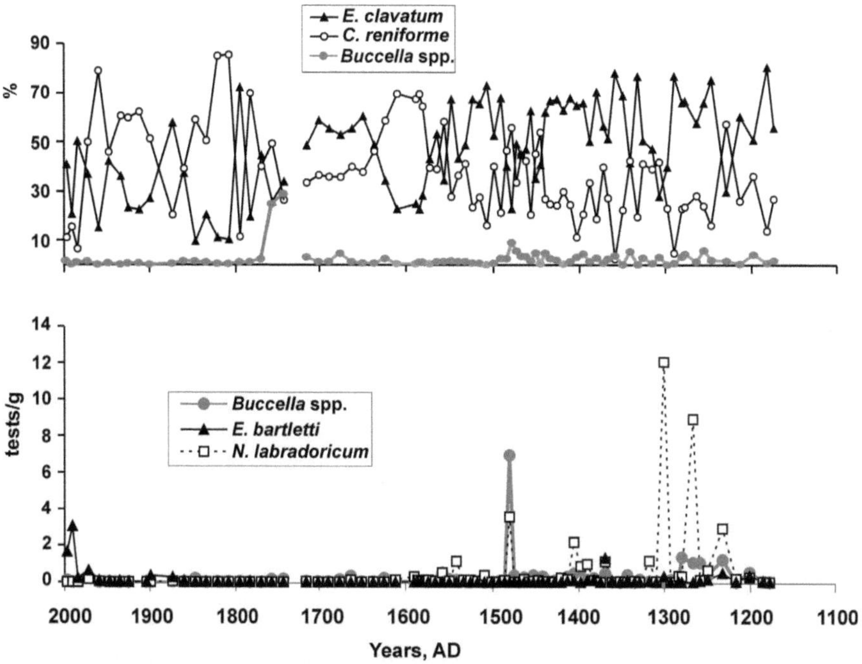

Fig. 4.20 Percentages (**a**) and abundance (**b**) of indicative benthic foraminiferal species in Core ASV-987 from the Russian Gavan' Fjord versus age

Hald and Korsun 1997; Korsun and Hald 1998), dominate throughout the Core ASV-987 section (Fig. 4.20). Description of the units and our interpretation of obtained proxy records are presented below (Color Plate 4.8, Murdmaa et al. 2003, 2004; Ivanova et al. 2003a; Polyak et al. 2004).

The end of MWP is characterized by maximum benthic foraminiferal abundance and their high species variability, relatively low sedimentation rates (6–8 mm/yr.), absence of ice or iceberg-rafted coarse debris (up to the beginning of fourteenth century), moderate bioturbation, mainly high values of oxygen and carbon isotope composition of *E. clavatum* tests. The maximum *Nonion labradoricum* content is also noted here that suggests an abundant fresh organic matter supply to the bottom (Hald and Korsun 1997; Polyak et al. 2002a), as a consequence of high biological productivity during longer ice-free season. Rather high content of polyhaeta tubes and mollusk shells supports this suggestion. Minor amounts of planktic foraminifera in some samples, especially from the interval of early fourteenth century, also possibly point to relatively warm and long ice-free seasons. Rare planktic foraminifera are mainly represented by the boreal species *Turborotalia quinqueloba,* as well as by rare tests of *Neogloboquadrina pachyderma* dex., *Globigerinita glutinata, Globigerina bulloides,* and of Arctic species *N. pachyderma* sin. Our assumption about relatively long ice-free seasons corresponds to existing data on a rather mild climate of the northern hemisphere during the MWP (Jones et al. 2001).

Decrease of marine salt ions in the Greenland ice core GISP2 indicates weakening of the Icelandic Low, that is negative NAO index, for that time (Meeker and Mayewski 2002; Color Plate 4.8). Introducing of boreal species into the Russian Gavan' Fjord was likely related to episodic intensification of its water exchange with the open Barents Sea.

Dating of a mollusk shell from the emerged moraine of the Shokal'sky glacier is an evidence of the glacier front advance into the fjord by several kilometers about 1300–1400 AD (Zeeberg 2001), that is at the beginning of the LIA. The glacier advance possibly led to a double acceleration of sedimentation rates due to progradation of the submarine accumulative apron toward the Core ASV-987 site, today located about 5 km apart from the glacier front. Downslope flows of sea ice freezing brines involving the near-bottom nepheloid layer might also contribute to the increase in sedimentation rates (Murdmaa et al. 2006). The linear sedimentation rate was 12–18 mm/yr. during 15th century. The rapid sedimentation must have been maintained by a sufficiently intense glacier melting. Such an assumption is consistent with the data on variable, but rather high average summer air temperatures in the Northern Hemisphere reconstructed by several authors (Bradley and Jones 1993, Jones et al. 2001).

The glacier front advance was likely related to an intensification of the cyclonic activity and snowfalls owing to reinforcement of the Icelandic Low at the beginning of LIA (Meeker and Mayewski 2002). According to our data, the glacier front most extended into the fjord to 1470 AD and leaved there up to 1600, possibly with a temporary retreat in 1500–1550 AD. Light oxygen and carbon isotope composition of *E. clavatum* tests in some samples, especially around 1470 AD, are evidence of a strong meltwater influence on the bottom environment in the fjord. Brines formation during autumn-winter sea ice freezing might result in a halocline subsidence and lightening of benthic oxygen isotope values (Bauch and Bauch 2001; Duplessy et al. 2001, 2005). Heavier oxygen isotope composition and occurrence of planktic foraminifera in few other samples suggest episodic penetration of the transformed Atlantic water into the fjord. On the other hand, presence of planktic foraminifera (note the peak of *N. pachyderma* sin. about 1430 AD, Fig. 4.10), along with a rather high *N. labradoricum* content and abundance of total benthic foraminifera, as well as a permanent occurrence of polychaeta tubes and mollusk shells, likely reflects a productivity increase. Short-term phytoplankton blooms might have been caused either by the effect of close position of the ice margin or to temperature increase and longer ice-free seasons. If the planktic foraminifera were brought by transformed cooled Atlantic water, which penetrated into the fjord via deep channels, we can assume an estuarine circulation with outflow of the freshened relatively warm surface water from the fjord into the open sea during summer. Abundance of benthic foraminifera in this unit is comparable to that in Early Holocene sediments from other Barents Sea sections studied by the author. The maximum values correspond to the highest abundance counted in the Core ASV-880 from the Franz Victoria Trough (Fig. 4.10). Along with the high productivity, such an abundant occurrence of benthic foraminifera in the Russian Gavan' Fjord also might result from a bacterial activity on the sediment surface (Murdmaa et al. 2004).

Summer air temperatures in the western Arctic region substantially decreased by 1520 AD (Briffa et al. 1995). However, they were still rather high in the Russian Gavan' Fjord and summer seasons were long enough for intense glacier melting, iceberg calving, and rapid accumulation of fine-grained sediments from the glacier meltwater load, as well as for a rather high productivity. Winter cyclonic activity with strong winds led to unstability of the sea ice cover and favored the sediment transport by brines-induced nepheloid flows (the polynya effect resulting in deposition of gray laminae in the laminated sequence). The lamination becomes thinner upward in the section (Fig. 4.19), the amount of macrobenthos remains and coarse fractions (rock fragments up to 13 mm in size) increase. These features point to stabilization of the likely floating glacier front that maintained the rather intense iceberg calving at the end of sixteenth century.

Cooling by 1–3°C in the Barents Sea region during the LIA is thought to correlate with the last glaciers advance on the surrounding archipelagoes. Based on the tree ring data from the Polar Ural, a cooling by 2°C is ascertained at the beginning of sixteenth century (Briffa et al. 1995). Historical data indicate a considerable southward shift of the summer sea ice margin in Barents and Norwegian seas (Vinje 1997). Similar changes are monitored for the Icelandic Low by marine salt ions content in the Greenland ice core GISP2 (Meeker and Mayewski 2002) and by variations of hydrological parameters over the Vøring Plateau, eastern Norwegian Sea (Jansen and Koç 2000).

The late LIA, after 1600 AD, including its culmination in seventeenth century and following global warming (Bradley and Jones 1993; Briffa et al. 1995; Jones et al. 2001), is worse documented in our section of Core ASV-987. It is characterized by less distinct lamination of sediments (Fig. 4.19), lower coarse fractions content, amount of micro- and macrobenthos, as well as by slowing down of sedimentation rates. Owing to a lack of carbonate material for the radiocarbon dating, we could not estimate the real sedimentation rates. However, dates from other intervals (Fig. 4.19) allow us to assume a three-fold drop of the sedimentation rate near the boundary between sixteenth and seventeenth centuries (at the depth 175 cm below sea floor). The above-listed changes likely reflect a considerable glacier front retreat from the fjord in seventeenth century, when the LIA cooling culminated (Color Plate 4.8).

The productivity drop at ~1600 AD possibly reflects cold summer seasons with restricted glacier melting and almost permanent sea ice cover in the fjord. The only peak of *Buccella* spp. content in the interval between 1740 and 1760 AD reflects rather a resistance of this genus against dissolution than increase in productivity, as shown by absence of the peak on the abundance record (Fig. 4.18). The following considerable glacier retreat likely started in seventeenth century.

Low species variability and small number of benthic foraminifera, strong dominance of *E. clavatum* and *C. reniforme,* low *N. labradoricum* content, as well as absence of macrobenthos remains and planktic foraminifera in the late LIA interval (Color Plate 4.8) suggest severe climatic conditions with long winters. Judging by low coarse IRD content and slow accumulation rates of fine-grained material, iceberg calving and glacier melting almost ceased (Murdmaa et al. 2004). Seasonal contrast likely decreased as summers became short and cold, whereas winters were milder than before (Ogilvie and Jonson 2001).

Weakening of the Atlantic water inflow during the LIA culmination likely led to a southward migration of the Polar Front and summer sea ice margin in the Barents Sea. Cold and freshened Arctic waters possibly field the Russian Gavan' Fjord below the almost permanent ice cover. This assumption corresponds to our data on sediment cores with low benthic foraminiferal number and absence of planktic foraminifera from Barents Sea shelf depressions (Duplessy et al. 2001; Ivanova et al. 2002).

Some authors believe that the Little Ice Age might develop as a response to rapid reorganization of the global thermohaline circulation: southward shift of the NADW formation owing to an Arctic water debacle to the North Atlantic (Monin and Shishkov 1998). G. Bond and co-authors (1997) supposed that LIA represents an element of the last 1500-year climatic cycle. However, many authors cast doubt on the reality of such cycles in the Holocene (e.g., Risebrobakken et al. 2003). Advance of the Shokal'sky glacier and other Novaya Zemlya glaciers at the beginning of LIA might result from both higher winter cyclonic activity in the North Atlantic (owing to positive NAO index and weak Arctic oscillation) and decrease in summer air temperatures (Chitzov et al. 1968; Zeeberg and Forman 2000). So far as paleoclimatic data fail to confirm any significant summer temperature decrease at the beginning of LIA (Briffa 2000; Briffa et al. 1995), we suggest that the glaciers advance was at least partially controlled by winter atmospheric circulation. The remarkable increase of marine salt ions concentration in the GISP2 ice core at ~1400 AD pointing to strengthening of the Icelandic Low in winter seasons, with corresponding cyclons penetration into the Barents Sea, supports such suggestion (Meeker and Mayewski 2002; Ivanova et al. 2003a; Polyak et al. 2004).

Foraminiferal assemblages became more variable in the eighteenth century. Percentage of epifaunal species *C. lobatulus* preferring bottom environment with high oxygen content and restricted sediment accumulation (Hald and Korsun 1997; Polyak et al. 2002a) increases in these assemblages (Color Plate 4.8). Such an environment might develop when the glacier front was remote and water exchange with the open sea intensified. Lightening of the oxygen and carbon isotope composition of benthic foraminiferal tests at the end of LIA, as well as increase in *Elphidium bartletti* content in the foraminiferal assemblage (Figs. 4.10 and 4.18), points to a more intense glacier melting and transfer of oxygen isotope-depleted water onto the bottom with sinking brines.

The timing of the transition from the late LIA unit to that of the recent warming, fixed at 1m below sea floor in Core ASV-987, is problematic owing to a lack of radiocarbon datings within both units other than the "post-bomb" age (that is the date of first nuclear weapon tests) at 45 cm assumed to be about 1950 AD. Because a notable warming occurred around the Barents Sea at the beginning of the twentieth century (e.g., Briffa et al. 1995), we consider the time level, ca. 1900 AD, as a possible age for the units boundary. The linear sedimentation rate of 1.05 cm/yr characterizes the interval 0.45–1 m, and that of 0.8 cm/yr is estimated for the post-bomb interval 0–0.45 m (Murdmaa et al. 2004). The acceleration of sediment accumulation at the beginning of twentieth century was related to a more intense material delivery by the glacier meltwater discharge during warmer and longer summer seasons. Further glacier front retreat from the fjord to the land led to cessation of the iceberg rafting. Considerable decrease in sedimentation rates after 1950 possibly reflects a decrease in the Shokalsky glacier melting intensity after the 1940s cooling.

The recommenced estuarine circulation and brines formation led to intensification of bottom currents, as shown by a significant portion of *C. lobatulus* in the foraminiferal assemblage. However, productivity was still rather low, judging by a small foraminiferal number in sediments. The maximum percentage (up to 50%) and abundance of *E. bartletti* at the end of twentieth century is an evidence of the bottom water freshening owing to intense glacier melting, even if we assume a partial influence of selective dissolution on this maximum (Color Plate 4.8, Fig. 4.20). Freshening of the surface water possibly inhibited the phytoplankton bloom, as follows from the lack of planktic foraminifera and small number of benthic foraminifera. Isotope data confirm the freshening (Color Plate 4.8, Murdmaa et al. 2004; Polyak et al. 2004).

Summer temperature increase likely controlled the glaciers retreat in the twentieth century (Dowdeswell et al. 1997). However, the retreat is not ascertained for most of Novaya Zemlya glaciers, likely because it was compensated by a more abundant winter snowfalls (Chizov et al. 1968; Zeeberg and Forman 2000). The twentieth-century climate warming at high latitudes of the Northern Hemisphere is inferred from a substantial (by 33%) reduction of the sea ice area in the Norwegian Sea and the northward retreat of their margin in April during last 135 years (Vinje 2001). It was related to a sea surface temperature rise by 1°C that started ca. 1880 AD, hence to intensification of the global thermohaline circulation.

The retreat of the Shokal'sky glacier likely continues, with minor oscillations, since seventeenth century till present. The glacier melting is also noted on Svalbard and Franz Josef Land archipelagoes (Briffa 2000). Lightening of the oxygen and carbon isotope composition of *E. clavatum* tests and rather high percentage of *E. bartletti* in the benthic foraminiferal assemblage confirm intense melting of the Shokal'sky glacier during the last 50 years (Color Plate 4.8).

4.6 Summary

The grounded ice sheet totally isolated the Barents Sea and northwestern Kara Sea shelf from the World Ocean during the Last Glacial Maximum.

The early deglaciation of the ice sheet was related to the global warming, sea level rise, and episodic penetration of Atlantic waters from the west into the Bear Island Trough and from the north into the Franz Victoria and St. Anna troughs likely since 18 cal. ka BP. The intense melting of the Barents Sea ice sheet at the beginning of deglaciation led to abundant icebergs delivery into the Norwegian Sea and, ultimately, to a rapid meltwater inflow to the North Atlantic during the H-1 (17.3–15.7 cal. ka BP) event. The persistent Atlantic water inflow into the Barents Sea from the west and its penetration to the Arctic along the Svalbard and Eurasian continental slope has to be admitted since 14 or even 16 cal. ka BP.

Deglaciation and the Holocene thermal maximum were apparently time transgressive suggesting stronger AMOC control. The deglaciation started earlier (prior to 16 cal. ka BP.) in the south, later in the northern troughs (since 16–15 cal. ka BP.) and central Barents Sea (15–13 cal. ka BP.), and was the latest within archipelagoes. Similarly, the THM is documented to start earlier in the southwestern part of the sea as compared to the northern troughs.

The onset of marine sedimentation in Barents and Kara seas in the Holocene was accompanied by a substantial surface water productivity rise and increase in abundance of micro- and macrobenthos. Pronounced spatiotemporal millennial-scale variability of postglacial environments in the Eurasian Arctic seas resulted from changes in insolation, sea level rise, both surface and subsurface Atlantic Water inflow, and shifts in seasonal sea ice margin. As the millennial-scale climate variability strongly depended on the AMOC control, the centennial-scale changes seem to be mainly affected by NAO/AO dynamics, notably during the Last Millennium.

Correlation of warming intervals in the unique Last Millennium multi-proxy record from the Russian Gavan' Fjord at Novaya Zemlya with the negative NAO index and cooling intervals with the positive NAO index points to transfer of climatic signals by the atmosphere. Meanwhile, a periodical penetration of the transformed Atlantic water into the fjord, for example, at the end of MWP and during the recent warming, was probably related to intensification of the global thermohaline circulation.

Chapter 5
Paleoceanography of the Northern Indian Ocean: Linkages to Monsoon and Global Thermohaline Paleocirculation

Abstract The foraminiferal assemblages and other marine and terrestrial indicators suggest the onset of the Indian monsoon in the middle or even early Miocene. A number of marine and terrestrial paleoarchives monitor the general pattern of the intensified SW (summer) Indian monsoon during warm stages and NE (winter) monsoon during cold stages at glacial–interglacial and millennial timescales. The numerous short-lasting events were superimposed on this pattern during the Pleistocene–Holocene. The primary forcing mechanisms of the monsoon variability are contrast in land–ocean sensible heating and tropospheric latent heating which in turn largely depend on the orbital parameters of the Earth. Remarkable monsoon-induced variations are ascertained in surface circulation, coastal and open-ocean upwellings, primary production, and intensity of the oxygen minimum zone. Besides, the hydrological parameters and microfossil assemblages in the northern Indian Ocean have been significantly affected by variations in the intensity of the return branch of the global thermohaline circulation as well as the rate of the NADW and Southern Ocean waters propagation into the tropical Indian Ocean during the high (interglacial) and low (glacial) sea level stands. The most pronounced environmental changes are associated with the reorganization of the THC over the terminations.

Keywords Indian monsoon · Tibetan Plateau · Sensible heat · Latent heat · Arabian Upwelling · Return branch · Subtropical gyre · Mechanisms · Climate variability

5.1 Introduction

Over the last ~15 years more and more researchers support the opinion about a significant or even leading role of low latitudes in the global climate change (Imbrie et al. 1992; Sirocko 1996; Sirocko et al. 1996; Charles et al. 1996; Beaufort et al. 1997, 1999, 2001; Wang et al. 1999b; Wang and Sarnthein 1999; Lea et al. 2000, 2007; Berger et al. 2006; Mix 2006). The low latitudes have gained a growing attention as their warm water and air masses can better transform any increase in the solar radiation into an increase in the atmospheric content of the water vapor, hence providing the greenhouse effect (Kerr 1994) along with carbon dioxide and methane.

E.V. Ivanova, *The Global Thermohaline Paleocirculation*,
DOI 10.1007/978-90-481-2415-2_5,

In the Indo-Pacific region, a significant part of the tropical area is largely affected by monsoons both in the ocean and on the surrounding continents. In the geological past, the monsoon dynamics were controlled by interaction of external (solar radiation, glacial-age boundary conditions, including land–ocean temperature contrast and continental albedo, ENSO) and internal forcing mechanisms (e.g., Clemens et al. 1991, 1996; Clemens and Prell 1991; de Garidel-Thoron et al. 2001; Ivanova 2002c; see futher discussion in Section 5.6).

Observations and modeling results show that the monsoon represents a very complicate system with feedbacks not fully understood up to date. The monsoon circulation in the atmosphere and ocean is primarily caused by seasonal redistribution of the sensible and latent heat fluxes (Clemens et al. 1991, 1996), and thus of the atmospheric lows and highs. Monsoons are sensitive to heat transport variations in the surface branch of the oceanic conveyor which may lead to short-term oscillations of the monsoon circulation (Zahn 1994). However, linkages between the monsoon and thermohaline circulations are still enigmatic.

According to the prevailing point of view, Indian and East Asian monsoons are considered as parts of the united Asian monsoon system that is controlled by the relationship between atmospheric pressure over the Tibetan Plateau and adjacent marine basins. Many authors emphasize a crucial importance not only of orographic barriers, especially of the Himalayas and Tibetan Plateau but also of the Southeastern Asian, Eastern African, Indonesian, and New Guinean mountains, in the onset and further evolution of the monsoon system (e.g., Ruddiman and Kutzbach 1989; Kutzbach et al. 1989, 1996; Ivanova and Ivanova 1996a, b; Ruddiman 1998; Webster et al. 1998). The Tibetan Plateau extends from 25° to 45°N and from 70° to 105°E. Its average height exceeds 4,000 m above the sea level. If the Tibetan Plateau did not exist there, the cold air from the west would penetrate much farther to the east, instead of turning to low latitudes (Webster et al. 1998).

5.2 Modern Surface Circulation in the Northern Indian Ocean

Seasonal contrasts of surface winds are nowhere so impressive as in the northern Indian Ocean (Ocean Circulation 1989; Hastenrath 1991; Fig. 5.1). The Indian monsoon represents a dynamic interaction between atmosphere, ocean, and continents. The sensible heating of the Tibetan Plateau and adjacent mountains depends on the solar radiation, Earth surface topography, snow line position in mountains, and water content in soils (Fig. 5.2; Clemens et al. 1991). These factors, in turn, are linked to the global heat budget (mainly to that of the Northern Hemisphere). The latent heat transport from the subtropical Indian Ocean to the Indian coast and the intensity of SW monsoon in the Arabian Sea are largely controlled by the Southern Hemisphere heat budget and by the size of Antarctic ice sheet.

The stable subtropical high-pressure cell over the Southern Indian Ocean is located at the sea level and centers at approximately 30°C (Neiman et al. 1997). In summer, a sharp atmospheric pressure gradient between this high and the South Asian Low causes the strong southwest monsoon over the northern Indian Ocean. In the winter, the South Asian low-pressure cell is replaced by the Tibetan High and

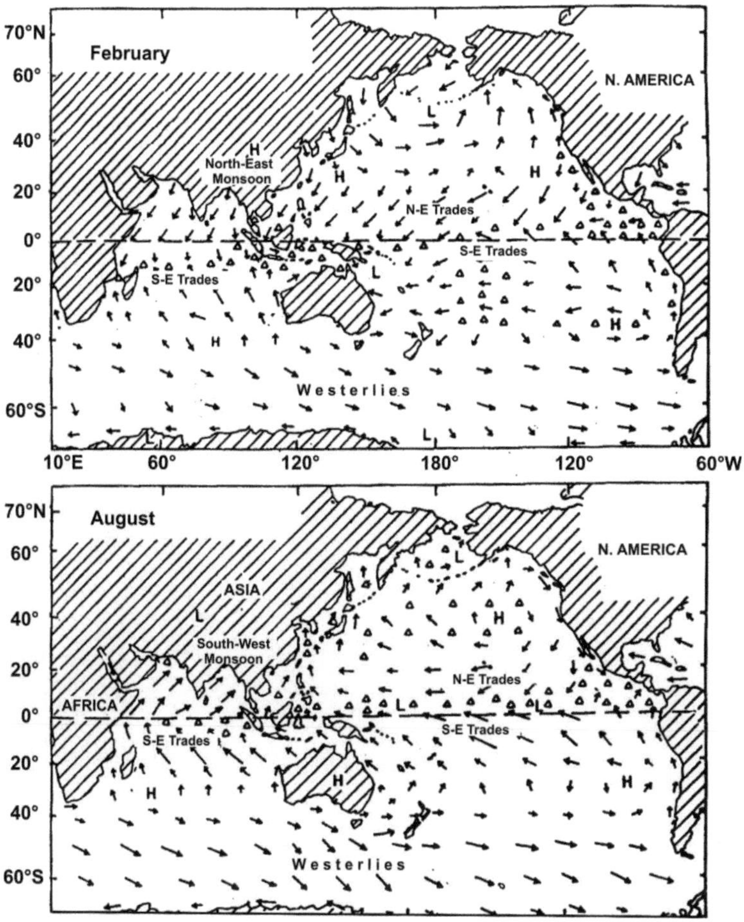

Fig. 5.1 Atmospheric circulation over Indian and Pacific oceans (after Pickard and Emery 1990). Printed with permission from Elsevier

a poorly manifested low-pressure cell is formed between the Tibetan and Southern Indian Ocean highs. Thus, northeast monsoon blows from Asia toward this low (Romanov 1994).

The strongest winds are associated with a low-level Findlater Jet that reaches its easternmost extension during the summer monsoon (Fig. 5.3; Anderson and Prell 1991; Romanov 1994; Madhupratap et al. 1996). Findlater Jet originates over the subtropical high-pressure cell, proceeds along the coast of Africa, and then crosses the Arabian Sea from the southwest to the northeast. It forces the southwest monsoon and supplies the latent heat to the southern coasts of Asia. The seasonal change in monsoon direction controls the corresponding seasonality of surface circulation in the northern Indian Ocean (Figs. 5.3 and 5.4).

During the SW monsoon (May through September) the South Equatorial Current intensifies, its northern branch turns northeast to form the Somali Current, as part of the anticyclonic summer circulation (Fig. 5.3). Intense coastal upwellings

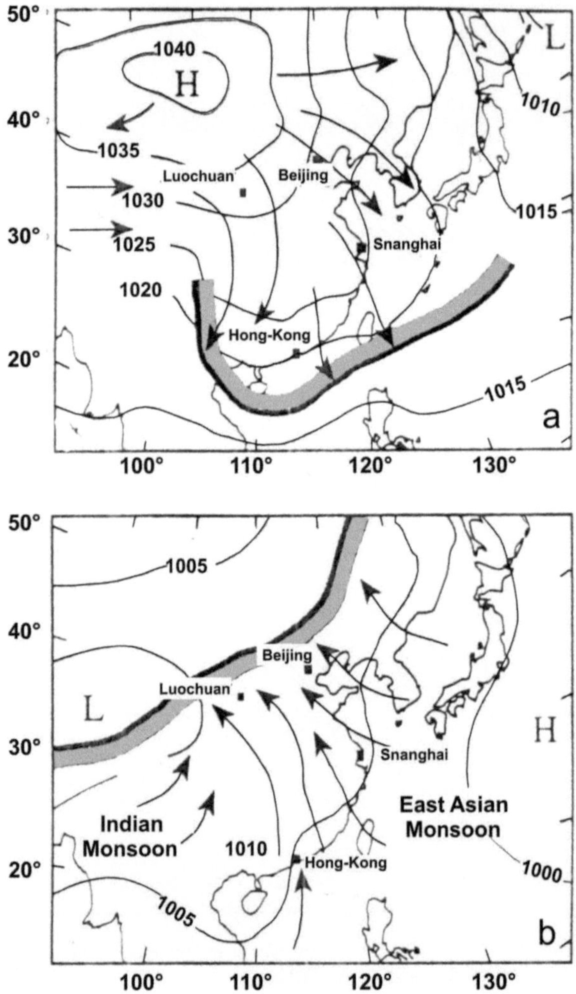

Fig. 5.2 Location of atmospheric pressure centers over the Tibetan Plateau and adjacent Pacific in winter (**a**) and summer (**b**) (after Porter and An 1995 with Nature permission). Printed with permission from AGU

along Somalia, Oman, and southwestern coast of India are associated with the SW monsoon, whereas an open-ocean upwelling is reported to be associated with the Findlater Jet. It occurs to the northwest of the Findlater Jet axis as a result of Ekman pumping (Anderson and Prell 1991; Brock et al. 1992; Rixen et al. 1996) and due to the filaments appearing off the Oman coast and extending to the southeast (Washburn and Armi 1988; Arnone et al. 1998; Manghnani et al. 1998). The Somali Current is also considerably controlled by the Findlater jet (Webster et al., 1998).

In contrast to the summer circulation pattern, cyclonic circulation prevails during the gentler NE monsoon (November through March). The NE monsoonal hydrography is characterized by sea surface cooling and convective overturning that cause the mixed-layer deepening (Filyushkin 1987; Bartolacci and Luther 1999). NE monsoonal winds are generally too weak to induce offshore Ekman pumping

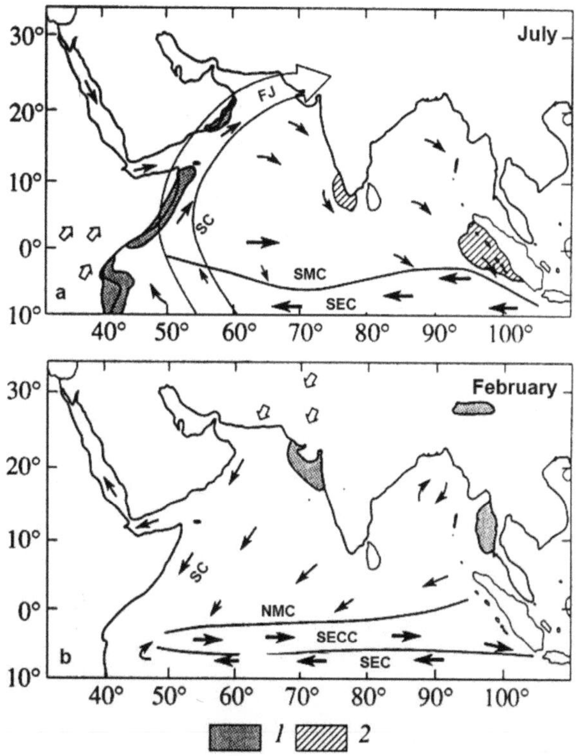

Fig. 5.3 Surface circulation of the northern Indian Ocean (after Clemens et al. 1991; Kroon 1991; Romanov 1994; Ivanova 2002c). Currents: (**a**) in July (summer monsoon maximum); (**b**) in February (winter monsoon maximum). *White arrows* show monsoon direction, *black arrows* indicate surface currents. (1) Seasonal upwellings; (2) episodic upwellings. The big *white arrow* shows the direction of latent heat transport from the high-pressure cell over the southern subtropical Indian Ocean to the Asian continent by the Findlater Jet

and upwelling takes place sporadically off Karachi and Malacca (Wyrtki 1973; Kroon 1991; Neiman et al. 1997). The winter circulation comprises the westward Northeast Monsoon Current, the southward Somali Current, and the South Equatorial Countercurrent (SEC). Unlike other oceans, in the Indian Ocean SEC is shifted to the Southern Hemisphere. In winter, the cyclonic circulation develops in the Arabian Sea, and the anticyclonic circulation occurs in the Bay of Bengal (Wyrtki 1973; Burkov and Neiman 1977; Burkov 1980). As shown in Section 1.1 (Fig. 1.5), warm Pacific waters feed the South Equatorial Current via the Indonesian Throughflow, and thus also penetrate westward to the Arabian Sea.

The reduced surface water temperatures, especially off the Oman coast, are also associated with the Ekman drift and upwellings (Levitus 1982; Brock et al. 1992; Rixen et al. 1996; Peeters 2000). Phosphate and nitrate content in the surface water is commonly higher in summer than in winter (Krey and Babenerd 1976; Ivanenkov 1979). In the Arabian Sea, it generally decreases from the west to the east. Nutrients are delivered to the northern part of the sea by filaments and eddies originating from the upwellings, or entrained into the photic zone from below during the winter convection (Gardner et al. 1999; Bartolacci and Luther 1999). The upwellings and associated vertical fluxes of nutrients caused by the Ekman pumping are responsible for the maxima of primary production (Brock et al. 1992; Prahl et al. 2000; Schulte and Müller 2001).

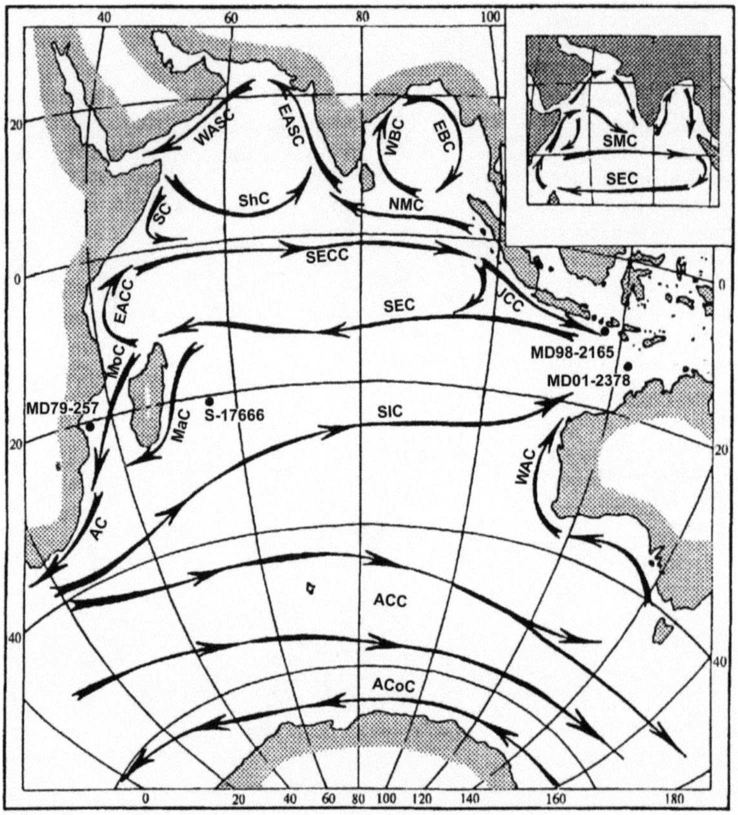

Fig. 5.4 Scheme of the surface circulation of the Indian Ocean during the boreal winter (during the boreal summer – in the insert) (modified after Neiman et al. 1997). WASC (EASC) = West (East) Arabian Sea Current, ShC = Shtokman Current, SC = Somali Current, NMC (SMC) = Northeast (Southwest) Monsoon Current, WBC (EBC) = West (East) Bengal Current, JCC = Java Coastal Current, SECC= South Equatorial Countercurrent, SEC= South Equatorial Current, MaC = Madagascar Current, MoC = Mozambique Current, AC = Agulhas Current, SIC = South Indian Ocean Current, WAC = West Australian Current, ACC= Antarctic Circumpolar Current, ACoC = Antarctic Coastal Current. *Black dots* indicate location of sediment cores considered in Section 5.7 after (Ivanova et al. 1995; Holbourn et al. 2005; Waelbroeck et al. 2006; Levi et al. 2007; Dürkop et al. 2008)

The mean annual and seasonal SST in the monsoon region restricted by the front at 10°N (Wyrtki 1973; Tchernia 1978; Gordon 2001) ranges within the narrow limits, from 25 to 30°C (Wyrtki et al. 1971; Levitus 1982; Levitus and Boyer 1994; Neiman et al. 1997). Thickness of the upper quasi-isothermal layer varies from 10 to 150 m (Filyushkin 1987). However, high vertical temperature gradients characterize the thermocline layer. High salinity, owing to a strong evaporation, in the Arabian and Red seas (35.5–37 and 37–41 psu, respectively) results in intense convection at the thermocline and intermediate water depths (Wyrtki et al. 1971;

Tchernia 1978; Gordon 2001). The saline intermediate waters penetrate to the south and are traced below the Agulhas Current (Gordon et al. 1992; see Section 1.1).

The intermediate water upwells in the equatorial divergence between 3° and 7°C (Neiman et al. 1997). The upwelling here is driven mostly by the westerlies (Walker circulation) as in other oceans. The equatorial westerlies are related to the Southern Oscillation Index that is to the pressure gradient between Tahiti and Darwin (see Section 1.22). They are especially strong between 7°N and 7°C in spring and autumn, but not in winter and summer as monsoon winds (Lappo et al. 1990; Beaufort et al. 1997, 1999).

The subtropical anticyclonic gyre in the southern Indian Ocean comprises the currents of permanent directions as in other oceans. The warm return surface branch of the global thermohaline circulation is associated with the SEC, Madagascar, and Mozambique Currents on the northwestern periphery of the anticyclonic gyre, whereas the Antarctic Circumpolar Current forms the southern periphery of the gyre. The warm waters of the return conveyor's branch are transported by the Agulhas Linkage from the Madagascar and Mozambique Currents to the Atlantic (see Section 1.1). The major salt, heat, and nutrients transport from the Atlantic to the Indian and Pacific oceans occur at intermediate and deep levels below the ACC. The AABW propagates northward from the convective cells in the Southern Ocean and is traced in the Arabian Sea at the depth of 3,800–4,000 m (Bruevich 1966; Lappo 1984; Lappo et al. 1990; Gordon 1986, 1989, 2001; Koshlyakov et al. 2001).

The well-expressed oxygen minimum layer at depths of 100–1,500 m characterizes the northwestern part of the Indian Ocean (Wyrtki et al. 1971; Wyrtki 1973; Qasim 1982; von Rad et al. 1999). It is formed due to the high biological productivity in the euphotic zone, denitrification, and microbial organic matter decay and isolation of the region in the north. An upwelling uplifts the oxygen minimum layer to the surface off southern India (Wyrtki 1973; Qasim 1982).

5.3 Onset and Development of the Monsoon Circulation in the Indian Ocean During the Neogene

The formation of Indian Ocean in its close to modern shape started after the opening of the ocean gateways separating Antarctica from the other continents in the Paleogene. Although the precise timing of circum-Antarctic gateways is still controversial, they certainly opened not later than in the early Oligocene as a result of the deep-sea spreading (see Section 1.3; Fig. 1.7; Lawver et al. 1992; Allen and Armstrong 2008, and references therein). The Indian Ocean became separated from the Thetys on the northwest in the early Oligocene after the collision between Arabian and Eurasian Plates (Allen and Armstrong 2008; see Section 1.3, Fig. 1.7). Some studies suggest that monsoon circulation existed near the southern Asian coast already in the Cretaceous, prior to the collision between Indian and Asian plates (Hemleben et al. 2001). However, this region was a part of Tethys Ocean in the Cretaceous, whereas in the proto Indian Ocean the monsoon circulation could not develop during the Mesozoic (Ivanova 2000 and references therein).

The analysis of published micropaleontological data obtained in Deep-Sea Drilling Legs 23, 24, and 115 (Vincent and Toumarkine 1990) shows that already at the beginning of the Miocene, 22.5–21 Ma ago, when equatorial–tropical and equatorial planktic foraminiferal assemblages spread over the northern part of the ocean, relatively cold-water species occurred at sites 709, 710, 714 near the equator (Ivanova 1994; Ivanova and Ivanova 1996a, b). This enables to assume that the equatorial divergence might appear at that time (Ivanova 1994; Ivanova and Ivanova 1996a). More cold-water assemblages, as compared to surrounding ones proliferated near the equator throughout the Miocene and Pliocene (DSDP/ODP sites 707, 709, 710, 712, 236), off the southern India at the end of early Miocene (sites 714, 715), at the end of late Miocene – beginning of the Pliocene and in the late Pliocene (Site 219), as well as in the Gulf of Aden at the end of middle Miocene and in the late Pliocene (sites 231, 233). These assemblages likely point to the appearance of monsoon-driven upwellings (Ivanova 1994; Ivanova and Ivanova 1996a, b; Fig. 5.5).

Diatom abundance and their assemblages dominated by *Thalassionema* group, oxygen isotope records on benthic foraminifera, increased uvigerinids content, and species diversity of bolivinids are evidence of intense upwelling accompanied by rather high primary production and development of the oxygen minimum zone off the southern Indostan 17–10 Ma ago (Boersma and Mikkelsen 1990). Appearance

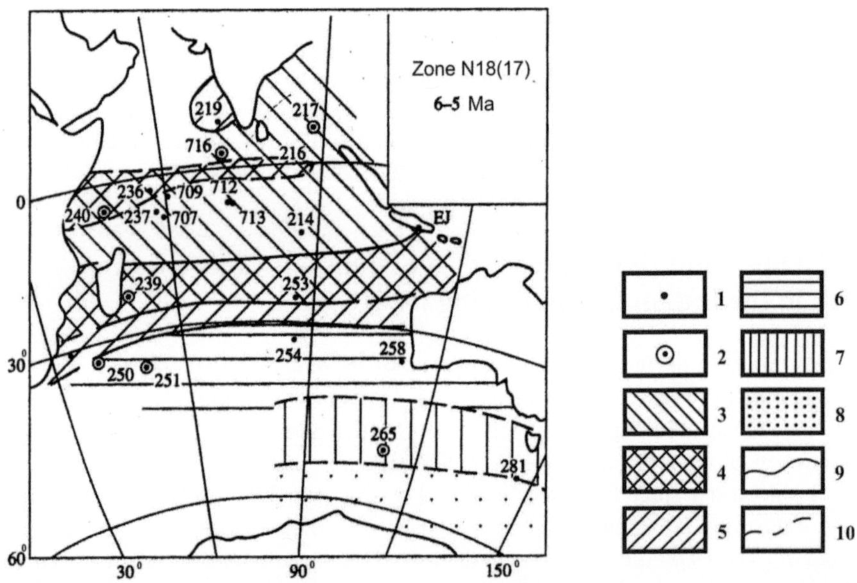

Fig. 5.5 Climatic zones of the Indian Ocean in the Late Miocene reconstructed based on planktic foraminiferal data from the DSDP/ODP reports (modified from Ivanova and Ivanova 1996a). Symbols: (1) deep-sea drilling sites with well-preserved foraminiferal assemblages; (2) deep-sea drilling sites with foraminiferal assemblages strongly affected by dissolution; (3–8) climatic zones: (3) Equatorial; (4) Equatorial–Tropical; (5) Tropical; (6) Subtropical; (7) Temperate; (8) Subpolar + Polar; (9, 10) boundaries between the zones: (9) established and (10) hypothetical

of the cold-water species *Coscinodiscus marginatus* in diatom assemblages points to the upwelling intensification off the Arabian and southern Indian coasts (Boersma and Mikkelsen 1990).

The preconditions of the onset of the monsoon circulation were closing of the Tethys Ocean in the east (Adams et al. 1983), uplift of the Himalayas and Tibet, and increase in the land–sea thermal gradient (Ruddiman and Kutzbach 1989; Prell and Kutzbach 1992; Klootwijk et al. 1992; Ruddiman 1998). Final isolation of the Antarctic Continent by the circumpolar current (Lawver et al. 1992), as well as the growth of its ice sheet, especially intensive in the Middle Miocene (Shackleton and Kennett 1975), also favored the development of monsoon circulation. Modeling experiments showed that the Tibetan uplift played an important role not only in the monsoon onset but also in changes of the Northern Hemisphere atmospheric circulation. When uplifted, Tibet became the barrier that intensified zonal westerlies and impedes the heat exchange between low and high latitudes (Ruddiman and Kutzbach 1989; Kutzbach et al. 1989). The cooling that involved the entire Indian Ocean at the end of middle Miocene (Ivanova 1994; Blyum et al. 1988; Ivanova et al. 1989; Ivanova and Ivanova 1996b) was also manifested over the southern subtropical gyre. This might lead to intensification of the latent heat transfer to the southern Asia, thus promoting the monsoon onset and development.

Several authors suggest that the monsoon circulation originated later, in the middle Miocene (Levitan and Seidov 1989) or the late Miocene (Prell et al. 1992), whereas upwellings off the Somali and Oman coasts appeared not later than in the late Miocene (Kroon et al. 1991). Geochemical data, coccolithophoride assemblages, and increased *Globigerina bulloides* content reveal an intensification of the Arabian upwelling in the late Miocene (Prell et al. 1989; Rahman and Roth 1990; Kroon et al. 1991). This is coeval with the Antarctic ice sheet growth (i.e., increase in the meridional thermal gradient) and with a large-scale plankton bloom in the tropical Indo-Pacific from 6 to 4 Ma ago. The latter was accompanied by an extension of the oxygen minimum zone in the Indian Ocean. The plankton bloom was possibly related to intensification of the monsoon circulation and decrease in oxygen content within the intermediate layer (Gupta and Thomas 1999).

Significant reduction of the circum-equatorial water exchange via the narrowing of Central American and Indonesian gateways about 4.6–4 Ma ago (Kennett et al. 1985; Haug and Tiedemann 1998) led to the dramatic reorganization of the global thermohaline circulation portrayed by the planktic foraminiferal assemblages (Blyum et al. 1987, 1988; Ivanova et al. 1989; Haug and Tiedemann 1998; Haug et al. 1999). Onset of the Northern Hemisphere glaciation in the late Pliocene, 2.73 Ma ago, also caused considerable changes in the THC and the water column structure of the Indian Ocean.

Spectral analysis of lithological indicators of the upwelling at the ODP Site 722, drilled on the Owen Ridge, shows that high-frequency oscillations of the monsoon circulation were mainly driven by precession from 3.4 to 2.5 Ma ago (Murray and Prell 1992). A subsequent shift toward longer-term oscillations characteristic for high latitudes was likely related to the onset of the Northern Hemisphere glaciation. 41-ka obliquity cycle dominated the paleoarchives of aeolian dust outblow from African and Arabian deserts during the late Pliocene and early Pleistocene (DeMenocal et al. 1991). A 100-ka eccentricity cycle became prevailing in the paleodata records since the mid-Pleistocene revolution established between oxygen isotope stages 25 and 22 (Farrell and Janecek 1991; Maslin and Ridgwell 2005). However, as it is shown below, the monsoonal climate in the northern Indian Ocean and on surrounding lands is primarily driven by the precession and obliquity cycles which are superimposed on the eccentricity cycles (e.g., Clemens et al. 1991, see Section 5.6).

5.4 Variations in Paleoenvironments and Monsoon Circulation During the Late Pleistocene–Holocene

Modeling experiments show that the onset of high land–sea thermal gradients triggering the intensification of monsoon circulation needs a strong global-scale cooling including the decrease in the North Atlantic SSTs, the buildup of the ice sheet in Europe and the extensive snow cover in Asia (Overpeck et al. 1996). Such conditions appeared only in the Pleistocene. Location of the deep-sea cores and DSDP/ODP sites used to study variations in different monsoon circulation proxies are shown in Fig. 5.6.

An analysis of numerous paleoceanographic proxies in the ODP Site 723 drilled on the Arabian continental slope, about 100-km offshore, within the Oman upwelling center (Anderson and Prell 1991, 1993; Niitsuma et al. 1991; Zahn and

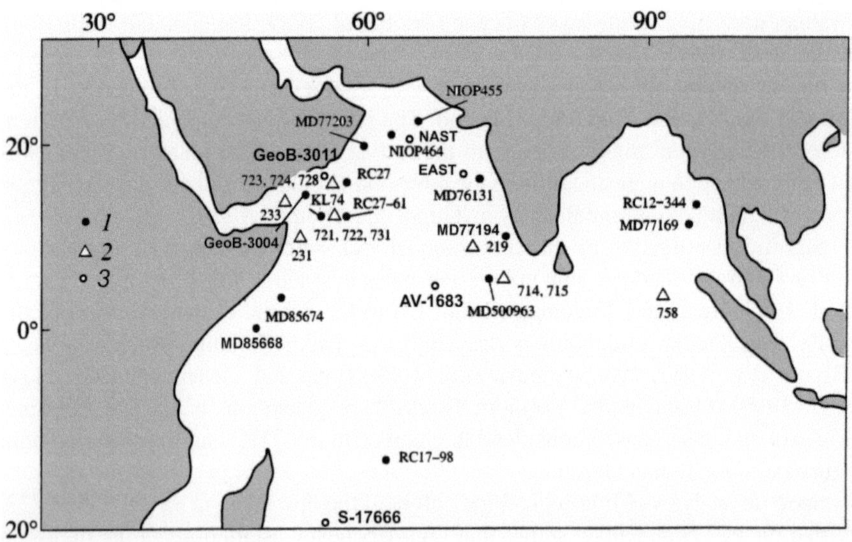

Fig. 5.6 Location of sediment cores and deep-sea drilling sites data from which are used in this chapter: (1) cores; (2) boreholes; (3) cores studied by the author (from Ivanova 2002c with additions). The full list of references is given in the paper cited and in the Sections 5.3, 5.4, 5.5, 5.6, and 5.7

Pedersen 1991; Emeis et al. 1995), revealed intensification of the SW summer monsoon and an increase in productivity mainly during interglacials. However, the U^k_{37}_SSTmeasurements carried out in the ODP sites 724 and 728 displayed the concurrent warmings. Some authors believe that the mean anuual SSTs were more affected by the winter cooling than by the summer monsoon weakening (Emeis et al. 1995; Rostek et al. 1997).

An indirect evidence for the monsoon intensification during the interglacial stages is provided by the elevated amount of foraminiferal fragments at Site 728 drilled on the continental slope of Arabia, at the water depth 1427.8 m. The high fragmentation suggests an enhanced dissolution due to the productivity increase in the euphotic zone and water undersaturation with calcium carbonate (Anderson and Prell 1991). Data on the adjacent shallower Site 723 (water depth 808, Niitsuma et al. 1991) and core Geo B-3004 from the middle bathyal (14°36.3N̆, 52°55.20Ĕ, water depth 1,803 m, Schmiedel and Leuschner 2005) confirm an increase in organic matter content and intensification of the monsoon-controlled upwelling during interglacial stages. During the glacial MIS 2, 4, and 6, the upwelling was weaker due to the lower snowline in the mountains and reduction of the thermal gradient between land and sea (Anderson and Prell 1993; Ivanova et al. 2003b). The similar pattern was ascertained near the Sokotra Island (10°N) where the high productivity intervals in MIS 3, 40–50 cal. ka BP, and after the LGM were interrupted by the weakening of the upwelling in MIS 2 (Ouahdi 1997).

However, not all the paleorecords point to the strong Indian summer monsoon during interglacials. Several geochemical, lithological, and micropaleontological proxy records display the events of intensified monsoons during both glacial and interglacial stages (Fontugne and Duplessy 1986; Clemens and Prell 1991; Emeis et al. 1995; Naidu and Malmgren 1996a, b; Rostek et al. 1997; Almogi-Labin et al. 2000). Therefore, they support the earlier assumption about the strengthening of winter monsoon during glacials (Duplessy 1982).

According to our data (Ivanova 1985, 1988; Ivanova et al. 1990), the mean annual SST fell by 2°C below the modern values in the equatorial divergence and by 3.5–4°C in the southern Arabian Sea during the stadials conceivably due to more intense equatorial westerlies, and also by 3–4°C in the Gulf of Aden. Similar cooling values during the last glacial were inferred from the alkenone measurements in the northern Arabian Sea (3.5–4°C, Schulte and Müller 2001) and in the tropical Indian Ocean (1.5–2.5°C, Sonzogni et al. 1998).

In the Red Sea, the oxygen minimum zone reduced during cold phases owing to a more intensive sinking of oxygen-rich waters in the northern part of the sea and weaker water advection from the Gulf of Aden because of the lower sea level stand (Ivanova 1985; Almogi-Labin et al. 1998). These events were responded by a higher number of mesopelagic pteropods during the coolings (Almogi-Labin et al. 1998), as well as by a dramatic decrease in planktic foraminiferal abundance. The species diversity of pteropods also decreased at the LGM, when the connection with the Indian Ocean ceased via the Bab el Mandeb Strait and the surface water salinity rose up to 50 psu (Ivanova 1985; Ivanova and Kiselev 1985). Input of the Red Sea water into the returning branch of the global thermohaline circulation in the Agulhas

Linkage area (see Section 1.1) must have been weakened correspondingly. On the contrary, the stronger oxygen minimum zone developed due to productivity increase in the Arabian Sea during interstadials (e.g., Schulz et al. 1998).

The Somalian upwelling became stronger at glacial–interglacial boundaries, about 130 and 15–10 cal. ka BP, as well as within the interval 65–25 ka BP (Caulet et al. 1992). The summer monsoon-driven variations of this upwelling were also affected by changes in the sea level and Pacific water inflow to the western Indian Ocean by the upper limb of the global thermohaline circulation (Vergnaud-Grazzini et al. 1995; Vénec-Peyré et al. 1995).

The faithful multiproxy analysis of several sediment cores from the Arabian Sea (Sirocko et al. 1993; Overpeck et al. 1996) provided timing for two short-term phases of monsoon strengthening within the Termination I. The first phase (13–12 ka BP, average 14.5 cal. ka BP) roughly coincides with Bølling warming and second phase (10–9.5 ka BP or about 11.4 cal. ka BP) corresponds to the beginning of Holocene (Fig. 5.7; Sirocko et al. 1993). Both phases were characterized

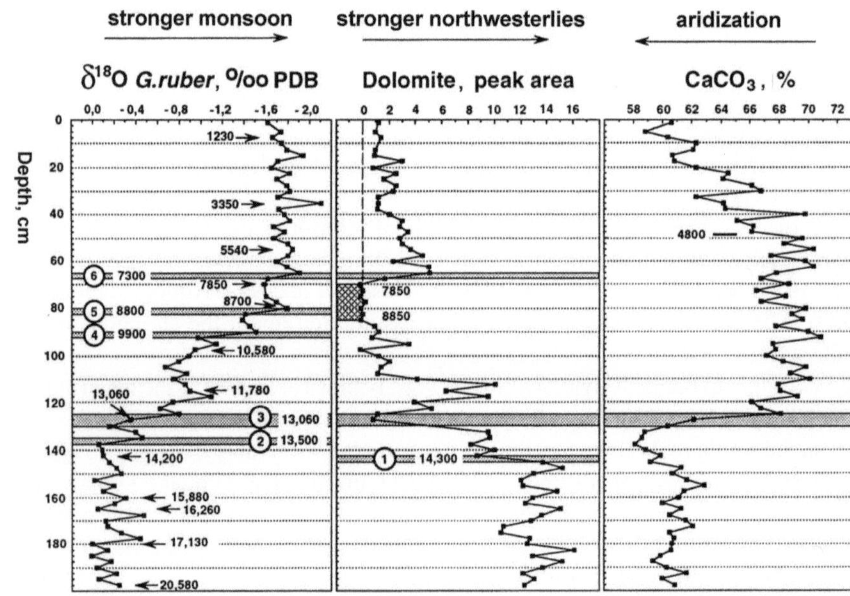

Fig. 5.7 Variations in indicators of the summer (southwest) monsoon-driven upwelling and northwesterlies during the last 24 cal ka in Core 74KL from the western Arabian Sea (for location see Fig. 5.6). (**a**) $\delta^{18}O$ record on *Globigerinoides ruber* indicating the intensity of Arabian upwelling; (**b**) dolomite indicating the influx of the Persian Gulf (Arabian) dust by the northwesterlies at site 74KL location; (**c**) $CaCO_3$record – indicator of the local aridization (modified after Sirocko et al. 1993 with Nature permission). The hatched area between 85 and 70 cm depth on the dolomite record indicates the depth interval of the summer monsoon maximum in the early Holocene. The short line at 50-cm depth on the $CaCO_3$ record indicates the knick point (change in profile) toward the late Holocene aridification. The *shaded belts* mark several short-term (300-years lasting) distinct events resulting from the rapid response of local climate to slow solar forcing. Radiocarbon ages are also shown

by the vigorous upwelling and more intense moisture advection over the Indian-Himalaya region, Arabia, and Africa. The palynological diagrams of sediment cores from the Arabian upwelling area (Overpeck et al. 1996) show an increase in taxones related to the NE winter monsoon. The sea surface salinity increased in the Bay of Bengal, near Site 758, approximately at the same time, likely as a result of the Arabian Sea water advection owing to the stronger SW monsoon and related Monsoon Current (Chen and Farrell 1991). In the northern Arabian Sea (Core KL 136; Schulte and Müller 2001), a fast warming by 2.5°C is ascertained during the first deglaciation phase (from H-1 to Bølling/Allerød) and a more gradual warming is fixed after the Younger Dryas (Fig. 5.8). The intensification of the summer monsoon is also evidenced by the SST and oxygen isotope data from the Andaman Sea core RC12-344 (12.46°N; 96.04°W; water depth 2,140 m) indicating the warming and increase in Irrawaddy River outflow during the Bølling–Allerød and low evaporation–precipitation (Rashid et al. 2007). The reversed conditions were documented for the Younger Dryas thus pointing to a weaker SW monsoon.

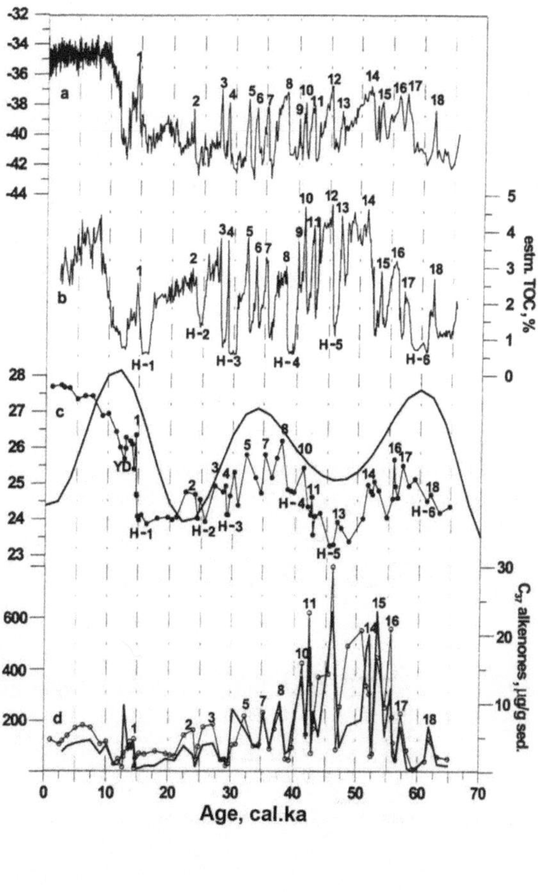

Fig. 5.8 Alkenone levels, sea surface temperature (SST), and gray-scale-based TOC record for Core 136 KL (23°7.34' N, 66° 29.83' E, water depth 568 m) from the northwestern Arabian Sea (**b–d**) compared to the oxygen isotope ($\delta^{18}O$) record for the GISP2 ice core (**a**) after (Schulte and Müller 2001 with Springer permission). Core 136 KL: (**b**) calculated gray-scale-based TOC record; (**c**) sea surface temperatures (*closed circles*) estimated from values of C_{37}–alkenones (the line without circles represents the insolation at 30°N); (**d**) content of C_{37}–alkenones versus age (*thin line with open circles*) and mass accumulation rate (MAR) of alkenones, $\mu g\ cm^{-2}$ 1,000 $year^{-1}$ (*thick line without circles*). Numbers(1–18) Daansgard-Oeschger events; (H-1–H-6) Heinrich events; YD = Younger Dryas

High humidity and lakes level rise are noted in Tibet, Arabia, and East Africa in the early Holocene, 11.5–5 cal. ka BP (Street and Grove 1979; van Campo et al. 1982). The pluvial was interrupted in Tibet and Ethiopia by an episode of drier climate 8–10 ka BP (Sirocko et al. 1993; Gasse and Van Campo 1994). Onset of pluvial conditions 11–8 cal. ka BP coincided with the suumer insolation maximum at 30–40°N (Sirocko et al. 1993; Zahn 1994; Gasse and Van Campo 1994; Overpeck et al. 1996). The increased river runoff into the ocean, particularly into the Bay of Bengal, led to a decrease of the surface water salinity in its northern part (Cullen 1981). The highest monsoon intensity is ascertained to occur about 3 ka after the insolation maximum and continued from 11 to 5 cal. ka BP (Overpeck et al. 1996). It was accompanied by the precipitation maximum and high lakes level stand in some regions of Arabia, India, and Africa (Street and Grove 1979; Ivanova 1988).

The data from ODP Site 723 off Arabia show that the summer monsoon-driven upwelling was the most vigorous 10.6–4.8 cal. ka BP (Naidu and Malmgren 1996a). However, the reduced summer and winter SST documented around the 8.2-cal. ka cold event (see also Section 3.6) are interpreted in terms of short-term decline of the upwelling (Naidu 2006). The study of planktic foraminiferal assemblages from the eastern Indian Ocean, near the Sunda Archipelago, provided evidence for weakening of the return branch of the global thermohaline circulation and strengthening of monsoonal upwelling off Java 10–8 ka BP (Ding et al. 2006). As today this upwelling is driven by the austral winter monsoon, these data seem to be in line with the intensification of the Northern Hemisphere summer (Southern Hemisphere winter) monsoon in the early Holocene.

Later on, the Indian monsoon and the Arabian upwelling weakened gradually owing to degradation of the Northern Hemisphere glaciation and the solar insolation decrease (Overpeck et al. 1996). The climate aridization related to weakening of the SW monsoon started at 5.3 cal. ka BP and reached the pessimum at 3.2–0.6 cal. ka BP (Sirocko et al. 1993; Overpeck et al. 1996). Data on Ethiopian lakes level, paleo-hydrology of the western Tibet, and palynological spectra of sections from the northwestern India and the eastern Arabian Sea support such suggestion (Ivanova 1988, 2002c; Gasse and Van Campo 1994). Several centennial-scale episodes of weaker monsoon and relatively arid climate are identified during deglaciation and early Holocene in several terrestrial and marine archives (Sirocko et al. 1993, 1999; Overpeck et al. 1996). Decadal-to-centennial-scale monsoon-controlled average summer SST variations of 2–2.5°C during the Holocene are inferred from the oxygen isotope record in Core 905 off Somalia (Jung et al. 2004).

Some recent studies (e.g., Fleitmann et al. 2007) suggest that the episodes of the summer monsoon strengthening were short living and superimposed on the gradual weakening trend of both Indian and East Asian monsoons and southward migration of the ITCZ during the Holocene owing to decreasing solar insolation. The centennial-to-decadal variations in the monsoonal precipitation inferred from stalagmites in Oman and Yemen caves are coherent with the temperature variations reconstructed in the Greenland ice cores.

5.5 Paleoproductivity Variations in the Arabian Sea Related to Monsoon and Thermohaline Circulation

As it was mentioned in the previous section, the primary production (PP) in the Oman coastal upwelling generally increased during the interglacials. We tried to reconstruct the quantitative variations in PP in eutrophic (western), mesotrophic (northern) and oligotrophic (eastern) parts of the Arabian Sea over the last climatic cycle with respect to the summer and winter monsoon oscillations (Ivanova et al. 2003b). The studied cores (Table 5.1, Fig. 5.9a) have been obtained in cruises 31/3 and 33/1 of the German R.V. *Meteor*. The oxygen isotope measurements on

Table 5.1 Location of the cores discussed in the Sections 5.5 and 5.7 (after Ivanova et al. 2003b; Fretzdorff et al. 2000)

Site	Location		Water depth (m)
	Latitude	Longitude	
EAST	15°35.5′ N	68°34.9′ E	3820
NAST	19°59.9′ N	65°41.0′ E	3167
GeoB 3011-1	16°32.1′ N	55°19.9′ E	2636
S-17666	20°36.5’ S	55°00.8’ E	3290

Planktic foraminiferal assemblages are studied by the author in all cores except for the EAST.

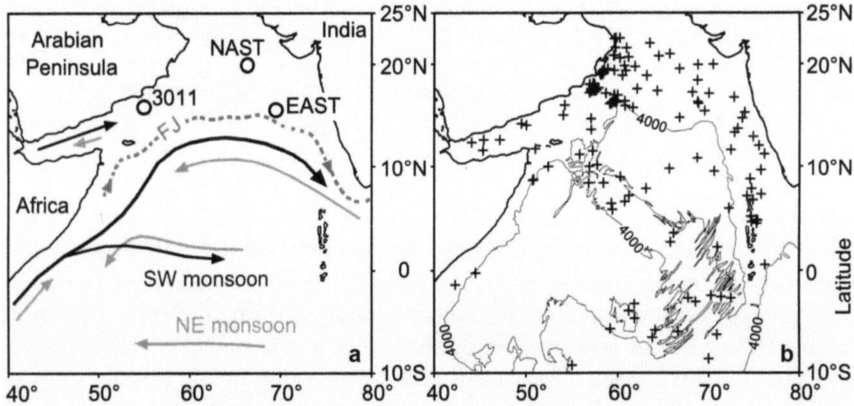

Fig. 5.9 The seasonal surface hydrography (**a**) and bathymetry (**b**) of the Arabian Sea with the location of the three sediment core (*circles*) GeoB 3011-1 (3011), NAST, EAST (**a**) and core tops (**b**) discussed in this section. (**a**) During the SW (summer) monsoons, a strong anticyclonic surface ocean circulation (*dark gray arrow*) is developed, which follows the direction of the Findlater Jet (FJ, *dashed gray arrow*). The cyclonic NE (winter) monsoonal circulation (*light gray arrows*) is weaker than the SW monsoonal circulation. Most of the 186 surface sediment samples (*crosses*; **b**) and the sediment cores (*circles*; **a**), which have been studied, are located above 4,000-m water depth, and foraminiferal faunas may be only slightly affected by dissolution (after Ivanova et al. 2003b with Elsevier permission)

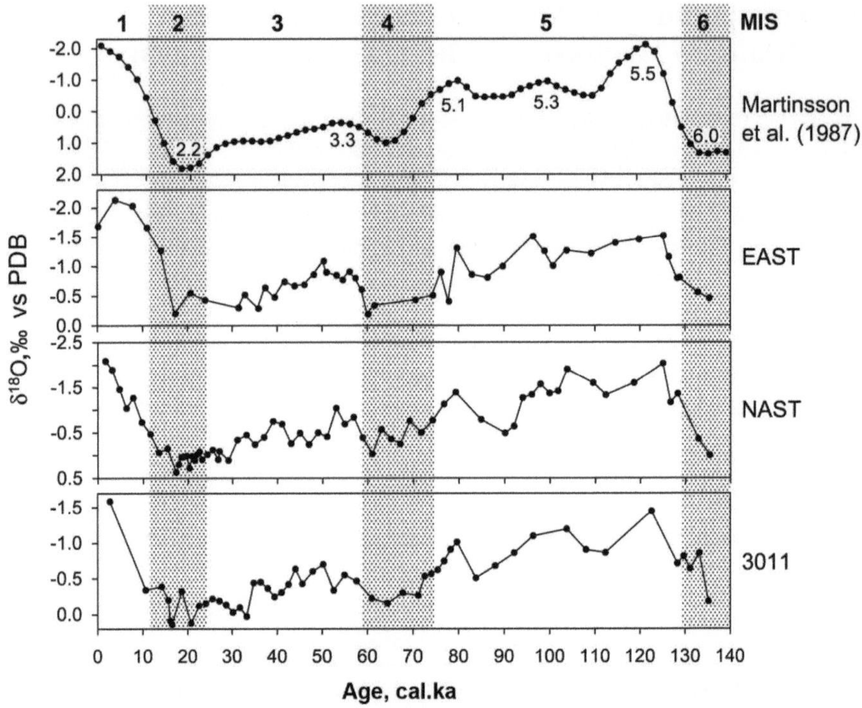

Fig. 5.10 Age model of the cores GeoB 3011-1 (3011), NAST, and EAST, according to the oxygen isotope record of *Globigerinoides ruber* ($\delta^{18}O$), ^{14}C data, and the last appearance datum level (LAD) of *G. ruber* (*pink*) (after Ivanova et al. 2003b with Elsevier permission). The age model was constructed by correlation with the SPECMAP stack of Martinson et al. (1987) using *AnalySeries* (Paillard et al. 1996). Glacial oxygen isotope stages (MIS) are indicated by stippling

Globigerinoides ruber and AMS-^{14}C datings on planktic foraminifera were carried out at the Leibniz Laboratory for Age Determination and Isotope Research at Kiel University, Germany, and at the Zurich Polytechnical School (ETH), Switzerland.

The oxygen isotope stratigraphy and age model of each core (Fig. 5.10) were established by graphic peak-to-peak correlation of planktic $\delta^{18}O$ curves versus the standard SPECMAP stack of Martinson et al. (1987) using the *AnalySeries* software (Paillard et al. 1996). AMS-^{14}C dates and the disappearance datum of *G. ruber* (pink) from the Indian Ocean, at Termination II, around 128 ka (Table 5.2) were used as calibration points. The AMS-^{14}C dates <24 ka are converted into calendar ages with the CALIB 4.3 software (Stuiver and Reimer 1993). Reservoir ages of 400 and 800 years are assumed to account for the reservoir effect of upwelling at all three sites (Cayre and Bard 1999; Naidu and Malmgren, 1996b). The AMS-^{14}C dates >24 ka are converted into calendar ages according to corrections suggested by Bard et al. (1990). Sedimentation rates (Fig. 5.11) are perceived by linear interpolation between tick points of the age model.

Table 5.2 Conventional ^{14}C ages and calendar ages for the cores at EAST, NAST, and GeoB 3011-1, and LAD for *Globigerinoides ruber* (pink) assumed as 128 cal ka BP (modified from Ivanova et al. 2003b).

Core	Depth in core, cm	Conventional ^{14}C age, years BP.	Calendar age (years) BP.	LAD *G. ruber* (pink), cm
EAST	90	28720 (+340/–330)	32070 (±1520)	466
NAST	140	24210 (+250/–240)	27560 (±1520)	550
3011-1	18	12740 (±70)	13563–13287	473
3011-1	33	14210 (±70)	15893–15419	
3011-1	53	14930 (±80)	16,742–16226	
3011-1	78	19570 (±130)	22209–21434	
3011-1	153	30260 (+350/–330)	33210 (±1520)	

Calendar ages younger than 24 ka are calibrated according to Stuiver and Reimer (1993; 1 sigma). Older samples are calibrated according to Bard et al. (1990). At EAST and NAST a reservoir effect of 400 years is assumed. At site 3011-1 a reservoir effect of 800 years is assumed due to strong upwelling in this area.

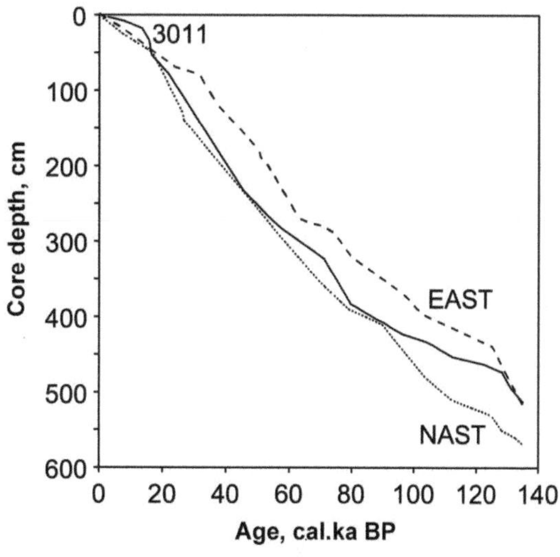

Fig. 5.11 Sedimentation rates in cores GeoB 3011-1 (3011), NAST, and EAST from the Arabian Sea, given as core depth (cm) versus age (ka BP), range between 1.0 and 24.2 cm ka^{-1} at GeoB 3011-1 and between 1.6 and 9.3 cm ka^{-1} at NAST and EAST (after Ivanova et al. 2003b with permission from Elsevier)

PP was estimated applying principal component analysis and transfer function to planktic foraminiferal assemblages. The quantitative foraminiferal analysis was performed by the author for cores GeoB 3011-1, NAST and by A.D. Singh for the core EAST (see Fig. 5.9a for sites location). Planktic foraminifera counts on the species level were performed on at least 300 specimens from aliquots of the size fraction >150 microns. The sampling interval 5–10 cm and relatively high sedimentation rates (Fig. 5.11) provide average time resolution for the oxygen isotope record and calculated paleoproductivity record from 2 to 2.7 ka.

We selected core top foraminiferal counts from 186 sites: 163 sites from Prell et al. (1999), 22 sites from Cayre et al. (1999a), and one sample from the EAST site

Table 5.3 Scaled varimax factor scores of 22 common foraminiferal species in 186 core tops from the Arabian Sea (E. Ivanova and R. Schiebel, unpublished; reproduced from (Ivanova 2003))

Species\Factors	F1	F2	F3	F4
G. conglobatus	−0.06	0.275	0.126	0.316
G. ruber	0.479	**4.197**	0.345	−0.532
G. tenellus	0.125	0.251	0.004	−0.125
G. sacculifer	−0.146	1.701	0.368	0.82
S. dehiscens	−0.013	0.015	0.032	0.181
G. aequilateralis	0.005	0.458	0.126	0.464
G. calida	0.103	0.244	0.041	0.083
G. bulloides	**4.529**	−0.447	0.027	0.258
G. falconensis	0.967	−0.049	−0.132	−0.301
G. digitata	−0.033	0.042	−0.025	0.119
G. rubescens	0.089	0.213	0.029	−0.099
*N. pachyderma*dex.	0.007	0.012	0.073	0.298
N. dutertrei	0.357	0.587	0.11	**2.407**
G. conglomerata	−0.046	0.256	0.106	0.23
G. hexagona	−0.038	0.246	−0.037	0.041
P. obliquiloculata	−0.244	0.162	−0.216	1.521
*G. truncatulinoides*dex.	0.001	0.002	0.009	0.028
G. crassaformis	−0.007	0.017	0.031	0.115
G. menardii	−0.254	−0.197	−0.713	**3.299**
G. tumida	−0.058	−0.161	−0.109	1.051
G. menardii flexuosa	−0.016	0.001	−0.057	0.065
G. glutinata	0.083	0.52	**-4.592**	−0.484

(Figs. 5.9b; Ivanova et al. 2003b) as modern analog data for calculation of transfer functions on the surface sediment foraminiferal assemblage. To address statistical significance, the rare species occurring at a frequency of <2% were excluded from a total dataset following the common practice (Prell 1985; Pflaumann et al. 1996; Waelbroeck et al. 1998; Barash and Yushina 1999a, b). The input dataset contains 22 planktic foraminiferal species. Data on the mean annual PP at the selected sites refer to Antoine et al. (1996). Q-mode principal component analysis (PCA), the regression, and the transfer function analysis were performed using the programs PaleoToolBox and WinTransfer (http://www.pangaea.de/ Software/PaleoTools/), which contain the original CABFAC program package of Imbrie and Kipp (1971).

The PCA of the planktic foraminiferal surface dataset revealed four factors (F1–F4), which explain 96.9% of the total variance, providing high communalities (>0.82; on average 0.97) for a total of 186 samples. The varimax factor scores computed by R. Schiebel for the common 22 species (Table 5.3) demonstrate the remarkable difference between the factors F1–F4, whereas mapping of the areas with maximum factor loading values reveals their relationship to ecological parameters (Fig. 5.12). The factors correlate with the trophic state of the surface ocean, the sea surface temperature, and calcite dissolution (Ivanova et al. 2003b).

The most significant species of Factor 1 that describes 36.5% of the total variance is *G. bulloides* (factor score 0.97). The regional distribution of high loadings in F1

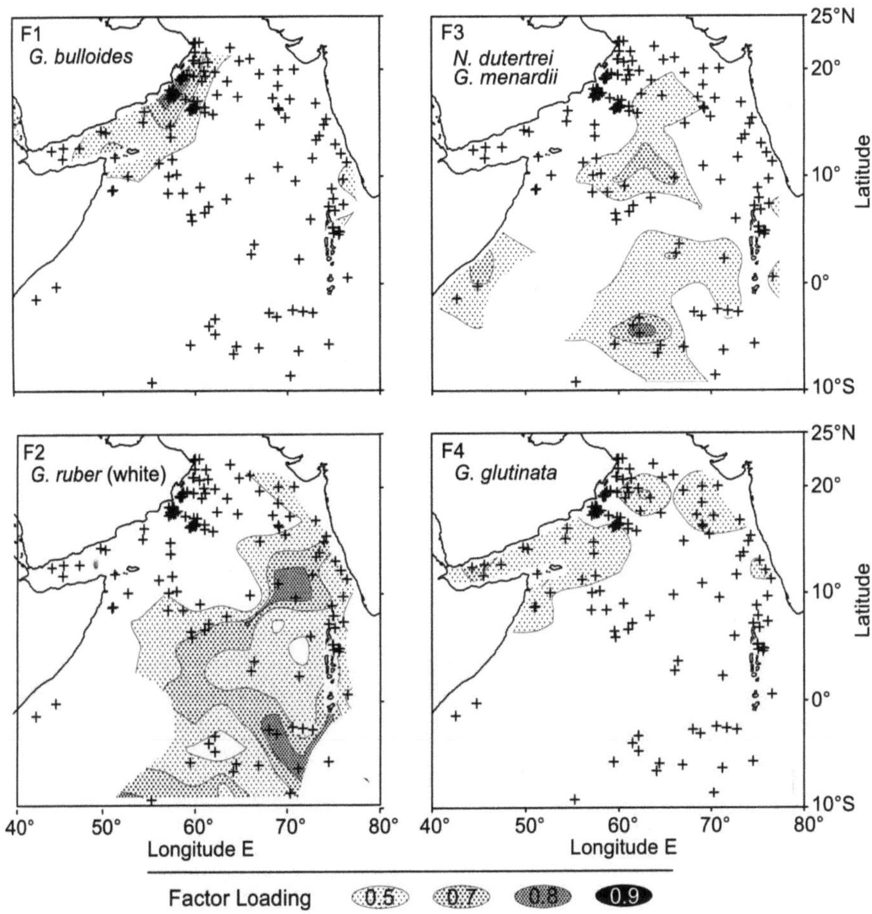

Fig. 5.12 Regional distribution of the four multivariate factors calculated from the surface sediment data set (186 samples, see Fig. 5.9 for location). Factor loadings are given below the four panels. Planktic foraminiferal species with the highest factor scores are *G. bulloides* (F1) *G. ruber (white)* (F2), *N. dutertrei* and *G. menardii* (F3), and *G. glutinata* (F4), indicating upwelling (F1), warm and oligotrophic surface waters (F2), calcite dissolution (F3), and diatom production following to nutrient entrainment (F4), respectively. For details see Table 5.3 (from Ivanova et al. 2003b with permission from Elsevier)

displays the area of strongest upwelling within the Arabian Sea (Fig. 5.12). This is in a good accordance with previous data on *G. bulloides* distribution in the water column and surface sediments (Ivanova 1988; Kroon and Ganssen 1989; Naidu and Malmgren 1996a, b; Peeters 2000). Factor 2 (variance = 21.4%) is dominated by *G. ruber* (white) and *Globigerinoides sacculifer* and mirrors regions of highest surface water temperatures and low primary productivity in the central and southern Arabian Sea. The highest factor score of factor 3 (27.9% of the total variance) is represented by thick-walled tests of *Globorotalia menardii, Pulleniatina obliquiloculata,* and

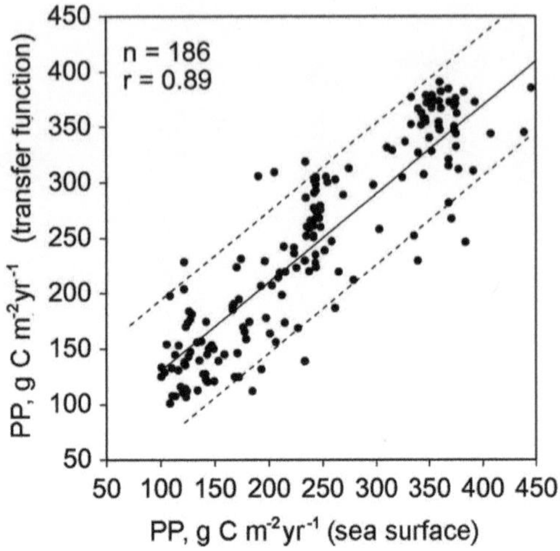

Fig. 5.13 The modern primary production (PP, sea surface; Antoine et al. 1996) plotted versus the estimated PP (transfer function) at the same sites (modified from Ivanova et al. 2003b). The coefficient of a linear regression for the whole set of 186 core top sites is equal to 0.89 ($y = a + bx$, with a = 39.4, $b = 0.84$). *Dashed lines* indicate the standard deviation of PP = 49.3 g C $m^{-2}year^{-1}$

Neogloboquadrina dutertrei that are species resistant to dissolution. The regional distribution of high loadings in F3 is rather patchy and primarily matches the deepest parts of the sea. Hence, this factor may display enhanced calcite dissolution. Factor 4 (variance = 11.2%) is made up of *Globigerinita glutinata* (factor loading = –0.98), the species associated with the increased but not the highest productivity within the open-ocean upwelling and filaments area of the northern Arabian Sea (Anderson and Prell 1993; Manghnani et al. 1998; see Section 5.1) and in the Gulf of Aden, between Somalian and Arabian upwellings (Fig. 5.12).

Factor loadings of F1 and estimated productivity rates are well positively correlated to modern primary production (PP) values (after Antoine et al. 1996) at the individual core top sites (Fig. 5.13; Table 5.4). The standard deviation of the derived transfer function is ± 32 g C m^{-2} $year^{-1}$for a PP calibration range from 100 to 450 g C m^{-2} $year^{-1}$.

Paleoproductivity rates of the three cores vary between 120 and 380 g C m^{-2} $year^{-1}$ (Color Plate 5.1). The average communality of core samples is 0.92 and >0.7 throughout the whole three cores dataset (Ivanova et al. 2003b). As the significant non-analogous conditions were not detected, all of the initial samples are considered in the following reconstruction.

The magnitude of variations in paleoproductivity during the last 135 ka appeared to be less pronounced in the Oman upwelling representing one of the most productive areas in the World Ocean, as compared to the lower productivity area in the northeastern Arabian Sea. The highest productivity values up to 250–320 g C m^{-2} $year^{-1}$ estimated in Core GeoB 3011-1 reflect intensified upwelling due to stronger SW monsoon mainly during interglacials and interstadials. The minimum values, 210–270 g C m^{-2} $year^{-1}$, characterize stadials (Fig. 5.14). Correlation of the productivity maximums with interstadials is confirmed by high concentrations and accumulation rates of benthic foraminiferal tests during MIS 3, as well as radiolarians in MIS 1, 3, 5 (Fig. 5.14).

Table 5.4 Results of the regression analysis: multiple regression coefficients and regression statistics for the transfer function of 186 samples, 22 taxa, and four factors (after Ivanova et al. 2003b with permission from Elsevier)

Variable	Regression coefficient	Standard error of regression coefficient	τ-value
$F1^2$	621.49	264.91	2.35
$F3^2$	−145.72	197.71	−0.74
F1F4	307.44	332.53	0.93
F2F3	549.94	267.37	2.06
F1F3	199.85	277.58	0.72
F3F4	81.92	311.54	0.26
F3	−561.08	433.28	−1.30
$F2^2$	−222.75	206.33	−1.08
F2	236.41	447.46	0.53
F2F4	178.75	321.26	0.56
$F4^2$	444.48	276.63	1.61
F1F2	158.78	284.05	0.56
F4	−603.29	519.20	−1.16
F1	−504.49	501.45	−1.01
Intercept	200.81		
Statistical value		PP	
Minimum of absolute residuals		0.3 g Cm^{-2} $year^{-1}$	
Maximum of absolute residuals		173.3 g Cm^{-2} $year^{-1}$	
Mean value of absolute residuals		34.2 g Cm^{-2} $year^{-1}$	
Standard deviation of absolute residuals		33.7 g Cm^{-2} $year^{-1}$	
Overestimated samples		105	
Underestimated samples		81	

PP = primary production.

At the NAST core site, within the mesotrophic northern Arabian Sea, estimates of the primary production reach values as high as 200–350 g C m^{-2} $year^{-1}$ in interstadial MIS 3.1, 3.3, 5.3, 5.5, as well as at the boundary MIS 5/MIS 4 (Fig. 5.15). They dropped down to 120–180 g C m^{-2} $year^{-1}$ during cold stadials, deglaciations (terminations I and II), and the Holocene. At the EAST core site, within the oligotrophic part of the sea, the productivity values generally exceeded 200 g C m^{-2} $year^{-1}$ during MIS 5, likely reaching maximum values of about 370 g C m^{-2} $year^{-1}$ around the boundary MIS 5/MIS 4 (Fig. 5.16). However, these maximum values should be considered with caution as they are obtained on the few foraminiferal assemblages that are close to non-anologous, yet the commonality is >0.9. The primary production decreased to values 120–210 g C m^{-2} $year^{-1}$ during MIS 4-1 (Ivanova et al. 2003b).

Reconstructed PP variations in three Arabian Sea cores significantly exceed the standard deviation of estimate (±49 g C m^{-2} $year^{-1}$, see Fig. 5.13) and mainly result from oscillations of the summer and winter monsoon circulation in the Indian Ocean. As it is shown in Section 5.3, the SW monsoon and related Oman upwelling, as well as the filaments and open-ocean upwelling in the northwestern Indian Ocean

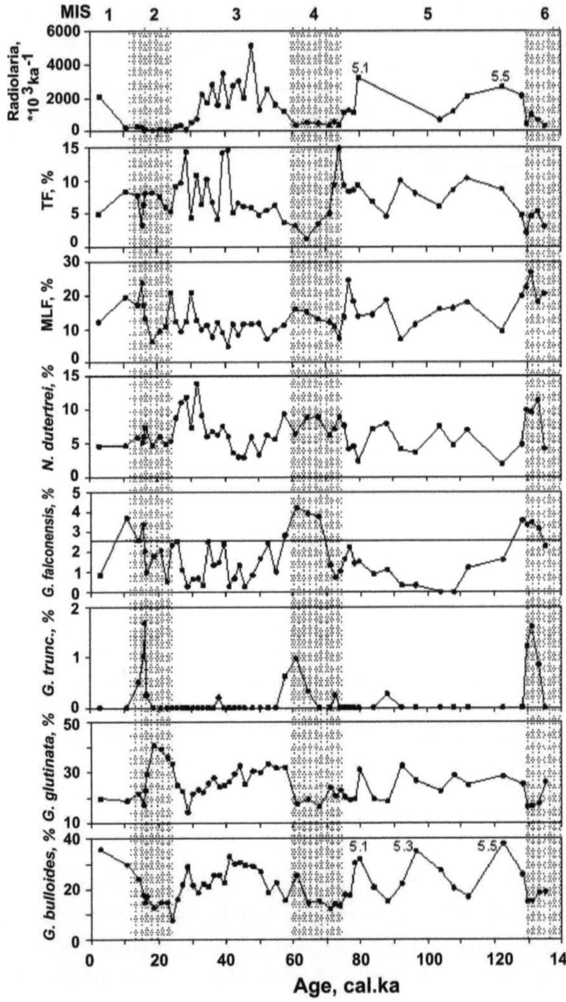

Fig. 5.14 Time series of several indicative planktic foraminiferal species, mixed layer foraminifera (MLF): *G. ruber* plus *G. sacculifer*; thermocline foraminifera (TF): *G. menardii, G. tumida, G. crassaformis, P. obliquiloculata*; and accumulation rates of radiolarians in Core GeoB-3011 from the western Arabian Sea showing distinct variations associated to climatic oscillations (modified from Ivanova et al. 2003b). For core location see Fig. 5.9 and Table 5.1. MIS = oxygen isotope stages. Glacial stages are indicated by shading. Maximum frequency of *G. bulloides* and radiolarians during substages MIS 5.5–5.1, and MIS 3, indicate enhanced surface water productivity. Enhanced faunal parts of *G. truncatulinoides* (*G. trunc.*) and *G. falconensis* during the glacial stages may point toward a substantial change in surface-to-subsurface water circulation

intensified over interglacials and interstadials due to the solar radiation increase during the boreal summer and stronger heating of the Tibetan Plateau (CLIMAP 1984; Ivanova 1988, 2002c; Anderson and Prell 1991, 1993; Clemens et al. 1991; Zahn and Pedersen 1991). The Findlater Jet axis might turn to the east, thus favoring mixing and surface water productivity increase in the eastern Arabian Sea near the EAST core site (Ivanova et al. 2003b).

The high primary production values characteristic for interglacials and interstadials, especially in the northwestern Arabian Sea (GeoB 3011-1 and NAST), are coherent with intensified summer monsoons (Anderson and Prell 1991, 1993; Clemens et al. 1991; Zahn and Pedersen 1991; Emeis et al. 1995). This is corroborated by the comparison of productivity and monsoon index record (Color Plate 5.1)

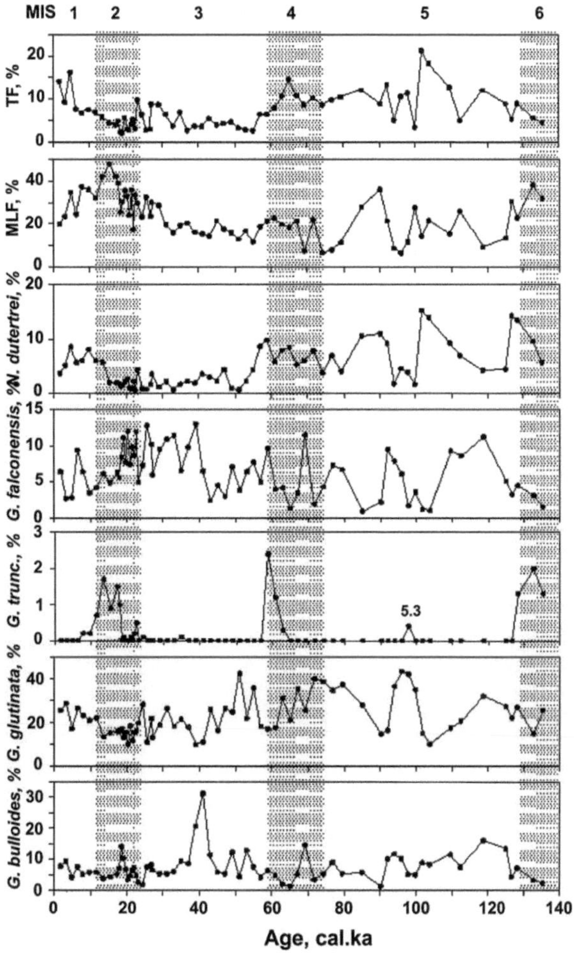

Fig. 5.15 Time series of several indicative planktic foraminiferal species, mixed-layer foraminifera (MLF), and thermocline foraminifera (TF) in NAST core from the northern Arabian Sea showing distinct variations associated to climatic oscillations (modified from Ivanova et al. 2003b with additions). See Fig. 5.9 and Table 5.1 for location and Fig. 5.14 for legend. The maximum frequency of *G. bulloides* at 40 ka may indicate enhanced upwelling during the early substage MIS 3.1. Maximum frequency of *G. glutinata* may indicate enhanced production of diatoms by nutrient entrainment into the surface water during the substages MIS 5.5, 5.3, and 5.1. Enhanced faunal parts of *G. truncatulinoides* (*G. trunc.*) during the glacial stages are similar to GeoB 3011-1 and may point toward an increased intermediate water input into the Arabian Sea

that reflects high summer insolation at low latitudes (Rossignol-Strick 1983). Yet, the comparison of PP record from the EAST core with the monsoon index record shows much worse correlation than for the sites GeoB 3011-1 and NAST. This observation points to the lower sensitivity of the oligotrophic EAST site to the monsoon-induced changes in productivity.

At sites GeoB 3011-1 and NAST productivity maxima follow high monsoon index values (i.e., precession minima) with a lag of several thousand years (Color Plate 5.1). Similarly, the lag of maxima in productivity and other monsoon tracers relative to maxima in precessional radiation was found by G.-J. Reichart and colleagues (1997, 1998; Fig. 5.17) and S. Clemens and colleagues (1991; Fig. 5.18).

The productivity maxima correlate with high faunal portion of *G. bulloides* and *G. glutinata* in foraminiferal assemblages (Figs. 5.14, 5.15, and 5.16; Appendix,

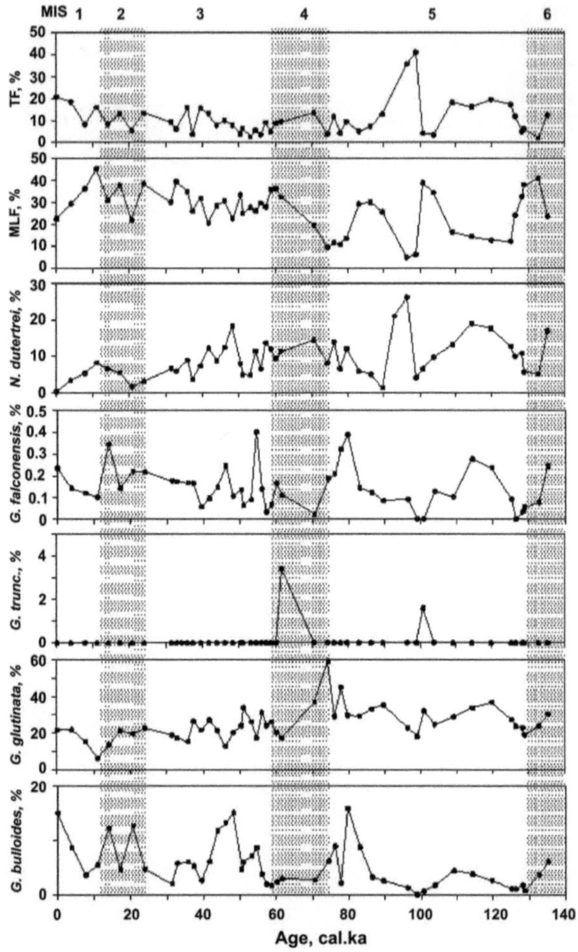

Fig. 5.16 Time series of several indicative planktic foraminiferal species, mixed-layer foraminifera (MLF) and thermocline foraminifera (TF) in EAST core from the eastern Arabian Sea (modified from Ivanova et al. 2003b). See Fig. 5.9 and Table 5.1 for location and Fig. 5.14 for legend. Faunal changes seem to be less correlated to the climatic cycles than at GeoB 3011-1 and NAST. The temporal distribution of *G. bulloides* and *G. glutinata* may point toward a change in the productivity pattern around 80 ka. *Globigerina falconensis* and *N. dutertrei* show no obvious correlation with climatic changes during the last 135 ka at EAST. The occurrence of *G. truncatulinoides* (*G. trunc.*) at the end of MIS 4 is similar to GeoB-3011-1 and NAST

Tables 8, 9), as well as with factors 1 and 4, where these species dominate respectively. As mentioned above, *G. bulloides*is known to be an indicator species of upwelling, especially characteristic for the Oman coastal area. *G. glutinata* is among dominant species in recent and Late Quaternary sediments of the Arabian Sea (Cullen and Prell 1984; Ivanova 1988; Kroon 1988; Anderson and Prell 1991, 1993; Prell et al. 1999). At the meso- to oligotrophic parts of the Arabian Sea around EAST and NAST, the distribution of *G. glutinata*is even more significantly associated with paleoproductivity than that of *G. bulloides* which can be explained by the diet of *G. glutinata* that preferentially consists of diatoms (Hemleben et al. 1989; Schiebel et al. 2001).

The abundance of diatoms in the northwestern part of the Arabian Sea, in turn, seems to depend on nutrients entrainment into the mixed layer from below (Schiebel et al. 2001) and by lateral advection via filaments associated with the Oman upwelling especially during the summer monsoon (Reichart et al. 1997). This

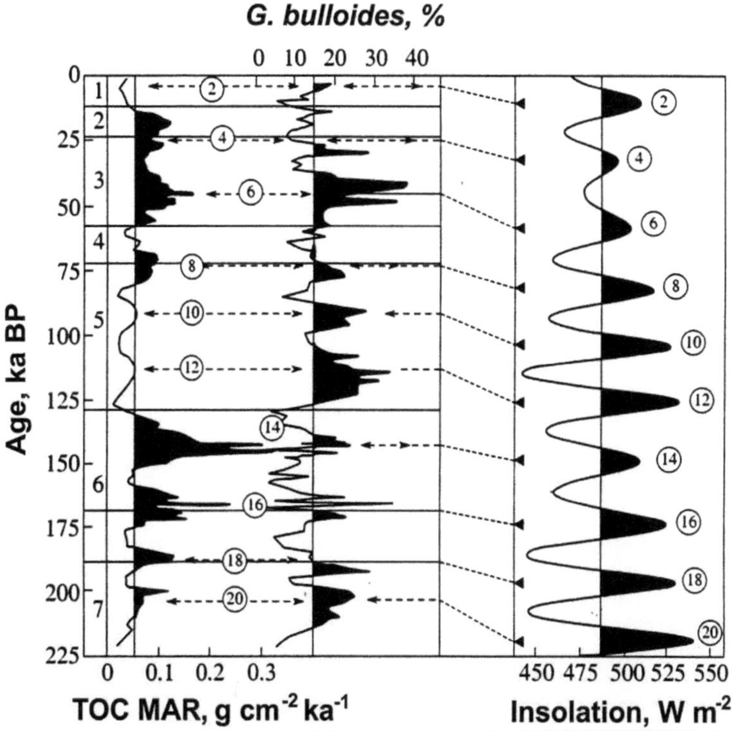

Fig. 5.17 Variations of monsoon intensity indicators in Core NIOP464 (2215.4'N, 6335.1'E water depth 1,470 m) from the northern Arabian Sea correlated with the June insolation at 35°N (modified from Reichart et al. 1997). The intervals of high productivity (shown in *black*) correlate to insolation maxima with a time lag of about 10 ka. Numbers in circles denote precession minima corresponding to the insolation maxima

is in line with the assumption of Anderson and Prell (1991) that *G. glutinata* reflects SW monsoon variations in the open-ocean upwelling area of the Arabian Sea.

However, not all productivity maxima in our sections, especially those in the EAST core, can be explained by the enhanced summer monsoon. The high values characteristic for the beginning of MIS 2 at site GeoB3011-1 and of the boundary MIS 5/MIS 4 at NAST and EAST likely reflect a stronger NE monsoon. An increase in the mixed-layer thickness owing to surface water cooling and deeper convective overturning in the northeastern Arabian Sea (NAST and EAST) seems to be related to the intensified winter monsoon (Reichart et al. 1997, 1998; Ivanova et al. 2003b). The deep convection was accompanied by nutrients entrainment into the photic zone from the intermediate layer.

The mixed-layer deepening led to an increase in the percentage of such relatively oligotrophic planktic foraminiferal species as *G. sacculifer* and *G. ruber* that generally correspond to stadial intervals in all three studied cores (Ivanova et al. 2003b). The total abundance of these species increases from the west to the east (Figs. 5.14, 5.15, and 5.16) in consistence with the today mixed-layer deepening in

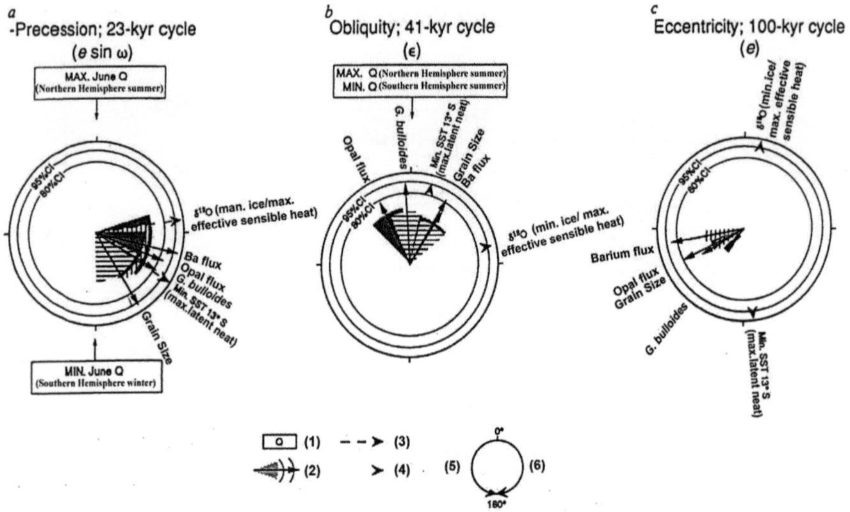

Fig. 5.18 The cross-spectral relationships between orbital variations and changes in climate proxies that occur over a given frequency band (after Clemens et al. 1991 with Nature permission). Zero phase on each wheel is set at (**a**) maximum precession (June 21 perihelion), (**b**) maximum obliquity, and (**c**) maximum eccentricity. Vectors represent the timing of proxy maxima relative to orbital maxima with positive angular distances from zero corresponding to the measured phase difference (+(counter clockwise) represents a lead; – (clockwise) represents a lag. The length of phase vectors indicates the degree of coherence with a given parameter. Significance level, at the 80 and 95% confidence intervals (CI) are given by the inner and middle *circles*. *Dashed arrows* indicate significant coherence between monsoon tracers and potential internal forcing mechanisms (=80% CI). Phase errors between cores are ~ ±2,000 years (±31° for precession, ±18° for obliquity, ±7° for eccentricity) shown as *shaded* areas centered on phase wheel vectors. Legend: 1 = phases of relevant radiation maxima (Q); 2 = monsoon tracers: vectors = phase of monsoon stress maxima, shading = phase error estimate, parenthesis = confidence intervals (CI, 80, 95%) relative to ETP (that is a frequency spectrum of a normalized, summed record of eccentricity, obliquity, and precession); 3 = potential internal forcing mechanisms, coherent with both ETP and monsoon indices (80% CI); 4 = potential internal forcing mechanisms, coherent with ETP but not coherent with monsoon indices; 5 = positive, phase lead; 6 = negative, phase lag

the same direction (Krey and Babenerd 1976; Rao et al. 1989). A similar pattern was found earlier in the Somalian upwelling (Vénec-Peyré et al. 1995, 1997). The highest abundances of *G. sacculifer* and *G. ruber* at the end of glacial MIS 2 and 6 were followed by a decrease to the lower interstadial values over both terminations. This trend coincides with the appearance episodes of the deep-dwelling planktic species *Globorotalia truncatulinoides,* also followed by the gradual disappearance of the species over terminations (Figs. 5.14, 5.15, and 5.16). Increased faunal portion of the subsurface species *Globigerina falconensis* corresponds to stadials in cores GeoB 3011-1 and NAST. The distribution of *G. falconensis* is decoupled from the productivity record. This is corroborated by F. Peeters (2000), who found *G. falconensis* in plankton tows mainly during non-upwelling autumn to winter season at water depth between 50 and 100 m. In recent sediments, this species is more frequent in the northern parts of the Arabian Sea and Bay of Bengal more strongly affected by the NE winter monsoon than the southern parts (Cullen and Prell 1984;

Ivanova 1988). However, at NAST and EAST, neither a positive nor a negative correlation of *G. falconensis* and productivity could be seen. The high abundance of *Neogloboquadrina pachyderma* dextral and *Neogloboquadrina dutertrei* proliferating in subsurface layer at low latitudes (Kroon 1988; Steens et al. 1992; Kincaid et al. 2000) probably reflects nutrients entrainment from below at sites GeoB-3011 and NAST areas due to the enhanced NE monsoon in MIS 3.1, 4, 5.2, and at the boundary MIS 3/MIS 2 (Ivanova et al. 2003b).

Some authors (e.g., Reichart et al. 1998) believe that appearance of the deep dwelling *G. truncatulinoides* in stadial intervals is an evidence for the deepening of convective mixing down to 500–800 m. However, such a suggestion hardly can be accepted, as the present winter mixed-layer depth does not exceed 100–200 m in this region (Krey and Babenerd 1976; Filyushkin 1987; Rao et al. 1989). *G. truncatulinoides*is a typical species of subtropical gyre of the Indian Ocean and practically absent in recent sediments of the Arabian Sea (Cullen and Prell 1984; Ivanova 1988). The rare expatriates of *G. truncatulinoides* have likely been transported into the Arabian Sea by the South Equatorial Current and Somali Current at intermediate depths (>200 m) and could complete their life cycle in conditions of the enhanced winter convection as proposed by G.-J. Reichart and colleagues (1998). These currents entrain a significant portion of the Pacific surface waters inflowing to the tropical Indian Ocean via the Indonesian Throughflow (Schott and McCreary 2001). Besides, the same authors ascertained the equatorward water transport within the modern subtropical gyre, the gyre margins being part of the anticyclonic circulation and westward flow along the SEC. A shift of the Southern Ocean fronts toward the Equator during MIS 2 (e.g., Gersonde et al. 2003; Peeters et al. 2004) may have favored a planktic foraminifera advection to the north and toward the Arabian Sea.

Yet, the species is rather deep dwelling, hence its appearance as well as the coeval maximum of the oligotrophic species pointing to the mixed-layer deepening might also indicate the intensified water input from the ITF area via the South Equatorial Current at the end of stadials – beginning of terminations owing to the reorganization of the THC. This suggestion seems to be in line with the resumption of the Agulhas Linkage at the end of glacials ascertained by F. Peeters and colleagues (2004) as both South Equatorial and Agulhas currents are largely fed by the ITF waters (see also Section 5.7). Later on, during the terminations, the intensified SW monsoon led to decreased winter convection depth, enhanced PP, stronger OMZ, and finally to the disappearance of *G. truncatulinoides* and decrease in the amount of mixed-layer foraminifera (Reichart et al. 1998; Ivanova et al. 2003b).

Our results on planktic foraminiferal species distribution in the studied cores, as well as the published data on the Somalian upwelling and the western equatorial Indian Ocean (Vénec-Peyré et al. 1995, 1997), suggest a reorganization of the upper water column during the terminations. It was likely related to the global sea level rise and enhanced Pacific water inflow into the Indian Ocean through the ITF. An intensification of the return branch of the THC at the transition from glacials to interglacials was demonstrated by several recent studies (e.g., Holbourn et al. 2005; Fig. 5.4). As shown in Section 1.1, this might result in a surface water temperature rise in the northern Indian Ocean that promoted a strengthening of the summer monsoon. Such a reorganization of the upper water column was responded by the abovementioned decreasing trends of oligotrophic species content in foraminiferal assemblages from the Arabian Sea cores.

The high-productivity interval within the MIS-3 (40–50 ka BP in our cores) identified by many authors (Zahn and Pedersen 1991; Caulet et al. 1992; den Dulk et al. 1998; Hermelin and Shimmield 1995; Almogi-Labin et al. 2000; Vénec-Peyré and Caulet 2000; Schmiedel and Leuschner 2005) cannot be explained only by intensified summer monsoon and related upwelling because PP values were considerably higher here, as compared to those in MIS 5.1, 5.3, 5.5, even far off the upwelling (Schulte and Müller 2001). This productivity maximum partially correlates with the Heinrich event H-5 (Fig. 5.8), but was likely longer. Episodes of enhanced southeastern or northwestern monsoon possibly alternated but might partially coincide. This is affirmed by detailed studies of laminated sediments

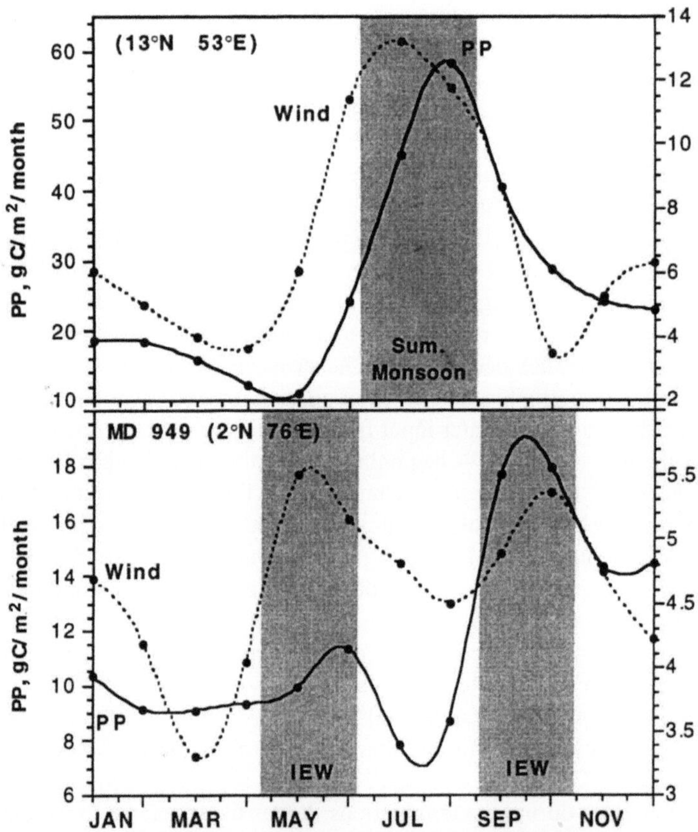

Fig. 5.19 Seasonal comparison of primary production (*solid lines*) and wind speed (*dotted lines*) at a location dominated by the boreal summer monsoon (Gulf of Aden, 13°N, 53°E) (*upper panel*) and another dominated by the Indian Ocean Equatorial Westerlies (IEW) (position of core MD900949: 2°N, 76°E – *lower panel*) after (Beaufort et al., 1999). Windy seasons are *shaded*. Satellite data on productivity are taken from Antoine et al. (1996), data on wind speed from COADS data base (1947–1991)

from the northern Arabian Sea (von Rad et al. 1999) where several phases of high, low, and moderate intensity of both winter and summer monsoon, in various combinations, are identified during the last 30 ka.

Similar low-productivity episodes characterize the beginning of interglacial MIS 1 and 5.5 in Core MD 77169 from the Bay of Bengal (see Fig. 5.6 for location), where they are thought to be related to weakening of the NE monsoon and/or to intensification of the SW monsoon promoting the inflow of freshened nutrient-depleted water from the Andaman Sea into the Bay of Bengal (Fontugne and Duplessy 1986).

Moderate productivity values calculated in three Arabian Sea cores during MIS 2 (150–250 g C m^{-2} $year^{-1}$) generally correspond to those obtained earlier for the LGM in the northern part of the Sea (den Dulk et al. 1998; von Rad et al. 1999). Our estimates for the Holocene in the EAST core (140–210 g C m^{-2} $year^{-1}$) are comparable with those computed for Core MD 77194 from the Laccadivian Islands area also using Q-mode factor analysis and transfer functions (Cayre and Bard 1999). However, the relatively low productivity values obtained for the LGM in the EAST core differ dramatically from high estimates for Core MD 77194. This might possibly result from a productivity increase due to the stronger westerlies in the equatorial zone (Fig. 5.19; Beaufort 1996; Beaufort et al. 1999; Cayre et al. 1999). The latter explanation seems more preferable than that proposed by O. Cayre and E. Bard (1999), who assumed an impact of the intensified winter monsoon on the productivity increase near the Laccadives. However, this effect should also appear in the EAST core, which is not the case. Hence, the two sites were conceivably affected by different mechanisms.

5.6 Mechanisms Controlling the Monsoon Intensity During the Pleistocene

Paleoclimatic observations and modeling results revealed several factors controlling the monsoon variability at millennial and centennial time scales. Land surface topography and orbital parameters of the Earth, as well as greenhouse gases content in the atmosphere are among the most important ones (Clemens et al. 1991, 1996; Clemens and Prell 1991, 1996; Hagelberg and Mix 1991; Prell and Kutzbach 1992; Overpeck et al. 1996; Anderson and Prell 1993; Zahn 1994). Distribution of atmospheric highs and lows, ocean and land surface temperatures, altitude of snow line in Asia, sea ice cover at high latitudes, and trans-equatorial latent heat transport from the southern subtropics to the monsoon region depend on these factors. The highest summer monsoon intensity during the last interglacial (MIS 5.5) is assumed to be related to the sharp seasonal ocean–land contrasts owing to the maximum summer and minimum winter insolation at 30°N (e.g., Leuschner and Sirocko 2003).

To examine the forcing mechanisms of the Indian summer monsoon, S. Clemens and co-authors (1991) applied cross spectral analysis to several paleoarchives. They compared variations of productivity proxies (contentof *G. bulloides*, opal, barium, and size of aeolian dust particles) in ODP Site 722 and Core RC 27-61 from the

Owen Ridge, Arabian Sea, with SST oscillations reconstructed by foraminiferal assemblages in Core RC 17-98 from the subtropical gyre of the Indian Ocean. The proxy records are correlated with the summer solar radiation record of the Northern Hemisphere for the last 350 ka calculated from the normalized, summed time series of the Earth's orbital parameters, eccentricity, obliquity, and precession. As a result, the authors ascertained that the SW summer monsoon was especially strong 9, 50, 123, 215, 261, and 329 ka BP (that is at MIS 1, 3, 5.5, 7.3, 8.3, 9.3 corresponding to interglacials and interstadials). On the contrary, it was weaker at the LGM, 21 ka BP.

This study clearly brings out that the increased summer radiation associated with the high precession and obliquity values forced strengthening of the summer monsoon circulation in the northern Indian Ocean area via the increase in the differential sensible heating (Clemens et al. 1991). The authors suggest the cross-equatorial transport of the latent heat and its further release over the Asian Plateau, as the second forcing mechanism. The release of latent heat originating from the southern subtropical Indian Ocean over the Asian Plateau creates a positive feedback amplifying the SW monsoon (Hagelberg and Mix 1991). S. Clemens and colleagues (1991) concluded that the monsoon proxy records display stronger coherence with precession (23-ka cycles) than with obliquity (41-ka cycles) and that the impact of the glacial–interglacial variability associated with the 100-ka eccentricity cycles on the monsoonal climate is rather negligible (Fig. 5.18). However, several modeling experiments and paleoceanographic studies (e.g., Prell and Kutzbach 1987, 1992; Anderson and Prell 1991, 1993; Clemens et al. 1996) demonstrate more significant influence of 100-ka cycles on the monsoon variability. They show that the glacial-age boundary conditions considerably accounted for the sensible heating of the Himalayas – Tibetan Plateau region through changes in snow line and thus in albedo, as well as for latent heat transport across the Equator through the decrease of SST and increase of wind strength hence increasing evaporation in the subtropical Indian Ocean. The influence of the boundary conditions on the climate of the monsoon domain is confirmed by the models predicting a 2–3°C temperature drop and a 10% decrease in precipitation over Equatorial Africa and South Asia during the LGM (Zahn 1994 and reference therein). These estimates are consistent with the palynological studies and other data (e.g., Gasse and Van Campo 1994). The very recent study of the benthic foraminiferal assemblages and stable isotopes in DSDP Site 219 off southwestern India (Gupta et al. 2008) also demonstrates the remarkable increase in SW monsoon-induced productivity coeval with the shift to increased amplitude of glacial–interglacial cycles 1.2–0.9 Ma ago.

In contrast to "zero-phase hypothesis" of W. Ruddiman (2006c) that postulates the near-zero phase of the timing of boreal summer monsoon strength relative to maxima of NH precession-driven radiation in the Northern Hemisphere, the recent study by S. Clemens and W. Prell (2007) confirms their above-mentioned previous finding of a ~8 ka lag in the timing of monsoon strength (Fig. 5.18; Clemens et al. 1991; Clemens and Prell 2003). This phase lag is interpreted in terms of a combined response of wind strength, upwelling, and other agents of the monsoon climate system to multiple forcing mechanisms including Northern and Southern hemispheres radiation and global ice volume.

The high-resolution studies ascertained that variations in the oxygen minimum zone in the northern Arabian Sea at orbital and sub-orbital timescales primarily depend on the three agents. They include surface water productivity and subsurface oxygen consumption controlled by monsoon-driven coastal and open-ocean upwellings (Reichart et al. 1997, 1998; Den Dulk et al. 1998, 2000; Schulz et al. 1998; von Rad et al. 1999; Altabet et al. 2002), on deep convective mixing related to strong winter monsoon (Reichart et al. 1998) and on changes in production and ventilation of intermediate and deep waters (e.g., Zahn and Pedersen 1991; Schmiedel and Leuschner 2005). For the productivity maxima, an approximately 10-ka lag with respect to the precession minima was recognized (Fig. 5.17; Reichart et al. 1997), which reflects the delay of the strongest summer monsoon-induced upwelling relative to the maximum insolation. The oxygen content in the bottom water commonly rises with the productivity decrease. Therefore, oxygen content also correlates with the monsoon intensity, though it depends as well on the advection of deep waters from the south (Reichart et al. 1997).

Besides, G. Schmiedel and D. Leuschner (2005) demonstrated that the periods of OMZ deepening inferred from the benthic foraminiferal assemblages are restricted to cold intervals and lag the SW monsoon proxies like relative abundance of *G. bulloides* by 4–5 ka (Figs. 5.20 and 5.21). Low abundance or absence of the high-oxygen indicator species in Core GeoB-3004 indicates the intervals of OMZ deepening and sluggish deep-water circulation. The tight correlation between the variability in abundance of the deep OMZ indicator species and the SW monsoon insolation forcing (Fig. 5.20) lends a strong support to the model of G. Schmiedel and D. Leuschner (2005). The model accounts for the duration of deep OMZ presence in the Arabian Sea in respect to the relative timing between changes in organic

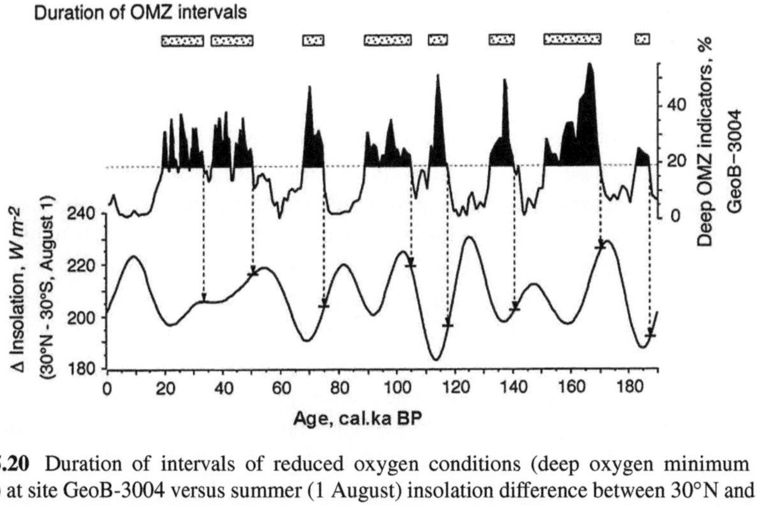

Fig. 5.20 Duration of intervals of reduced oxygen conditions (deep oxygen minimum zone, OMZ) at site GeoB-3004 versus summer (1 August) insolation difference between 30°N and 30°C (Leuschner and Sirocko 2003) at onset of deep OMZ presence (after Schmiedel and Leuschner 2005). Printed with permission from AGU

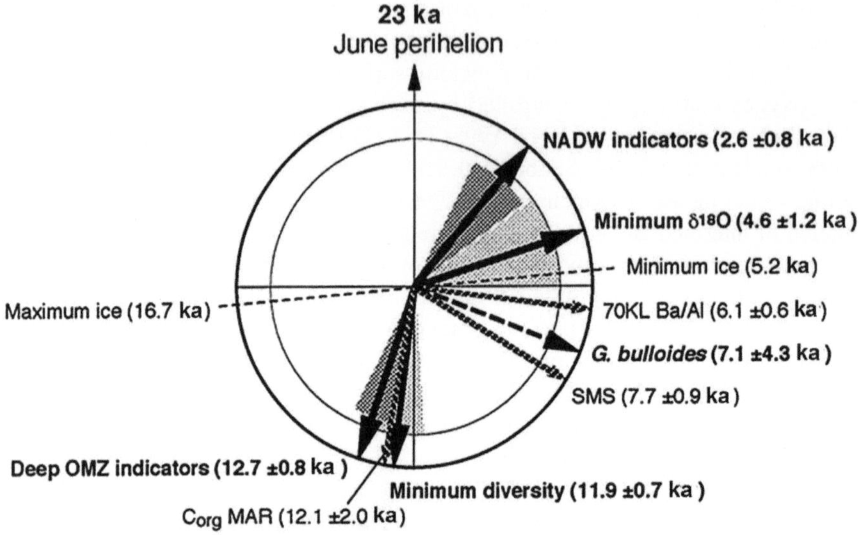

Fig. 5.21 Precessional phase wheel showing different faunal proxies and the benthic $\delta^{18}O$ record of core GeoB-3004 relative to maximum boreal insolation of 21 June. *Shaded* sectors indicate 80% confidence intervals. Phase relationships of the southwest monsoon stack (SMS) and organic carbon mass accumulation rate (Corg MAR) of Clemens and Prell (2003) and the *Ba/Al* record of Leuschner and Sirocko (2003) are shown for comparison (after Schmiedel and Leuschner 2005). Printed with permission from AGU

matter fluxes and deep-water circulation. Thus, this model offers a new insight into the interaction between the thermohaline circulation manifested in the rate of NADW transport to the Arabian Sea and intensity of SW monsoon.

The modeling experiments demonstrate that the warming in the North Atlantic 14.7–14.1 cal. ka BP promoted a warming over the Tibetan Plateau conceivably sufficient to reduce the snow cover (and hence, the albedo and moisture content in the soil) in the spring and at the beginning of the summer. This warming probably resulted in an earlier and stronger summer monsoon (Overpeck et al. 1996). The warming might also be favored by the enhanced greenhouse gases concentration in the atmosphere. As a result, the positive feedback appeared with the further intensification of the monsoon due to the terrestrial vegetation extension. Owing to the absence of influence of glacial boundary conditions on the thermal regime over the Tibetan Plateau, the decrease in the monsoon intensity during the late Holocene became more gradual than its intensification during the termination.

Some studies suggest the strengthening of both Indian and East Asian summer monsoons during the DO interstadials due to low pressure above the Tibetan Plateau and Asian continent (e.g. Cai et al. 2006). The coherent millennial-scale changes in the East Asian and Indian monsoons are documented by high-resolution studies of stalagmites in the Hulu and Xiaobailong caves in the Eastern and Southwestern China, respectively (Wang et al. 2001; Cai et al. 2006). The prominent resemblance of dry (wet) episodes (especially remarkable in the Hulu Cave) with the Antarctic warmings (coolings) on paleotemperature record of Byrd ice core provides evidence of the link between high Southern latitudes and Asian monsoon. Y. Cai and colleagues (2006) speculated that the summer monsoons strengthened by the pressure decrease above the Tibetan Plateau and Asian continent were further intensified by a stronger trans-equatorial air flow caused by high temperature gradient resulted from the Southern Hemisphere cooling.

Air temperature over Tibetan Plateau may also rise during El Niño events causing a subsequent intensification of the summer monsoon. On the contrary, thick snow cover and corresponding high albedo led to an increase in water content of soil, delay and weakening of the summer monsoon (Lappo et al. 1990; Overpeck et al. 1996). Some authors assume that the summer monsoon weakening owing to reduction of the westerlies at the equator might result in a situation similar to ENSO in the equatorial Pacific (Webster et al. 1998).

The weakening of summer monsoon and more persistent La Niñas are suggested by L. Stott and colleagues for the DO interstadials in contrast to "Super El Niño" mode associated with vigorous winter monsoon during the stadials (see Section 7.7).

On the other hand, several studies of the present relationships between the Indian Ocean climate and ENSO demonstrated that the air–sea interaction in this basin manifested in east–west sea surface temperature asymmetry and precipitation anomalies (so-called Indian Ocean Dipole) is rather independent of El Niño–Southern Oscillation (Saji et al. 1999; Webster et al. 1999). This internal mode of the Indian Ocean climate dynamics, although difficult to ascertain, should have been operating in the past. However, several authors demonstrated contradictory observations of the ENSO–Indian Ocean Dipole interaction in the 20th century (e.g., Zinke et al. 2004). Thus, the possible influence of ENSO and Indian Ocean Dipole on the monsoon variability is still strongly debatable.

Recently, S. Jung and colleagues (2004) proposed two potential scenarios to explain the observed coherent centennial to decadal changes in Holocene monsoonal records from the Arabian Sea, continental China, and South China Sea. This coherency enables the authors to imply the joint forcing mechanism for the entire region dominated by the Asian monsoon. The first scenario suggests wind-induced temperature variations in the Western Pacific Warm Pool to be the best candidates to driving force. The second scenario alternatively involves the process controlling the simultaneity of changes in the trade wind strength in the Indian and Pacific oceans and in turn forced by the decadal-scale variations in solar insolation or linked to ENSO phenomenon.

5.7 Paleoceanographic Changes in the Tropical Indian Ocean Linked to Variations in the Return Branch of Global Thermohaline Circulation

Variations in the upper limb of the global thermohaline circulation are studied in Core S-17666 (20°36S, 55°55E, water depth 3,292 m) retrieved by the German R.V. *Sonne* nearby the Reunion Island, at the western margin of subtropical gyre

Table 5.5 Age-depth relationships in core S17-666 after (Fretzdorff et al. 2000 with permission from Springer-Verlag)

Isotope Events	Depth, cm	Age, ka	Sedimentation rate, cm/ka
2.0	20	12	
2.2	50	14.5	12.00
3.0	100	26	4.35
4.0	190	62	2.50
4.2	200	65	3.33
5.0	210	71	1.67
6.0	320	128	1.93
7.0	560	186	4.14

The isotope events from the SPECMAP time scale (Imbrie et al. 1984) were used as tie points to calculate the sedimentation rates.

(Ivanova et al. 1995; Fretzdorff et al. 2000, see Fig. 5.4 for site location). The core site is situated on the passage of the Southern Equatorial Current which carries the warm Pacific waters from the ITF to the Mozambique and Madagaskar Currents and then to the Agulhas Current. The present SST at the core site is 27°C in summer and 23°C in winter, the mean annual salinity is 35 psu (Levitus 1982).

The age model of the core S-17666 is based on the oxygen isotope record of surface-dwelling planktic foraminifera *G. ruber* (white) correlated to the SPECMAP curve (Imbrie et al. 1984; Table 5.5). The oxygen isotope measurements are performed at CFR (now LSCE) CNRS–CEA, Gif-sur-Yvette, France. Disappearance of *G. ruber* (pink)at Termination II was used as an independent tie point (Fretzdorff et al. 2000).

According to several studies (Climap Project Members 1981; Ivanova et al. 1985; Ivanova 1988; Barrows and Juggins 2005), during the LGM the surface waters of the subtropical gyre cooled by 0–1°C in the centre and 3–4°C at the periphery as compared to the modern seasonal and mean annual SST. We estimated foraminiferal-based summer and winter SST in Core S-17666 during the last 128 ka BP applying the modern analog technique (Fig. 5.22, Appendix, Table 10; Ivanova et al. 1995; see also Chapter 2) and using the database of 800 planktic foraminiferal assemblages from surface sediments of the tropical Indo-Pacific mainly from Prell (1985). Modern summer and winter SST are derived from (Levitus 1982). The MAT-derived SST display negligible oscillations during the last climatic cycle, being close to 23°C in winter (decreasing to 21–22°C only at two short-term events) and within the limits 26.3–27.6°C in summer, with maximum values at interglacial MIS 5.5 and 5.1 (Ivanova et al. 1995; Fig. 5.20). These oscillations are comparable to the error of modern SST estimates by MAT equal to ±1.5°C (Prell 1985) but this is consistent with the above-mentioned stability of the subtropical gyre. The dissimilarity values (=0.25 throughout the record) are acceptable for the tropics (except for two samples excluded from the paleotemperature estimates). Thus, the SST estimates are quite robust inspite of abundant fragments pointing to dissolution of foraminiferal tests at the end of MIS 6–beginning of MIS 5, as well

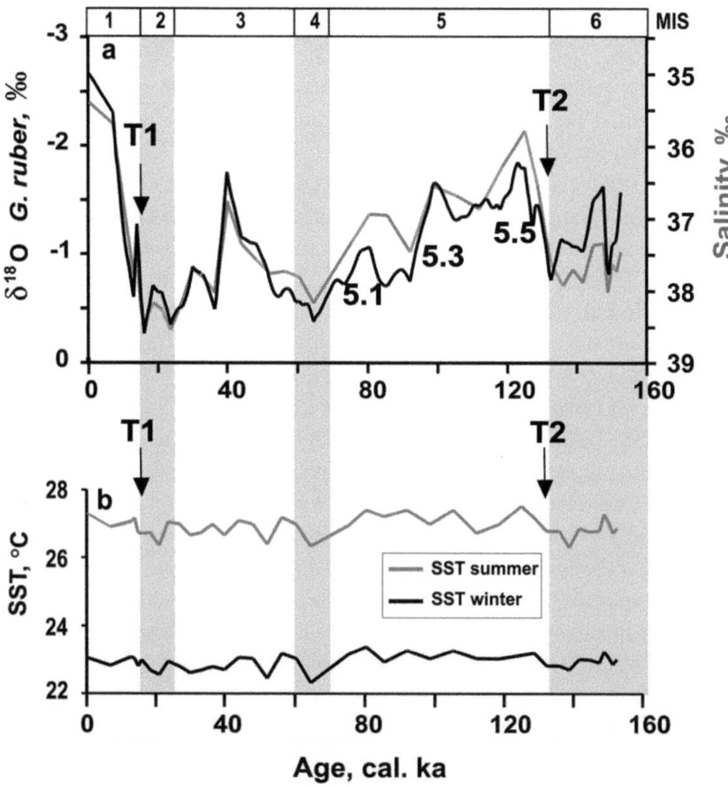

Fig. 5.22 Time seies of (**a**) oxygen isotopes ($\delta^{18}O$) measured on *Globigerinoides ruber* (*white*) (*black line*), mean annual sea surface salinity (*gray line*), (**b**) foraminiferal MAT-derived winter (*black line*), and summer (*gray line*) sea surface temperature (SST) during the last 160 ka in Core S-17666 from the western margin of the subtropical gyre, Indian Ocean. T1 and T2 indicate terminations, T2 also marks the disappearance of *G. ruber* (*pink*) MIS = oxygen isotope stages. $\delta^{18}O$ record and age model from (Fretzdorff et al. 2000, see also Table 5.5)

as at the boundaries MIS 5/MIS 4 and MIS 3/MIS 2. Location of the core at the depth close to the present foraminiferal lysocline estimated at 3,300 m (Belyaeva and Burmistrova 1984) explains the dissolution. The lysocline depth varied considerably during the Pleistocene (e.g., Murdmaa and Ivanova 1987; Ivanova and Shackleton 1993) but the MAT-derived SST in core S-17666 seems not to be considerably biased by dissolution. The very recent foraminiferal-based SST estimates by MAT in Core MD79-257 (Levi et al. 2007) collected from Mozambique Channel, westward of our Core S-17666, display variations of the same order of magnitude (about 1–2°C) for the austral summer over the last 30 ka. Somewhat higher magnitude of winter SST variability, up to 5–6°C, can be explained by a more marginal location of Core MD79-257 as compared to Core S-17666.

Sea surface salinity (SSS) variations through time are calculated based on the MAT-derived SSTs and $\delta^{18}O$ record of surface-dwelling species *G. ruber*

(Fig. 5.20). SSS is generally expressed as local $\delta^{18}O$ anomaly of seawater (Duplessy et al. 1991b, 1992, 2001; Rostek et al. 1993; Wang et al. 1999a). We extracted $\delta^{18}O$ of seawater from the standard paleotemperature equation (Shackleton 1974) and then computed the local water $\delta^{18}O$ changes by subtraction of the effect of global continental ice volume on mean ocean $\delta^{18}O$ during the last glacial cycle (after Labeyrie et al. 1987). Local salinity anomalies due to changes in hydrological budget (precipitation–evaporation) and lateral surface water advection are calculated using the present local $\delta^{18}O$ seawater/salinity relationship. According to F. Rostek and co-authors (1993), $\delta^{18}O$ increase by 1‰ corresponds to 2.7 psu salinity rise in the tropical Indian Ocean. The obtained local anomalies in salinity are summarized with present salinity value at the core site and with salinity variations owing to glacio-eustatic sea level changes in the Pleistocene: the $\delta^{18}O$ increase by 1.17 ‰ corresponds to 1.11 psu salinity rise (Labeyrie et al. 1987).

The lowest SSS values occur in MIS 5 and at the end of MIS 3 hence characterizing interglacials and interstadials. This might result from an evaporation decrease in the subtropical gyre owing to trade winds weakening (Ivanova 1988; Clemens et al. 1991), as well as from an increased inflow of the lower salinity water from the ITF due to intensified thermohaline circulation during interglacials (see Section 3.1; e.g. Seidov and Haupt 1999; Sarnthein et al. 2000a). Both assumptions are in line with the slight interglacial warming at core site (Fig. 5.20). Evaporation decrease leads to the surface water temperature rise, whereas the advection of the warm Pacific water to the Madagascar region by the return surface branch of the THC results in the same effect. New data by F. Peeters and colleagues (2004) about the enhanced warm surface water advection from the Indian Ocean to the Atlantic by the Agulhas Current at interglacials during the past 500 ka support this suggestion, as well as the data by A. Holbourn and co-authors (2005; Color Plate 5.2) about intensified return flow through the straits between the Australia and New Guinea due to sea level rise.

On the contrary, the salinity increased at core location by 3.6 psu at the LGM, likely related to weakening of the THC and intensified evaporation owing to enhanced trade winds. The SSS decrease during interstadials and increase during stadials is ascertained in the Indonesian gateway area (Müller and Opdyke 2000; De Deckker et al. 2003). The salinity rise along the passage of the warm tropical Pacific waters across the Indian Ocean is consistent with their reduced transport to the Atlantic via the Agulhas leakage suggested by some studies (e.g., Berger and Wefer 1996; Peeters et al. 2004). High SSS at the margin of the subtropical gyre in turn might result in a more intense convection in the subtropical convergence and thus in an enhanced intermediate water formation (Ivanova et al. 1995).

According to the data by M. Spooner and co-authors (2005), a higher content of deep-dwelling species in the Banda Sea sediments of MIS 2 and 3 points to a reduced thickness of the low-salinity mixed layer. Similar data on Core MD01-2378 (13°04.95' S and 121°47.27' E, 1,783 m water depth) from the Indian Ocean side of the Indonesian gateway support this assumption (Holbourn et al. 2005; Color Plate 5.2). M.-T. Vénec-Peyré and colleagues (1995) speculated that the conveyor return branch weakening resulted from the World Ocean level drop, shallowing of the Indonesian straits and the South China Sea. However, the deep passage between emerged Sunda and Sahul shelves remained open, hence maintaining penetration of Pacific waters into the Indian Ocean (Hantoro 1997). The overall glacial weakening of the thermohaline circulation owing to reduced NADW formation during MIS 2 (see Section 3.3) might be another cause of the salinity rise at site S-17666.

Similar pattern of SSS variability in the tropical Indian Ocean over the last 30 ka was recently demonstrated by C. Levi and colleagues (2007) at the millennial time scale. The authors established two periods of increased salinity at around Henrich event H-1, 18 –14.5 cal ka BP, and at the Younger Dryas in two AMS-^{14}C-dated sediment cores along the route of warm surface water transport from the ITF to the Atlantic across the Indian Ocean. The cores were collected nearby the Pacific waterway out of the ITF (IMAGES core MD98-2165, 9°38.96'S, 118°20.31'E, water depth 2,100 m) and from the Mozambique Channel, on the further pathway of this water to the Agulhas Current (core MD79-257, 20°24'S, 36°20'E, water depth 1,262 m) (Fig. 5.4). The primary implication of the documented SSS increase during H-1 and YD concerns the resulting accumulation of salt in the tropical Atlantic which favors an abrupt resumption of the thermohaline circulation and northern hemisphere warming after these cooling events. C. Levi and colleagues (2007) assume that the observed millennial-scale oscillations of sea surface salinity resulted from the interaction between the relatively slow expansion and decay of the Northern Hemisphere ice sheets and ENSO variability manifested in latitudinal migration of the Intertropical Convergence Zone (see also Section 7.7). The sufficient ice sheets extent in the Northern Hemisphere is suggested to be a precondition of these oscillations as it forms a positive feedback between high and low latitudes.

H-events and YD were characterized by the increased intermediate water production due to brines release in the North Atlantic and Nordic seas (Vidal et al. 1997; Dokken and Jansen 1999; see also Section 3.4). C. Waelbroeck and colleagues (2006) demonstrated the depletion in oxygen isotope values and increase in carbon isotope values of epibenthic foraminifera *Cibicides* in the same core MD98-2165 after the LGM with further oscillations within Termination I, 18–10 cal. ka BP (Fig. 5.23). These findings supported by the study of two cores from the tropical and mid-latitudes are interpreted in terms of more active circulation at about 2,100–1,200 m water depth in the Indian Ocean during H-1 and likely YD comparing to the last glacial and Bølling–Allerød. Correlation of the Indian Ocean stable isotope records with those from the North and South Atlantic enables C. Waelbroeck and colleagues (2006) to speculate about the well-ventilated intermediate water propagation from the North Atlantic via the Southern Ocean to the site MD98-2165 in the Indian Ocean, at the outflow of the Indonesian Throughflow.

In contrast, based on the benthic foraminifera and stable isotope study of the nearby IMAGES Core MD01-2378 (see Fig. 5.4 for core location) A. Dürkop and colleagues (2008) call on the transmission of the Southern Ocean signal into the eastern Indian Ocean and Timor Sea through the transport of deep and intermediate waters during MIS 3. These authors also argue that the tight coupling between Asian and Australian monsoon systems might enable the transferring of

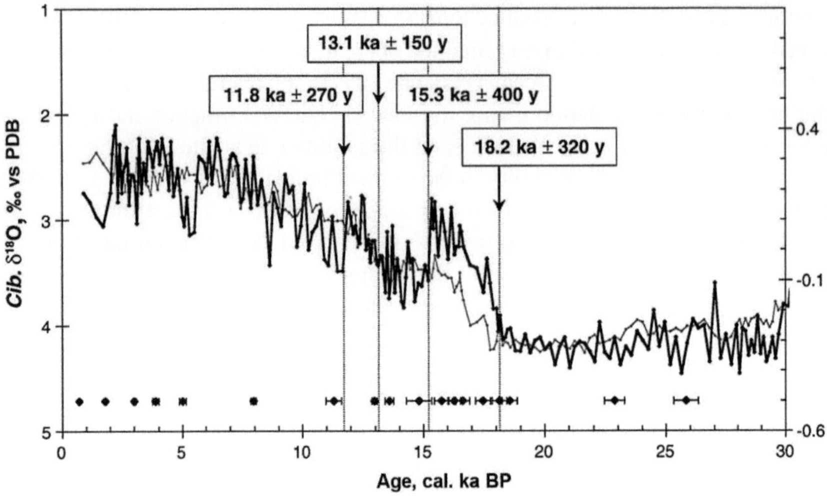

Fig. 5.23 Epibenthic *Cibicides* $\delta^{18}O$ (*fine line*) and $\delta^{13}C$ (*bold line*) records versus calendar age in Core MD98-2165 at the way out of the ITF reflecting the variations in the intensity of intermediate water circulation (after Waelbroeck et al. 2006 with Elsevier permission). *Solid diamonds* indicate dated levels

Northern Hemisphere signal established in the planktic oxygen isotope record of Core MD01-2378 through the atmospheric teleconnection.

5.8 Summary

The onset of the Indian monsoon in the middle or even early Miocene was forced by changes in the basin configuration resulting from several geodynamic events during the Paleogene and the beginning of Neogene and by reorganization in the global thermohaline circulation. The most significant events include the India–Asia collision, uplift of Tibet and Himalayas, closure of the Tethys Ocean gateway, final isolation of the Antarctic continent and its ice sheet growth.

The well-established correlation of a number of proxy time series to the solar radiation record points to the direct orbital (primarily precession) control of the monsoon circulation rather independent of climate changes in the high latitudes. Besides dominating Milankovich cycles, millennial and shorter-lasting cyclic monsoon variations are ascertained which point to non-linear relationships between the internal dynamics of monsoon-forming mechanisms and multiple external factors. The strengthening of both Indian and East Asian summer monsoons during the DO interstadials in Greenland and coeval coolings in Antarctica provide evidence of the high northern and southern latitudes involvement in the tropical temperature and precipitation changes. The latter in turn may strongly affect the heat and moisture transport to the high northern latitudes and finally the global ice volume.

Intensification of the upper limb of thermohaline circulation during interglacials and weakening during stadials were expressed by sea surface salinity decrease or increase, respectively, along the route of the warm waters of the return THC branch across the Indian Ocean, from the ITF to the Atlantic. The salinity increase during the glacials might promote a more intense convection and enhanced intermediate water formation in the southern subtropical convergence zone. Besides, it might enable the accumulation of salt in the tropical Atlantic, thus favoring the abrupt AMOC resumption and warming in the Nordic seas area after the cold episodes. Very small SST oscillations during the last glacial cycle support the earlier suggestions about stability of the Indian Ocean subtropical gyre (Climap Project Members 1981; Ivanova 1988).

Variations of the primary production in the northwestern Arabian Sea, inferred from planktic foraminiferal assemblages, correlate with the monsoon index record (that is with the summer monsoon intensification during interglacials) showing a lag of several thousand years. Several events of the prevailing winter monsoon are identified in MIS 2, 4, 5.2, at the end of MIS 6, and at the transition MIS 5/MIS 4. Enhanced convection and entrainment of nutrients to the photic zone are related to these events in the northeastern part of the sea, whereas summer monsoon had a rather weak influence on this area. The productivity maximum at MIS 3 fixed in many sections from the northern Indian Ocean possibly reflects an intensification of both summer and winter monsoon. The peak abundance of oligotrophic foraminiferal species and short-term unusual occurrence of deep-dwelling *G. truncatulinoides* in the Arabian Sea are associated with the end of glacials and beginning of terminations. These events are likely related both to the reorganization of the monsoon circulation and to the enhanced Pacific water inflow into the tropical Indian Ocean due to intensification of the return branch of the global thermohaline circulation.

Chapter 6
Influence of the Thermohaline Circulation on Paleoceanographic Events in the South China Sea

Abstract During the last climatic cycle, the monsoon-governed marginal South China Sea and surrounding lands experienced dramatic climate changes on millennial-to-decadal scales. At glacial intervals, the influence of the upper limb of the THC on the South China Sea decreased due to the sea level drop and emergence of the Sunda subcontinent that cut off the inflow of warm Indo-Pacific waters via the Borneo Strait. The sea became a semi-closed basin with the estuarine circulation, oxygen-minimum layer, and the only passageway to the open western Pacific via the Luzon Strait in the northeast. Unlike interglacials, the amplification of climatic signal in the South China Sea due to its sharp isolation from the THC during the glacials along with the strengthening of winter monsoon and weakening of summer monsoon was manifested by remarkable decrease in winter SST and surface water salinity, and increase in seasonality, mixed-layer, and thermocline depths over the major part of the basin, except for the upwelling area off Luzon tip. The short-term variability of hydrological parameters superimposed on the glacial–interglacial cyclicity suggests the global teleconnections as several climatic events in the South China Sea are coeval with DO cycles and changes in the Indian monsoon.

Keywords East Asian monsoon · Tibetan Plateau · Borneo Strait · Sunda shelf · Luzon Strait · Upwelling · Indo-Pacific waters · Surface salinity · Climate variability

6.1 Introduction

As shown in Chapter 5, the Asian monsoon system includes East Asian monsoon along with the above-described Indian Ocean monsoon (Fig. 5.1). The East Asian monsoon is mainly controlled by seasonal variations in heating of the Tibetan Plateau and surrounding southeastern Asian mountain chains. During winter, a strong high atmospheric pressure appears over Tibet, and the monsoon wind is directed from the land to the sea in the western Pacific region. In contrast, the summer monsoon wind blows from the sea to the land, toward the Tibetan low (Fig. 5.2). The strong seasonality in wind directions, temperature, and precipitation lead to extensive moisture transport from the sea to the land during summer and

E.V. Ivanova, *The Global Thermohaline Paleocirculation*,
DOI 10.1007/978-90-481-2415-2_6,

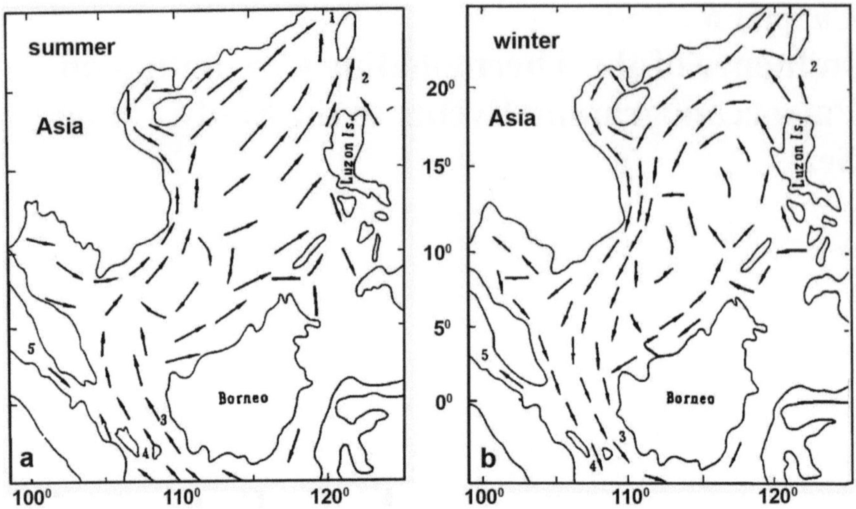

Fig. 6.1 Present (**a**) summer and (**b**) winter surface circulation pattern in the SCS (after Wyrtki 1961; Wang and Wang 1990). (1) Taiwan Strait; (2) Bashi Strait (northern channel of Luzon Strait); (3) Karlmata Strait; (4) Gaspar Strait; (5) Malacca Strait

finally to heat and moisture transfer from low to high latitudes. Thus, the East Asian monsoon represents a basic element in the atmospheric heat budget of the Northern Hemisphere and exerts a significant control on the global climate change (Wang and Wang 1990; Wang et al. 1999a). In particular, the climate of the SCS and the surrounding lands is governed by the South Asian monsoon (Tchernia 1980). As a result of the seasonally reversing winds, the-present day surface circulation of the South China Sea (SCS) is generally clockwise in summer and counter-clockwise in winter (Wyrtki 1961; Fig. 6.1). During the summer monsoon, the warm waters of the return branch of the THC enter the sea from the tropical Indo-Pacific via the Borneo Strait. Enhanced concentrations of chlorophyll, silicate, and phosphate indicate an upwelling off Vietnam (Levitus and Boyer 1994), also predicted by climatology-driven circulation model (Wiesner et al. 1996). During the winter monsoon, the flow patterns are reversed, the relatively cold waters flow into the sea primarily from the subtropical western Pacific via the Luzon (Bashi) Strait, and upwelling occurs off the Luzon tip (Wyrtki 1961; Wang et al. 1999a; Huang et al. 2002).

Both the East Asian and the Indian Ocean monsoon are shown to display clear orbital-scale variability and depend on the glacial boundary conditions (Morley and Heusser 1997). Variations in the East Asian monsoon are extensively studied in the North China loess profiles (Kukla et al. 1988; Liu and Ding 1993; Banerjee 1995; Porter and An 1995). Phases of climate aridization and enhanced dust outblow by winter monsoon to the SCS mainly correlate with glacials, Heinrich events, and Younger Dryas. Interglacials are characterized by humid climate and soil formation. These phases are also described in core sections from the Sulu Sea (Kudrass et al. 1991). The Early Holocene humidity maximum in China (Pye and Zhou 1989; Shi et al. 1993; Porter and An 1995) corresponds to the summer monsoon maximum in the Arabian Sea (Sirocko et al. 1993, 1996). Therefore, one may expect that similar events of enhanced precipitation and river runoff in the South China and Indonesia led to a strong sea surface salinity decrease off Hong Kong and hence, can be traced in sediment sections from the northern and southeastern SCS, as well as in those from Indonesian seas.

6.2 Paleoceanographic Events in the South China Sea from the Last Glaciation to Holocene

The multi-proxy study of the past changes in the monsoon circulation was fulfilled in 11 sediment cores retrieved from isolated submarine rises at intermediate depths along three transects within the framework of the "Monitor Monsoon" Project (Sarnthein et al. 1994a). The cores were collected during the Cruise 95 of the German R.V. *Sonne* in 1994 from the areas with high or extremely high sedimentation rates (7–85 cm/ka). The cores recovered different stratigraphic range and consist of undisturbed hemipelagic mud (Wang et al. 1999a). The northern transect extends over 400–500 km southeast of Hong Kong. This transect is located offshore the freshwater plume in front of the Pearl River mouth, within the subtropical zone (Fig. 6.2, Table 6.1). The western transect lies in the upwelling area off South Vietnam. The southern transect is located in the tropical part of the sea, off the mouth of the "Molengraaff" paleoriver that drained the emerged Sunda shelf during the glacial low sea level stand. One additional core S-17927 documents the sea surface temperature and salinity regime to the west of Luzon Strait (Wang et al. 1999a). The history of winter upwelling 100 km offshore the Philippines was also investigated in nearby core S-17928 (Huang et al. 2002).

Table 6.1 Location, water depth, and core recovery of the cores discussed in this chapter (modified after Wang et al. 1999a)

Site	Longitude (E)	Latitude (N)	Water depth (m)	Recovery (m)
S-17937-2	117°39.9′	19°30.0′	3,428	12.92
S-17938-2	117°32.3′	19°47.2′	2,840	11.78
S-17939-2	117°27.3′	19°58.2′	2,474	12.74
S-17940-1	117°23.0′	19°07.0′	1,728	0.56
S-17940-2	117°23.0′	19°07.0′	1,727	13.30
S-17954-2	111°31.5′	14°47.8′	1,520	11.52
S-17955-2	112°10.6′	14°07.3′	2,393	11.66
S-17956-2	112°35.3′	13°50.9′	3,388	13.56
S-17961-2	112°19.9′	08°30.4′	1,968	10.30
S-17964-2	112°12.8′	06°09.5′	1,556	13.04
S-17961-3	112°12.8′	06°09.5′	1,556	9.12
S-17927-2	119°27.2′	17°15.1′	2,804	5.58

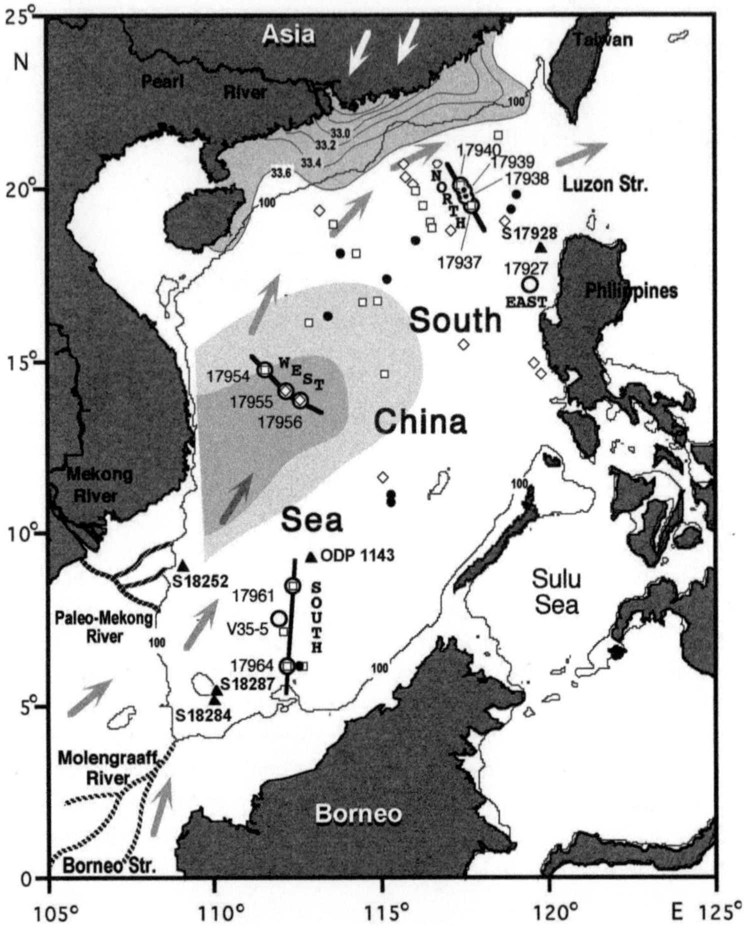

Fig. 6.2 Locations of 11 core sections (*large open circles*) on 3 transects and of 44 core-top samples studied by L. Wang and co-authors (1999a) and core sections (*black triangles*) studied by (Kienast et al. 2001; Huang et al. 2002; Tian et al. 2005; Steinke et al. 2008) from the SCS *Gray arrows* show modern surface currents during summer; modern upwelling region southeast of Vietnam during summer is stippled; *white arrows* show winter monsoon. *Shaded area* with contour lines shows the modern freshwater plume with reduced sea surface salinity in front of the Pearl River mouth during summer. Present 100-m isobath shows approximate position of coastline during glacial low sea level. Molengraaff and Mekong rivers on the (emerged) glacial Sunda shelf are indicated in the southwestern South China Sea (after Wang et al. 1999a). Symbols of core-top samples: dots = aeolian sediment supply; diamonds = sediment winnowing; squares = fluvial sediment supply

The monsoon variations are traced in subtropical and tropical parts of the sea during two last glacial cycles involving the last 250 ka (Ivanova et al. 1997; Wang et al. 1999a, b). The special emphasis was engaged to the moisture transport that is to the summer precipitation; to the winter dust transport depending on the wind strength and climate aridity; and to vegetation and climate changes on the Sunda subcontinent emerged during glaciations. The detailed AMS-^{14}C dating of the upper parts of the studied cores enables tracking the short-term (bidecadal and centennial) monsoon variations, to precise their phase relationships and linkages to global climate changes such as Heinrich and DO events defined in the North Atlantic realm (see Section 3.4 and 3.5).

Sea surface temperatures were reconstructed both by U^k_{37} and foraminiferal transfer functions (Wang et al. 1999a). The biomarkers U^k_{37} index allows estimating the mean annual temperature of the surface water (0–30 m) with the error of estimates about $\pm$ 0.15°C for the modern dataset (Pelejero and Grimalt 1997; Pelejero et al. 1999). Winter and summer SSTs were calculated based on downcore planktic foraminifera census counts using transfer functions FP-12E developed by Thompson (1981) for the northwestern Pacific. The error of estimate is $\pm$ 1.46°C for summer and $\pm$ 2.48°C for winter (Wang et al. 1999a).

Local sea surface salinity changes are calculated by subtracting the correction for the global ice volume (Labeyrie et al. 1987; Fairbanks 1989; Vogelsang 1990) and U^k_{37}-SST from oxygen isotope values of the surface-dwelling foraminifera *Globigerinoides ruber* using the equation by L. Wang and co-authors (1995b) for the tropical Atlantic and summer SST. As *G. ruber* calcifies its test mainly within the upper 50 m of the water column (Bé and Tolderlund 1971; Tolderlund and Bé 1971; Hemleben et al. 1989), and the alkenone-derived temperature is pertained to the upper 30 m, the reconstructed salinity values really characterize the surface water layer. Unlike the tropical Atlantic (Wang et al. 1995b), the U^k_{37}-SSTs in the SCS are considered to correspond to mean annual values as the highest *G. ruber* contents are found in sediment traps both at the end of summer and at the beginning of winter (Wiesner et al. 1996). According to our data (Wang et al. 1999a), the U^k_{37}-SSTs are well comparable with the foraminiferal-based mean annual SSTs and to hydrological data (Levitus and Boyer 1994). The standard error of the mean annual salinity estimates is 0.68 psu (Wang et al. 1999a). For comparison, salinity was also computed using the equation by F. Rostek and colleagues (1993) for the tropical Indian Ocean. The results of both calculations were similar and the error of estimate was within the same limits (see Section 5.7). To establish the same time resolution for temperature and salinity records in Core S-17940, where the deglaciation interval was studied in detail, the smoothing spline interpolation of the alkenone paleotemperature record was carried out (Wang et al. 1999a). All results of the study are available at the web site PANGAEA (http://www.pangaea.de).

Besides, about 1.7‰ amplitude of glacial–interglacial changes planktic and benthic oxygen isotope records demonstrate short-term events within the intervals of deglaciation and MIS 3 (Fig. 6.3). They are especially prominent in cores S-17940 and S-17939 with high sedimentation rates. In particular, Termination I is well traced in the northern part of the sea. Rapid climate changes during the Holocene are expressed by δ^{18}O oscillations with amplitudes 0.7–1‰. DO events are well defined in the northern transect, as well as the Younger Dryas cooling with δ^{18}O enrichment by 0.5–0.7‰ and characteristic double peak ascertained in the Greenland ice core (Alley et al. 1993; Stuiver et al. 1995; Figs. 6.3 and 6.4). The YD is less distinctly expressed in cores from the western and southern transects.

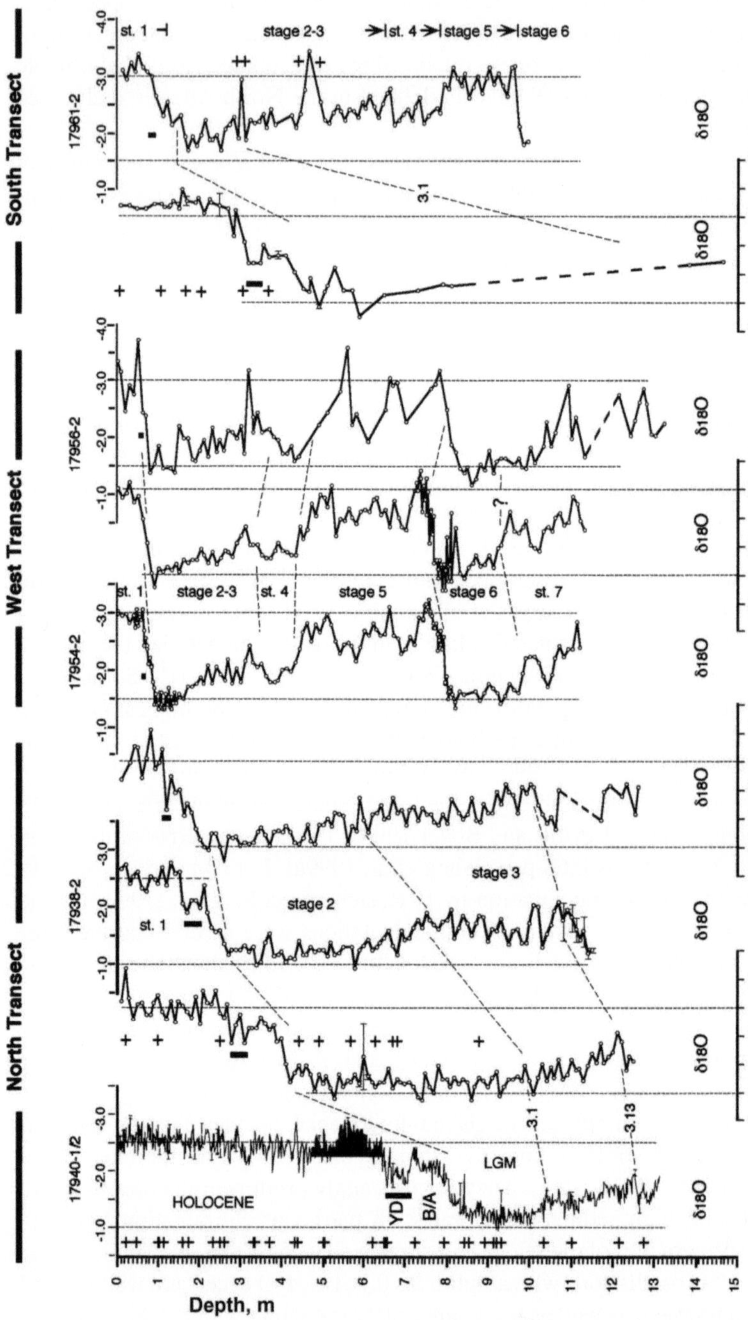

Fig. 6.3 Oxygen isotope records measured on planktic species *Globigerinoides ruber* in sediment cores from the South China Sea (for location of cores see Fig. 6.1) (modified from Wang et al. 1999a). Crosses show AMS-^{14}C-dated levels. Marine isotope stages are labeled and stage boundaries are connected by *dashed lines* between the different records for each transect B/A= Bølling–Allerød. LGM= Last Glacial Maximum. YD = Younger Dryas

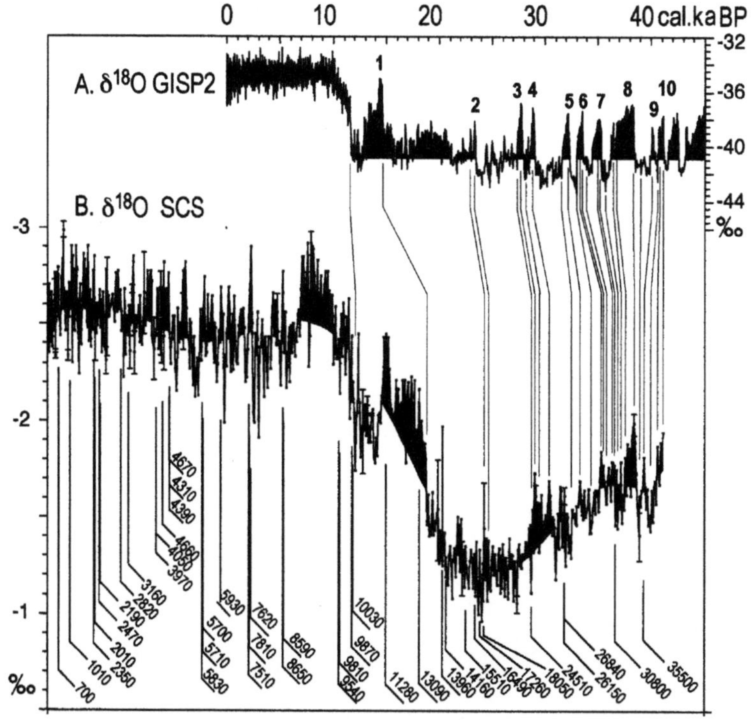

Fig. 6.4 Age models comparison: (**A**) of the Greenland ice-core GISP2 (Grootes and Stuiver 1997) and (**B**) of Core 17940 (for location see Fig. 6.1), based on AMS-^{14}C ages labeled below the planktic δ^{18}O record (modified from Wang et al. 1999a)

The correlation of δ^{18}O minima with warm interstadials of the DO cycles is based on the assumption that the δ^{18}O minima in the South China Sea correspond to the summer rainfall maxima (that is to low δ^{18}O values of local surface water), and that the latter, in turn, correspond to higher air temperature and stronger snowfalls in Greenland (Wang et al. 1999a). The δ^{18}O minima in the Bølling–Allerød, Preboreal, and Holocene result from monsoon precipitation maxima, as the local sea surface temperature remain rather stable (Fig. 6.5). The difference between ages of δ^{18}O excursions within MIS 2 and 3 in the core S-17940 from the SCS and the Greenland ice core, which are correlated in Fig. 6.4, is up to 3–5 ka. This is explained by a constant 3.5 ka deviation between radiocarbon and uranium–thorium (calendar) dates (Winn et al. 1991; Bard et al. 1998). Along with assumed increase of the reservoir effect (Duplessy et al. 1989; Adkins and Boyle 1997), the age difference >3.5 ka possibly results from a more intense radiocarbon production in the atmosphere during the last glacial cycle (Duplessy et al. 1989; Stuiver and Braziunas 1993; Laj et al. 1996), as it is ascertained in Norwegian and North seas (Vőlker et al. 1998). Below the MIS 3 interval, the age model of every core was constraint by correlation of its oxygen isotope record with the SPECMAP stack (Martinson et al. 1987).

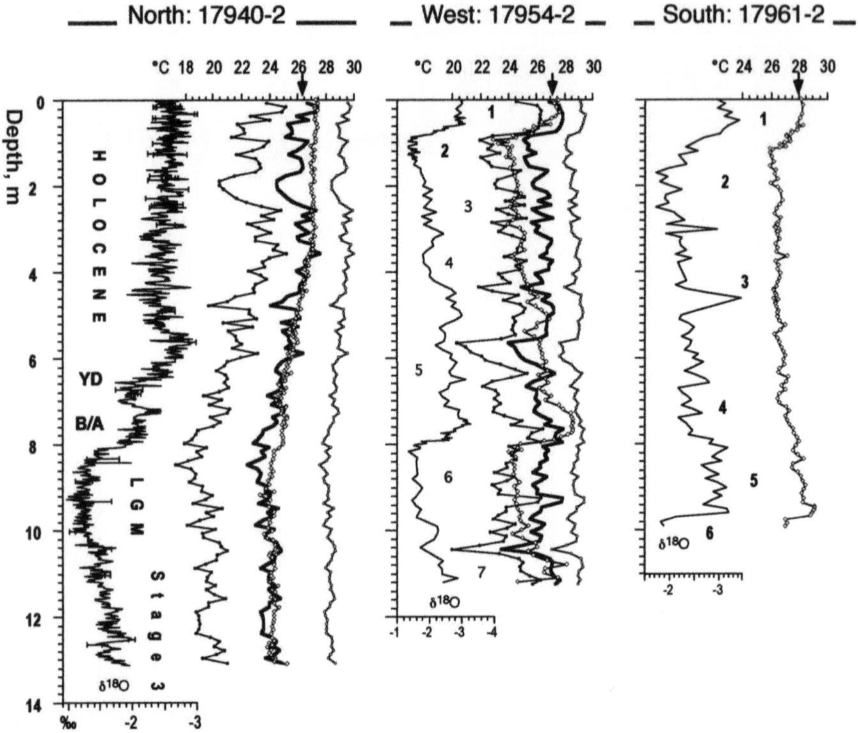

Fig. 6.5 Sea surface temperature (SST) records from the northern (S-17940), western (S-17954), and southern (S-17961) SCS (after Wang et al. 1999a with Elsevier permission). *Arrows mark* modern annual mean SST values at 0–30 m. Summer (*left*), annual mean (*central*), and winter (*right*) SST estimates are given for cores S-17940 and S-17954, based on planktic foraminifera counts and calculated using FP-12E transfer function of (Thompson 1981).YD= Younger Dryas; B/A= Bølling–Allerød; LGM= Last Glacial Maximum; SST = sea surface temperature

Foraminiferal-based SST variations in the northern, western, and southern parts of the sea demonstrate remarkable seasonal contrasts, stronger than those in the adjacent part of the Pacific (Fig. 6.5; Wang and Wang 1990; Thunell et al. 1994). Summer temperatures were rather stable in the northeastern part of the sea, especially during MIS 2 and 3. On the contrary, winter temperature record in Core S-17940 from the northern SCS shows considerable variations with amplitude up to 2°C during MIS 3. The winter temperature minimum at the end of LGM was followed by a gradual warming trend up to the mid-Holocene, about 6 cal. ka BP. Several cold episodes are superimposed on this trend in the Bølling, Younger Dryas, and early Holocene (about 8.2 cal. ka BP, see Section 3.6 and Section 4.5.4), as well as in the late Holocene (5.2–3.4 and 2–1 cal. ka BP). The winter temperature minimum correlates with the seasonality maximum (Fig. 6.5; Wang et al. 1999a). Temperature variations in the west, off the Vietnam coast, were quite similar to those reconstructed in the north, off Hong Kong. The U^k_{37}-SSTs are 1°C higher over the interglacial optimum than during the Holocene both in northern and in southern parts of the sea (Fig. 6.5). SST variations in the southern part of the sea generally did not exceed 3°C (Wang et al. 1999a).

During glacial winters, the northeastern monsoon might lead to intense anti-clockwise surface circulation, enhanced inflow of colder waters via the Luzon Strait, and a strong temperature decrease in the northern transect area, off the South China. The weak summer monsoon possibly resulted in a slow clockwise circulation that assumes close to present SST off the South China, strong seasonality, and reduced upwelling off Vietnam. Whereas summer temperatures were almost constant, winter temperatures fell by 4–7°C as compared to the modern values. The stability of summer temperatures is also documented in the southeastern part of the sea (Miao and Thunell 1994). The U^k_{37}-derived mean annual SSTs (Fig. 6.5) demonstrate glacial values lower than the interglacial ones by 3.5°C (24 and 27.5°C, respectively) in the north, by 3.5–4.5°C in the west (24 and 27.5–28.5°C), and by 2–3°C in the south (26 and 28–29°C, respectively). Therefore, the difference between the stadial and Holocene values of annual mean SST decreases from 4 to 5°C in the north and west to <3°C in the south (Wang et al. 1999a).

During interglacials, the warm Indo-Pacific water inflow into the southern part of the sea via the Borneo Strait was strongest in summer times, as today (Figs. 6.1 and 6.6; Wang and Wang 1990; Wang et al. 1999a; Huang et al. 2002; Steinke et al. 2008). This inflow might suppress the signal of coldwater signal entering from the subtropical Pacific via the Luzon Strait in winter. Such an effect may explain the shift of the U^k_{37}-derived mean annual SST toward summer values of foraminifera-based SST in MIS 5.5 (Wang et al. 1999a). The Borneo Strait reached its maximum

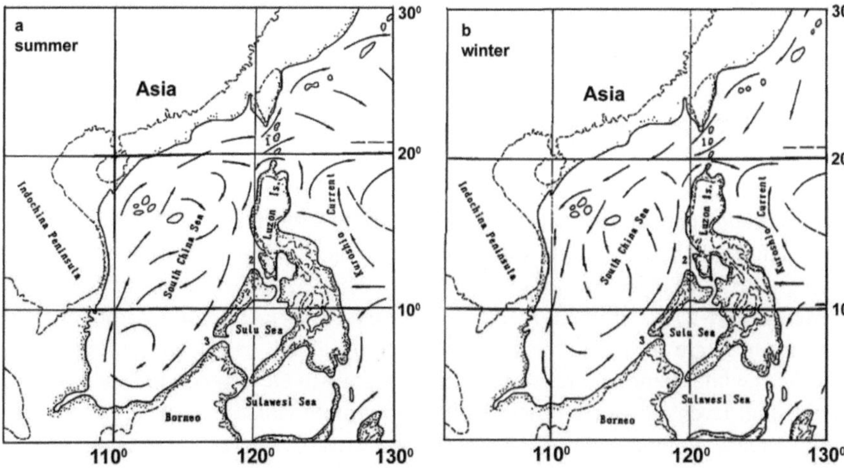

Fig. 6.6 Inferred glacial surface circulation patterns of the South China Sea, (**a**) summer and (**b**) winter (after Wang and Wang 1990). The present-day coastlines are drawn in the thin dotted line and the glacial coastlines are drawn along the present-day 100-m isobath. The *heavy dotted lines* indicate boundaries of different surface water masses, and *arrows* indicate surface current direction. (1) Bashi Strait (northern channel of Luzon Strait); (2) Mindoro Strait; (3) Balabac Strait. Printed with permission from AGU

width at that time, as the sea level was few meters higher than today (Labeyrie et al. 1987; Fairbanks 1989). The rate of the water flow in the surface return branch of the thermohaline circulation in the region conceivably displayed maximum values as follows from the paleodata on the ITF area and Indian Ocean (Ivanova et al. 1995; Holbourn et al. 2005; see Section 5.7). The intensified summer monsoon induced stronger mixing, shallowing of thermocline, more vigorous upwelling of Vietnam documented by changes in foraminiferal population (Huang et al. 2002), depleted carbon isotope values of surface-dwelling planktic foraminifera *G. ruber* in core S-17954 (Wang et al. 1999a; Fig. 6.7). In contrast, high amplitude of $\delta^{18}O$ values measured on the relatively deeper-dwelling planktic species *Pulleniatina obliquiloculata* and surface-dweller *G. ruber* in ODP Site 1143 suggest decreased mixed layer depths and thus weaker winter monsoon (Tian et al. 2005).

On the contrary, mean annual U^k_{37}-SST was close to winter foraminiferal SST during MIS 2–4, when the strait floor emerged and water was colder throughout the sea. The cold water was perhaps partially formed in the upwelling near the Luzon Strait (Pflaumann and Jian 1999; Wang et al. 1995; Steinke et al. 2008). Hence, the long-term glacial–interglacial variations of U^k_{37}-derived mean annual SST are related to sea level oscillations which result in closure or opening of the Borneo Strait and hence, control the inflow of warm water of the upper branch of the THC from the tropical Pacific. Enhanced cold water advection via the Luzon (Bashi) Strait forced by the intensified winter monsoon was superimposed on these variations (Wang et al. 1999a; Fig. 6.6). Along with the strengthening of winter monsoon, the intensified cold Pacific water inflow was associated with a southward shift of the polar front and the eastward migration of the Kuroshio Current during the stadials (Steinke et al. 2008). The winter monsoon induced stronger mixing, more vigorous upwelling off Luzon tip, and shallowing of thermocline as follows from decreased ratio of mixed-layer to thermocline-dwelling species (Huang et al. 2002; Steinke et al. 2008) and enhanced nutrients content deduced from the carbon isotope values of surface-dwelling species *G. ruber* in core S-17927 (Wang et al. 1999a). On the contrary, the upwelling off Vietnam was suppressed and the mixed layer deepened over the major part of the sea, except for the northeastern corner, as evidenced by lower amplitude between $\delta^{18}O$ values measured on *P. obliquiloculata* and *G. ruber* in ODP Site 1143 (Tian et al. 2005).

More short-term, millenium-scale, temperature variations show contradictive trends, which are difficult to explain. According to the data from Core S-17940 (Fig. 6.5), winter temperatures decreased during warm DO interstadials and Bølling interstadial. Even the warm Preboreal was not accompanied by any strong temperature increase. On the contrary, the temperature rose at the cold H-4 event, as well as during the warm DO interstadials 3, 4, and in the Allerød interstadial. However, these contradiction results seem at least partially related to errors of the transfer functions (Pflaumann and Jian 1999). Alternatively, Kienast and co-authors (2001) established an abrupt U^k_{37}-SST increase of at least 1°C, a 14.6 cal. ka BP coeval to the beginning of Bølling warming in Greenland ice core GISP2.

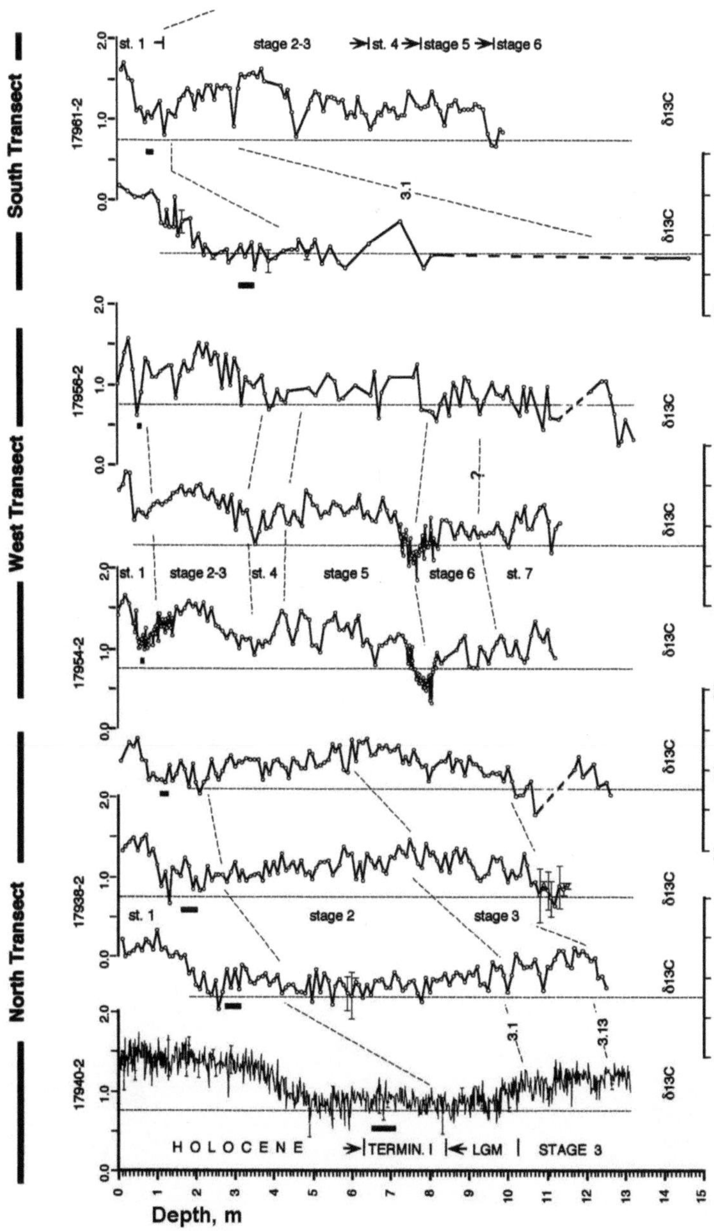

Fig. 6.7 Carbon isotope records measured on planktic species *G. ruber* in sediment cores from the SCS (modified from Wang et al. 1999a). Crosses show AMS-^{14}C-dated levels. Marine isotope stages are labeled and stage boundaries are connected by *dashed lines* between the different records for each transect. B/A= Bølling–Allerød. LGM= Last Glacial Maximum. Solid bars indicate Younger Dryas event

Grain-size distribution of the carbonate free and organic carbon free fraction of hemipelagic sediments is processed by means of silt mode and "Koopman index" (Koopman 1981) in order to distinguish between wind-supplied (dust) and river-borne (mud) sediments (Wang et al. 1999a). As a result, the pronounced aridity and humidity signals are identified. Variations in the intensity of wet summer and dry winter monsoon are deduced from the history of continental aridity in South China, which in turn is controlling the fluvial and/or aeolian sediment supply to the continental margin in front of the Pearl River mouth, southeast of Hong Kong. Here the hemipelagic sediments in cores S-17939 and S-17940 contain a detailed record of past changes in the aeolian and fluvial sediment input. In a first approximation, two different sources were distinguished by a simple empirical relationship between the percentage of clay fraction (<6 mm) and the primary modal grain size of siliciclastic silt (>6 mm). The obtained records (Fig. 6.8) of past changes in aridity and wetness on the mainland are supported by sea surface salinity reconstructions in the same cores and by the independent evidence from pollen spectra (Sun and Li 1999). Silt modal grain sizes also form a measure of past wind strengths (Sarnthein et al. 1981), in our case, of the winter monsoon that carries the aeolian loess from the South China to the sea.

The grain-size distribution data suggest that unlike the last 14.6 cal. ka, the beginning of deglaciation, as well as MIS 2 and 3, is characterized by prevailing of aeolian

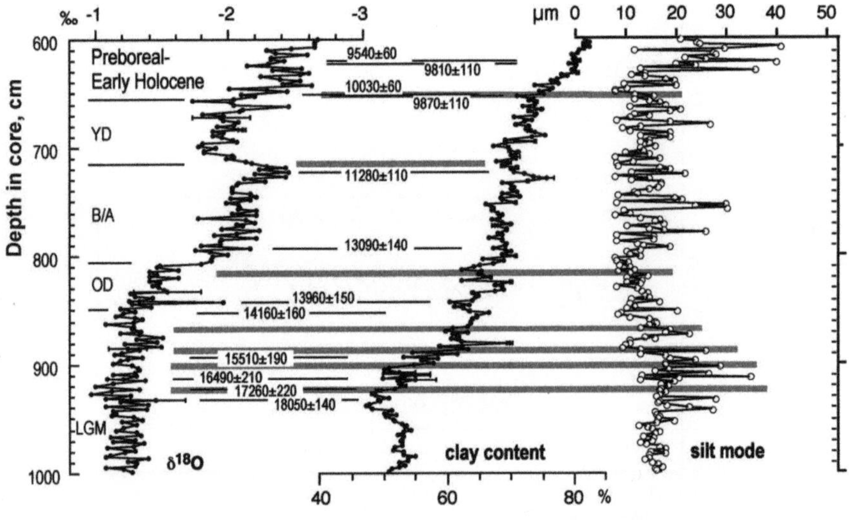

Fig. 6.8 Close-up of variations in terrigenic sediment input across the last deglaciation (600–1,000 cm-core depth) in Core S-17940 as compared to the planktic oxygen isotope record (modified after Wang et al. 1999a). Siliciclastic modal grain sizes (>6 mm) of dust stand for paleowind strength during winter, clay contents reflect intensity of fluvial sediment input. *Thin lines* mark levels with AMS-^{14}C dates, broad bars display major abrupt changes in the terrigenic sediment input and their linkage to δ^{18}O stratigraphy. LGM = Last Glacial Maximum; *OD*= Oldest Dryas; *B-A*= Bølling–Allerød; *YD*= Younger Dryas. The maximum in summer monsoon is established in Preboreal–early Holocene

dust and winnowing of sediments. The downslope reworking is also reflected by an increased lateral supply of shallow-water ostracods (Zhou and Zhao 1999), and possibly by the extremely low clay content within the area off Hong Kong. In part, the absence of riverine sediments of Hong Kong during the LGM may result from the local morphology of the continental slope. Here the glacial low sea level has led to a significant downcutting of the Pearl River mouth. Accordingly, any potential fluvial sediment discharge may have bypassed our core transect via canyons across the slope farther to the west and may be better encountered in the cores near the base of the slope. Nevertheless, the joint lack of fluvial sediments and massive aeolian-dust supply at core site S-17940 during the LGM together ascertain the continental aridity in South China. This suggestion is corroborated by further grain-size data. During MIS 2 and 3, the grain-size modes of silt amounted to 15–20 mm, values that are characteristic of loess deposits in Central China (Porter and An 1995). At the solar insolation minimum within the LGM, about 21 cal. ka BP, the grain-size modes reach 25–30 mm and depict a clear maximum in the strength of winter monsoon winds.

Domination of dust-borne pollen derived from semiarid and Alpine regions at site S-17940, the northern SCS, points to an arid climate in the continental South China (Sun and Li 1999). During the LGM, the sea surface salinity values at site S-17940 were 1–1.5 psu lower than today, suggesting an increased freshwater input. This may be explained by the glacial Pearl River mouth advance close to site S-17940 across the emerged broad shelf at the low sea level. However, glacial salinity values were >1 psu higher than during deglaciation, 14.6–9.5 cal. ka BP (Fig. 6.9), when the increase in precipitation was associated with the summer monsoon.

During MIS 3,centennial coherent oscillations of sea surface salinity and the clay fraction in core S-17940 demonstrate variations in the monsoon climate corresponding to those over South China, reconstructed by the loess record (Porter and An 1995). The high concentration of aeolian dust, coarsened modal grain size, and a salient clay minimum match a marked sea surface salinity maximum, hence document a peak in aridity. It was dated to about 35 ^{14}C ka and tentatively assigned to the cold Heinrich event H-4 in the North Atlantic (Bond and Lotti 1995; Vidal et al. 1997). Directly subsequent lack of aeolian-dust samples, increased clay, and markedly reduced SST record a phase of humid climate linked to the warm DO interstadial 8 (Figs. 6.4 and 6.9). Similar but less conspicuous climatic oscillations occur above in the core section and may be linked to Heinrich event 3 and DO interstadial 4 (about 29–30 cal. ka BP, Fig. 6.9). These results are interpreted in terms of the correspondence of short-term events of continental aridity in China to global cooling events whereas the phases of continental humidity are related to global warmings on millennial and centennial timescales (Wang et al. 1999a).

The discharge of fluvial sediments is mainly restricted to the Holocene and the Bølling. However, the earliest intensification of the summer monsoon accompanied by an increase in river runoff and monsoon rainfall is noted at about 21.5 cal. ka BP. The rapid (during several decades) increase in summer monsoon intensity occurred just after the Younger Dryas. The monsoon rainfall reached its maximum in the Preboreal and then reduced gradually. The Early Holocene river runoff and rainfall

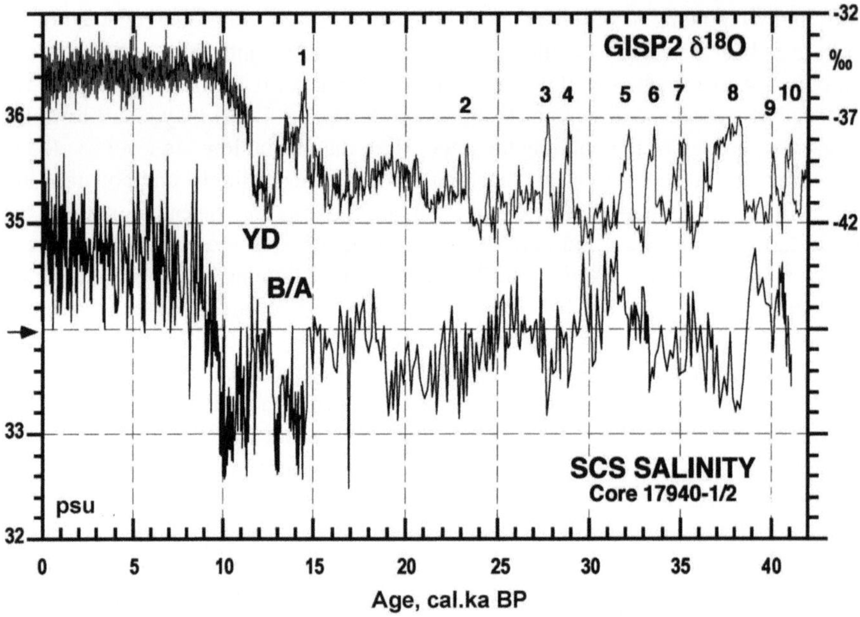

Fig. 6.9 Sea surface salinity variations at site S-17940 from the northern South China Sea, which are generally reversed to the $\delta^{18}O$-temperature oscillations in the GISP2 Greenland ice core record (Grootes and Stuiver 1997). Numbers are Dansgaard–Oeschger events (after Wang et al. 1999a with Elsevier permission)

maximum in China was reflected by sea surface salinity minimum in the western and especially, in the northern South China Sea, as well as by $\delta^{18}O$ and U^{k}_{37}-SST minima. So far as the sea surface salinity decreased to 32.6 psu in the northern South China Sea (Figs. 6.9 and 6.10), it cannot be explained by the freshened water inflow from the ocean. Only a substantially enhanced precipitation might result in such a freshening. This assumes a considerable increase in the intensity of summer monsoon that transported the moisture from the sea to the subtropical South China. The timing of this event is coeval with an early Holocene maximum in monsoon precipitation, which occurred in the countries encompassing the Arabian Sea (Sirocko et al. 1993). This synchroneity suggests the atmospheric forcing of long-term (and short-term) changes in the hydrological cycle of the Asian monsoon (Wang et al. 1999a). In the western and southern SCS, three cores (S-17954, S-17956, S-17961) show a distinct $\delta^{18}O$-freshwater signal during the earliest Holocene, possibly linked to an enhanced runoff of the Mekong River (Figs. 6.2 and 6.3). In other cores, the signal is less significant, partly because of lower sampling resolution, partly owing to core locations that are too distant from the major river mouths.

In the tropical southern SCS, the interglacial-to-glacial changes in paleoceanography and terrigenic sediment supply were equally affected by changes in sea level and climate. The glacial lowering of the sea level down to about 120 m (Labeyrie et al. 1987) led to an emergence of Sunda shelf, to a shift of its shoreline toward

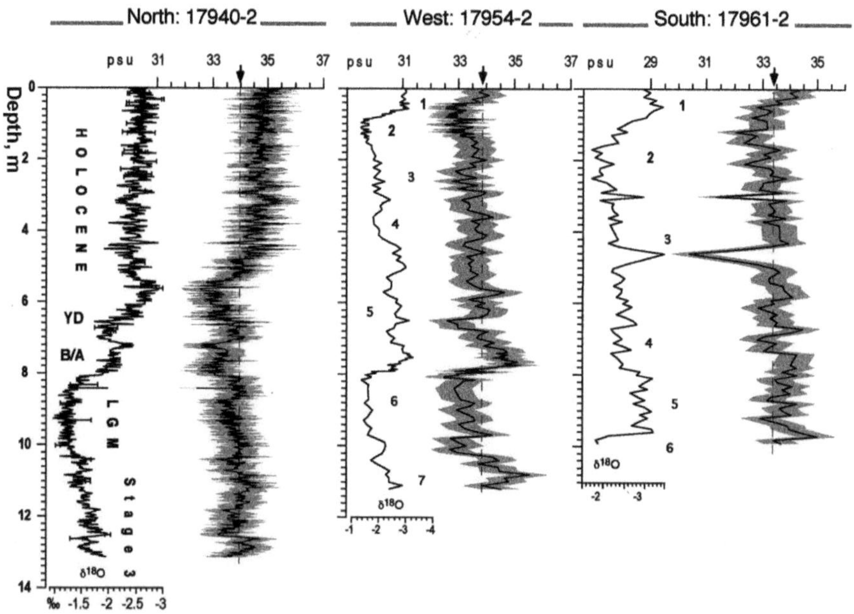

Fig. 6.10 Sea surface salinity (SSS) records covering up to 220,000 years in the SCS (modified from Wang et al., 1999a). Using the transfer equation of Wang et al. (1995b), SSS values are calculated from $\delta^{18}O$ of *G. ruber* after subtracting the global ice effect and - SST for 0–30 m (see Fig. 6.4). $\delta^{18}O$ curves serve for stratigraphic reference. Numbers refer to $\delta^{18}O$ stages; YD= Younger Dryas; B/A= Bølling–Allerød; LGM = Last Glacial Maximum. *Vertical arrows* mark modern mean surface salinity (0–30 m)

the site S-17964 (Fig. 6.1), and to the development of Mekong and Molengraaff river deltas at the shelf margin (Evans et al. 1995; Stattegger et al. 1997). These great shifts in shoreline and fluvial discharge may readily explain an abrupt, but long-lasting increase in sedimentation rates and high clay content at MIS 5/4 boundary, as well as the abrupt decrease in clay percentage and sedimentation rates (by a factor of 2) at both Termination II and the end of the YD (cores S-17961 and S-17964). This uniformity entails that fluvial runoff and possibly tropical conditions continued during the LGM on the emerged Sunda subcontinent, which became the major glacial refuges of tropical forest (Prentice and Sarnthein 1993). The model of a tropical rainforest on the glacial Sunda shelf is strongly supported by the pollen assemblages from the glacial section of core S-17964 (Wang et al., 1999a).

However, the simple model of solely sea level-induced changes in freshwater discharge hardly can explain oscillations in paleosalinity found in core 17961 from the southern SCS (Fig. 6.10). At this more distal site, sea surface salinity values have decreased by up to 2 psu only during and subsequent to the LGM. These low salinity values document a freshwater lid that culminated during deglaciation and the earliest Holocene, just in contrast to the coeval sea level rise and flooding of the Sunda shelf. Likewise, the significant sea level fall during early glacial MIS 4 did not induce the expected salinity reduction, which occurred later, in late MIS 4 and MIS 3.3. On the other hand, the sea surface salinity indeed increased during Termination II and reached the expected maximum along with the high sea level stand of MIS 5.5, such as in the late Holocene. Therefore, the long-term glacial-to-interglacial variations in tropical moisture budget are not yet fully understood.

The same problem concerns some prominent but very short-lasting changes in the tropical freshwater budget documented by the paleosalinity record of core S-17961. Here, the $\delta^{18}O$ record defines two major salinity reductions by almost 4 psu near 25 and 34–40 ^{14}C ka BP (Fig. 6.10). Unfortunately, the dating precision is still insufficient because of the unknown local ^{14}C-reservoir effect in the glacial South China Sea. Hence it's still unclear whether these events were linked to cold Heinrich events H-3 and H-4 or, conceivably, to the subsequent warm interstadials of the DO events 4 and 8. Meanwhile, the answer will be crucial to assess the budget of tropical moisture in the Asian monsoon region, and also to extend or confine the early suggestions by (Duplessy et al. 1981a) that the monsoon precipitation area was limited to the equatorial Indian Ocean at the LGM.

As soon as the sea level dropped by more than 40 m, the input of warm waters of the upper limb of the THC from the tropical Indo-Pacific ceased owing to the closing of the Borneo Strait and SCS turned into a semi-closed estuarine basin since MIS 4 to the LGM (Chappell et al. 1996). An estuarine circulation also characterizes the modern and early Holocene South China Sea, as shown by low epibenthic $\delta^{13}C$ values (about 0‰) in the southern part of the sea, at 1,550 m water depth (Core S-17964; Fig. 6.7). These values are similar to the δ^{13} signature of the lower intermediate water (Sarnthein et al. 1988) that enters from the western Pacific into the SCS. During MIS 2, the epibenthic $\delta^{13}C$ values dropped to –1.6‰ directly in front of the emerged Sunda shelf (Core S-17964). This $\delta^{13}C$ range was lower than almost everywhere else in the glacial ocean (Sarnthein et al. 1988, 1994b). So strongly depleted $\delta^{13}C$ values indicate well-developed oxygen-minimum layer near the southern margin of the SCS and a stable estuarine stratification of the surface water in conjunction with high fluxes of organic matter. This model is in line with the coeval extensive tropical runoff deduced from both the ongoing deposition of hemipelagic fluvial sediments and low salinity discussed above (Wang et al. 1999a).

Owing to the development of estuarine circulation, the isolated SCS was highly susceptible to oxygen depletion during the glacials. Even a minor increase in freshwater influx might cause the oxygen-depleted conditions below the mixed surface layer. Geomorphology of the Sunla shelf promoted the runoff delivery by the Mekong River from the eastern Himalayas into the SCS. The estuarine circulation was maintained by the inflow of old low-oxygen intermediate and deep waters from the western Pacific that is from the "end" of the THC (Gordon 1986; L. Wang 1992; P. Wang 1992) whereas the young oxygen-rich surface waters of the tropical Indo-Pacific could not enter the sea from the southwest, unlike today (Fig. 6.6).

In modern sediments of the South China Sea, the nutrient content of surface water is reflected in the planktic $\delta^{13}C$ record of *G. ruber* (white), while the nutrient level of subsurface water is mirrored in the $\delta^{13}C$ record of the subsurface dweller *P.*

obliquiloculata (Broecker and Peng 1982; Pflaumann and Jian 1999). Low values of these parameters(<0.9‰ and <0.4‰, respectively) trace the inflow of nutrient-enriched surface and subsurface Pacific water via the Luzon Strait into the northern SCS (Metzger and Hurlbert 1996). The rich nutrients of the flow originate from an upwelling cell near the northern tip of Luzon, whereas the Kuroshio Current is nutrient depleted. Another "tongue" of low $\delta^{13}C$ values of *G. ruber* (<1.2–<0.9‰) occurs south of Vietnam and depicts the nutrient enriched surface water in the local upwelling, induced by the summer monsoon (Wiesner et al. 1996). The $\delta^{13}C$ minimum in the far southwest may be also linked to the prominent nutrient discharge of Mekong and other tropical rivers. High $\delta^{13}C$ *G. ruber* (>1.3‰) characterizes the surface water in the remaining parts of the sea as largely nutrient depleted. Rather coarse grid of modern $\delta^{13}C$ data cannot detect any other local nutrient sources (Wang et al. 1999a).

However, as soon as the sea level dropped and the shoreline of China moved closer to shelf break, the variable local input of fluvial and aeolian nutrients probably played a more important role in the nutrient budget of surface water. Very low planktic $\delta^{13}C$ values at the proximal site S-17940 depict extremely enriched nutrients during deglaciation and the early Holocene. This fertility has triggered also a short-lasting extreme, organic-fluff-induced benthic $\delta^{13}C$ minimum of –1.4‰. Based on the siliciclastic grain-size data this Preboreal event was ascribed to outstanding runoff from the Pearl River (Fig. 6.8; Wang et al. 1999a). The surface water productivity in the northern SCS decreased later, as evidenced by rather high $\delta^{13}C$ values measured on *G. ruber* ~1.3–1.6‰. The same is true for MIS 3, except for some short-term excursions during warm DO events 3 and 8 (Fig. 6.7; Wang et al. 1999a).

In the cores from the western SCS, southeast of Vietnam, the high $\delta^{13}C$ values (1.0–1.7‰) demonstrate low nutrients content during cold stages 2, 3, 5.2, and 5.4 which indicates a weak upwelling intensity and hence low productivity. In contrast, extremely low $\delta^{13}C$ values of 0.4–0.6‰ document an enhanced abundance of nutrients, stronger upwelling, and summer monsoon during Termination I – the early Holocene – and during Termination II – early MIS 5.5 (Fig. 6.7). These results are in line with the reconstructed increase in productivity over both terminations – beginning of interglacials in Core GeoB-3011 from the Arabian upwelling (see Section 5.5; Ivanova et al. 2003b). Besides the upwelling, the enhanced nutrients content in the southwestern part of the SCS may reflect a river runoff increase due to more intense monsoon rainfall or snow melting in Himalayas during the deglaciation when the sea level was still low, and Mekong and Molengraaff rivers discharged near to the break of the Sunda shelf (Fig. 6.2). This concept is indeed supported by short-lasting significant drops in sea surface salinity (by 1.5 psu) found at site S-17954 during terminations I and II, MIS 5.3 and 5.1 (Fig. 6.10). The high nutrient supply by rivers from the glacial "Sunda Land" is documented more directly by extremely low glacial and early deglacial planktic and benthic $\delta^{13}C$ values at sites S-17961 and S-17964 (Fig. 6.7), then lying proximal to the Sunda coastline (Wang et al. 1999a).

The revealed monsoon-driven maxima in fluvial nutrient supply and intensified summer upwelling during deglaciations are consistent with the assumption introduced in Sections 5.5 and 6.1. It postulates an influence of the orbital periodicities on the intensity of the Indian and East Asian monsoons, deduced from the study of Arabian Sea cores (Clemens et al. 1991; Sirocko et al. 1996) and northwest Pacific (Morley and Heusser 1997) – Sulu Sea cores (de Garidel et al. 2001), respectively, and from the atmospheric climate modeling (Prell and Kutzbach 1987; Kutzbach et al. 1996).

6.3 Variations of the East Asian Monsoon During the Holocene

Alternating short-term (about 100 years) episodes with domination of either summer or winter monsoon are superimposed on the increasing trend in the summer monsoon intensity both in the South China Sea and in the Indian Ocean over the transition from the deglaciation to the Holocene (Wang et al. 1999b).

Distinct minima of foraminiferal SST in winter, and to the less extent in summer in Core S-17940 point to climate change at 8.2, 5.2–3.4, and 2–1 cal. ka BP, that is to the periodicity of about 3 ka (Figs. 6.3 and 6.4). The temperature minimum about 4.5 cal ka BP is noted in several sections from the West Pacific marginal seas (Oda and Takemoto 1992; Jian et al. 1996; Wei et al. 1998). Three relatively cold events match a disappearance or reduction of the deep-dwelling tropical species *P. obliquiloculata* (Pflaumann and Jian 1999). Judging by distribution and oxygen isotope composition of this species in recent sediments, *P. obliquiloculata*can trace the outlined nutrient-rich inflow of upwelled water from the Luzon Strait into the South China Sea during winter. Accordingly, the above-mentioned intra-Holocene episodes may reflect a short-lasting southward deflection of this inflow, probably linked to an increase in winter monsoon (Wang et al. 1999a). The high-resolution (20–40 years) salinity record of Core S-17940, adjacent to the Pearl River mouth (Fig. 6.9), distinctly reflects the variability of monsoonal rainfall in the South China and corresponding changes in the river runoff. According to geomorphological data, the sea level rise by 20 m in this region during the last 10 ka thus corresponded to that of the World Ocean (Fairbanks 1989). Hence, the distance from the Pearl River mouth has not changed.

The sea surface salinity gradually reached the average Holocene value 34.5 psu only after the well-expressed early Holocene maximum of freshwater runoff. A short-term double salinity peak of 35.5 psu at about 8.3–8.1 cal. ka BP was followed by a short-term decrease to 34.4 psu. These events match centennial oscillations from extremal arid to extremal humid conditions. The salinity maxima correlate with remarkable coolings in Greenland (Fig. 6.8) and Antarctic (Ciais et al. 1992). Upward in the section, salinity varies by $\pm$ 0.7 psu relative to average Holocene values. During the last 600 years, salinity increased to the modern value of 35.2 psu, possibly because of reduction in the monsoonal rainfall during the Little Ice Age. Judging by the cyclicity of the salinity record, one may expect a rapid return to stronger freshwater runoff rather soon (Wang et al. 1999a).

The fluctuations in temperature and salinity over the last 10 ka reflect cycles of 775 years and 102/84 years in the East Asian monsoon regime (Fig. 6.11). Each of these periodicities exceeds the upper limit of red noise at 80% confidence level.

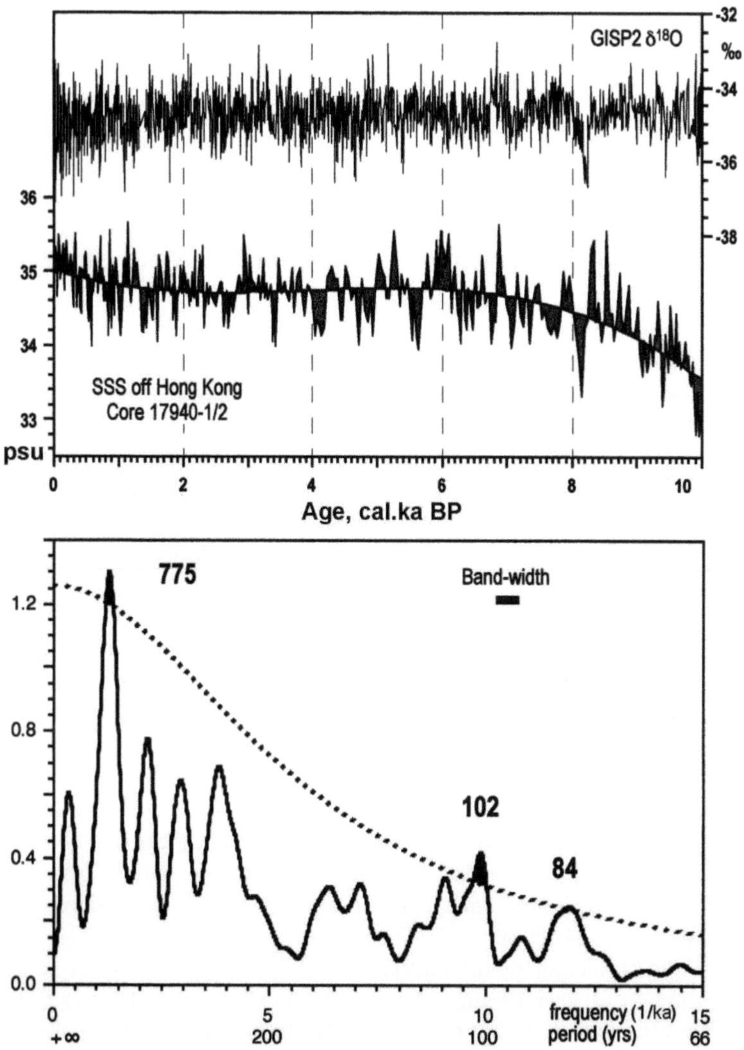

Fig. 6.11 Holocene changes in $\delta^{18}O$-based temperature record of the Greenland ice-core GISP2 (Grootes and Stuiver 1997) versus sea surface salinity (SSS) variations at site S-17940 off Hong Kong (northern South China Sea) over the last 10,000 years (after Wang et al., 1999a with Elsevier permission). Data are presented in the time (*upper panel*) and frequency domain (*lower panel*). Major temperature lows ($\delta^{18}O$ minima) on Greenland summit were coeval with SSS maxima, equal to lows and subsequent highs in monsoon precipitation in subtropical South China (that is at the end of the Preboreal maximum in summer monsoon, about 8,250 year BP)

Similar periodicities were also found during MIS 2 and 3, but were not defined precisely because of insufficient time control in this section of Core S-17940. The 775-year period amounts approximately to half of the 1,500-year cyclicity, found in monsoon-controlled sediment records from the Arabian Sea (Sirocko et al. 1996), North Atlantic, and Greenland (DO cycles, see Section 3.4). According to suggestion by Wang et al. (1999a), the 775-year periodicity may represent a harmonics of the 1,500-year cycle. The 102/84-year periodicities may possibly reflect the Gleissberg cycle of solar activity, frequently described in high-resolution records such as tree-ring and the GISP2 ice core (Stuiver and Braziunas 1993; Grootes and Stuiver 1997).

6.4 Linkages of the Monsoon Circulation to Global Climate Change

The glacial–interglacial variations in the East Asian monsoon, as well as more short-term, up to millennial-scale, changes are primarily controlled by temperature and pressure contrasts between the northwestern Pacific and Tibet with surrounding highlands. Modeling experiments and paleoclimatic reconstructions (Kutzbach and Gallimore 1988; An et al. 1991; Kutzbach et al. 1996; Morley and Heusser 1997) ascertained that variations in insolation anomalies forced by orbital factors played a crucial role in the onset of these contrasts, and therefore, in monsoon climate changes.

To reveal the linkages of the intensity of East Asian monsoon to the extension of the Antarctic sea ice and to growth and melting of the Northern Hemisphere ice sheets, we tried to test, as precisely as possible, the timing (leads and lags) of past variations in winter and summer monsoon versus changes in the most prominent marine and ice-core records of global climate and sea level change (Wang et al. 1999a). The special attention was focused on a few major events of monsoon evolution during the last glacial cycle. The GISP2 and VOSTOK $\delta^{18}O$-temperature records from the Greenland and Antarctic ice sheets, respectively (Grootes and Stuiver 1997; Sowers and Bender 1995) were used for the detailed turning of the well AMS-^{14}C-dated benthic and planktic oxygen isotope records from the SCS (Fig. 6.3; Wang et al. 1999a).

Core S-17940 documents the mid-Holocene decrease in the mean annual SST and the inflow of western Pacific subsurface water southward of Hong Kong since ~5.2–4.4 cal. ka BP (Figs. 6.4, 6.5, and 6.12) This event precisely correlates with a short-term, but marked weakening of NADW ventilation off the northwest Africa 4.8–4.6 cal. ka BP (Maslin et al. 1996) and thus suggests an atmospheric and/or thermohaline influence on the ocean circulation.

The major early Holocene fluctuations in sea surface salinity and monsoon intensity off Hong Kong, lasting about 300 years, during the 8.2 cal. ka BP are demonstated by Core S-17940 (see also Section 3.6 and 4.5.4). These fluctuations appear to coincide with the early Holocene cooling in Greenland (Fig. 6.11; Grootes

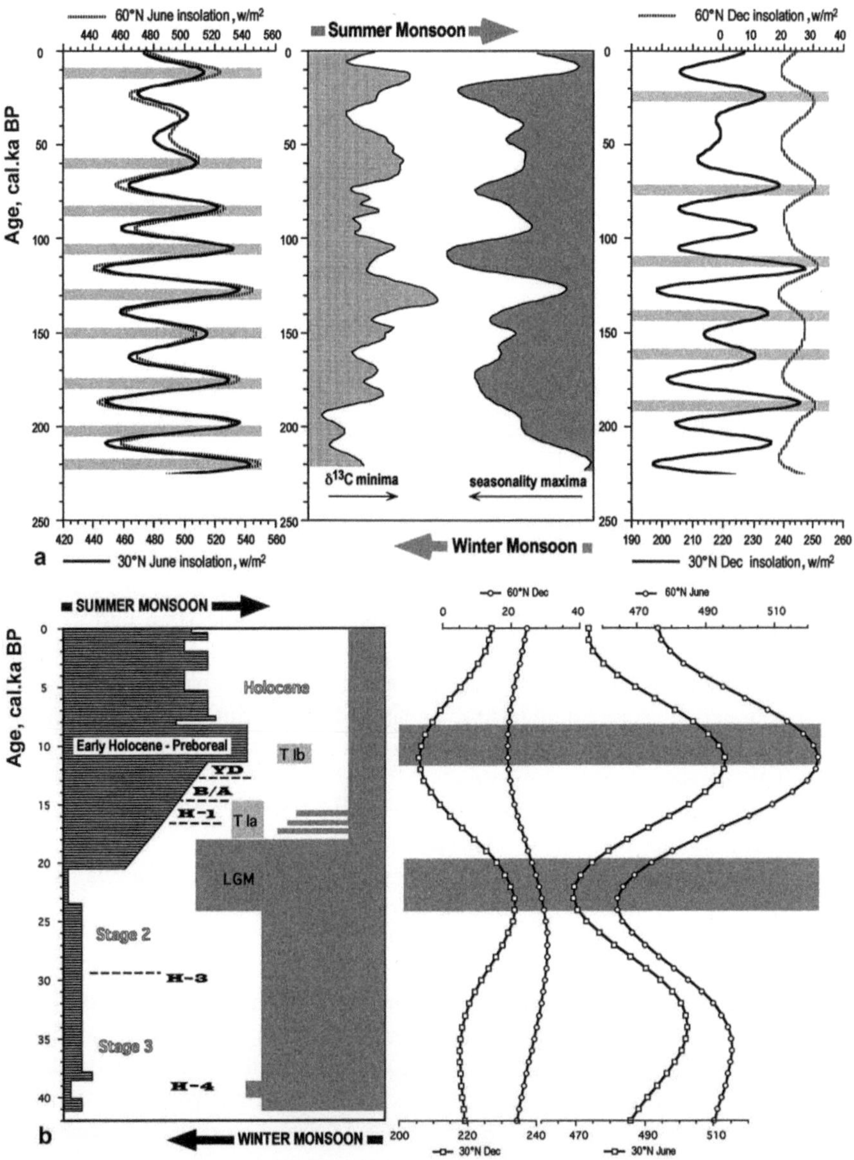

Fig. 6.12 Summary schemes of variations in the East Asian monsoon over the last glacial cycle versus changes in solar insolation at 60°N and 30°N (after Wang et al. 1999a with Elsevier permission). These latitudes represent the two end-member regions of the Asian winter and summer monsoon on land and on sea, characterized by totally different surface conditions. *Arrows* on top and bottom of the panel with the monsoon scheme point at increasing monsoon intensity. (**a**) Monsoon cycles over the last 220,000 years as inferred from a smoothed nutrient-upwelling record of planktic $\delta^{13}C$ minima (*G. ruber*) and the SST-seasonality record of planktic foraminifera counts in core S-17954. Peak glacial seasonality and winter monsoon largely correspond to minimum insolation during summer and a reduced insolation anomaly between land and sea. Vice versa,

and Stuiver 1997), Norwegian, and Barents seas (see Sections 3.6 and Section 4.5.4; Duplessy et al. 2001; Hald and Korsun 2008), with a short-lasting reduction (or even shutdown) in NADW ventilation (Maslin et al. 1996; Knaack 1997) and with the start of the mid-Holocene dust discharge from Arabia into the Arabian Sea (Sirocko et al. 1993). However, a more precise definition of the decadal- to- centennial scale leads and lags is necessary to understand the actual linkages between these climate signals. Owing to limited precision of radiocarbon dates, we cannot ascertain yet, whether the sea surface salinity maximum 8.3 cal. ka BP in the SCS indeed correlates with the distinct temperature minimum in Greenland, and whether the subsequent short salinity minimum, that is the maximum in monsoon rain 8.15 cal. ka BP really followed the cooling episode. If such remote events actually coincide, we can correlate cold phases in Greenland with arid events in the SCS and warm phases with humid episodes, respectively.

The less than 10-year long transition from the cold Younger Dryas to the warm Preboreal about 11.6 cal. ka BP, documented in the Greenland ice cores (Dansgaard et al. 1989; Grootes et al. 1993; Fig. 6.4) correlates with a rise in monsoon precipitation over less than 50–100-year off Hong Kong, as displayed by a decrease in salinity and onset of fluvial mud deposition in the sediment cores S-17940 and S-17939 (Fig. 6.9). Moreover, a rapid coeval rise in winter and mean annual SST occurred throughout the SCS (Fig. 6.5), including its southeastern part (Thunell and Miao 1996). The subsequent Preboreal to early Holocene culmination in the East Asian monsoon intensity well corresponds to a comparable maximum of the Indian summer monsoon found in Arabian Sea sediments (Sirocko et al. 1993). The solar insolation and summer temperatures in Greenland (Koerner and Fisher 1990) reached the maximum at the same time.

If the suggestion about the reservoir effect up to 1,500 years in the SCS (Wang et al. 1999a) is correct, and the $\delta^{18}O$ maximum with the age of 35.5 ^{14}C ka BP really corresponds to the date of 34 ^{14}C ka BP measured in the Norwegian Sea (Adkins and Boyle 1997), then the short-term monsoon oscillations during MIS 3, based on grain-size and salinity data from Core S-17940, may be considered as coeval with Heinrich events H-4 and H-3 (Figs. 6.3 and 6.5). The above data on the Holocene and Younger Dryas abrupt climate fluctuations, as well as a new record of the Indian monsoon variations during MIS 3 (Schulz et al. 1998) suggest that any short-term changes in the Asian monsoon system were closely linked to high-latitude climate forcing such as major iceberg surges in the North Atlantic (Broecker et al. 1990; Bond and Lotti 1995). These changes possibly occurred on decadal-to-centennial

Fig. 6.12 (continued) interglacial culminations in summer monsoon, such as about 10,000–11,000 year BP, parallel maximum insolation and maximum insolation anomalies between land and sea during summer (shaded bands). (**b**) Blow-up of monsoon variations over the last 40,000 years, based on variations in the differential input of siliciclastic sediments and sea surface salinity at site S-17940 (see Figs. 6.8 and 6.10). T Ia and T Ib = Termination Ia and Ib, YD=Younger Dryas, B/A = Bølling–Allerød, H-1, H-3, H-4 = Heinrich events 1, 3, and 4

timescales mainly via atmospheric climate forcing. Any leads or lags have not been fixed (Wang et al. 1999a).

In contrast, significant leads and lags mark the onset of glacial Termination IA when considerable age deviations are ascertained. The clay fraction content in sediments of Core S-17940, roughly corresponding to the fluvial runoff, began to rise from <50% up to an almost Holocene level of 60–70% as early as 17.0–14.7 ^{14}C ka BP, just after the solar insolation minimum at 18 ^{14}C ka BP (21 cal. ka BP). The increase in clay content preceded the onset of postglacial $\delta^{18}O$ signal of sea-level rise by 3–4 ka. The sea-level rise, in turn, slowly started at 14 ^{14}C ka BP and accelerated at ~13.2 ^{14}C ka BP. At this time the rise in fluvial clay input had finally reached the more distal site S-17939 on the northern transect. By contrast, the modal grain sizes of dust show the postglacial decrease from 15 to 13.4 ^{14}C ka BP.

The above timing relationships between different paleoclimatic signals related to the onset of Termination IA lead us to the following conclusions: (1) The clay supply and hence, the fluvial sediment supply from South China, increased irrespective of the postglacial sea level rise. (2) This early signal of enhanced summer monsoon rains appeared simultaneously with the start of Antarctic ice melting (Sowers and Bender 1995) and likely responded to climate forcing by southern high latitudes. (3) The delayed decrease in strength of winter-monsoon winds was coeval with a major melting of northern high-latitude ice sheets directly prior to Bølling and DO interstadial 1 (see GISP2 record in Fig. 6.4). The maximum in winter monsoon winds slightly lagged the minimum in northern insolation, whereas the peak in summer monsoon precipitation was in phase with maximum northern insolation (Fig. 6.12). The same relationship is true for the Indian monsoon system.

In most cores from the SCS, the records of monsoon variations are tied to global climate change via well-defined events on planktic oxygen isotope curves. The revealed up to 1,000-year age lag of paleoceanographic events in the SCS relative to the North Atlantic $\delta^{18}O$ records represents the time span needed for an ocean-wide spreading of the early deglacial $\delta^{18}O$ signal and confirms previous estimates of the ocean overturning time (800–1,000 years; Duplessy et al. 1991a). However, the actual lag may be larger in view of the high ^{14}C-reservoir ages in the SCS during Heinrich events (Wang et al. 1999a). If to accept the idea that the water overturning in the global conveyor belt is only 200–300 years (Lappo 1984; Anisimov et al. 2002), the age lag of the deglaciation signal likely does not exceed 800–1,000 years.

6.5 Summary

The alternation of two regimes of monsoon circulation in the SCS associated with the glacial–interglacial cycles was primarily driven by variations of solar insolation at 30–60°N. Both long-term (thousands of years) and short-term (century scale) variations in the East Asian monsoon are coeval with those of the Indian monsoon. This suggests a common, probably atmospheric, mechanism of their intensification.

A strongly intensified winter monsoon and weakened summer monsoon were typical of glacial periods. These cold stages involved a strongly reduced monsoon precipitation during summer that led to continental aridity in subtropical South China, a noticeable discharge of loess dust, markedly lowered SST during winter, generally enhanced seasonality over the SCS, and intensified upwelling off Luzon tip. In contrast, the Holocene and other interglacial regimes were characterized by a strengthened summer monsoon circulation and weak winter-monsoon winds. This change in monsoonal circulation led to extreme continental wetness in South China, to high winter SST and a low seasonality in the whole SCS, and to decreased thermocline depth and high upwelling-induced productivity southeast of Vietnam.

Glacial-to-interglacial sea level changes had a significant impact on the paleoceanography of the SCS via the closure and opening of the Borneo Strait at a threshold water depth of about 40 m. During interglacial periods with high sea level, the lateral advection of warm waters of the upper THC limb from the tropical Indo-Pacific led to a significant general rise in annual mean SST (according to U^k_{37} data). Vice versa, the glacial closure of the Borneo Strait promoted to an intensified inflow of cool and nutrient-rich subsurface water from the northeast, which was upwelled near the northern tip of Luzon during winter. This inflow is traced in SST, salinity, and planktic $\delta^{13}C$ records. On the contrary, the mixed-layer deepening off Vietnam is portrayed by increased abundance of oligothrophic surface dwelling species in foraminiferal population.

Salinity variations during the last glacial cycle were determined by the relationship between precipitation, river runoff, and advection of Pacific water from the east, and Indo-Pacific water from the southwest. So far as the precipitation was controlled by the monsoon intensity, the lateral advection via the straits by the sea level oscillations, and the river runoff by both of these factors, amplitude and timing of the salinity changes varied in different parts of the sea.

Intervals of the excess fluvial clay accumulation and extended periods of reduced salinity indicate an ongoing fluvial sediment supply along the southern margin of the SCS during MIS 2–4. Based on these data, the emerged Sunda shelf was humid and drained by tropical rivers. As a result, this area served a glacial refuge for tropical forests. Isolation of the SCS in the southwest during glacials was accompanied by intensification of the estuarine circulation along with a significant increase in river runoff. This led to an extreme oxygen depletion of the intermediate water along the southern margin of the sea.

Sea surface temperature, salinity, and siliciclastic grain-size records from the northern South China Sea reflect centennial to millennial-scale paleoclimatic events. Cold events re-occurred with intervals about 3,000 years during the Holocene. Arid episodes corresponded here to the Younger Dryas and possibly to the Heinrich event H-4. Short-term low-salinity anomalies in the southern and western parts of the South-China Sea during Terminations II and IA and in late MIS 3 were related to intensified summer monsoon rainfalls and conceivably correlate with the warm DO interstadials 8 and 4–3.

Chapter 7
The Role of Thermohaline Circulation in Global Teleconnections

Abstract Evidence for global teleconnections via the ocean and atmosphere is found throughout marine and terrestrial climate records at glacial–interglacial, millennial, centennial, and shorter time scales; however, their mechanisms are still enigmatic. The examples of linkages between the North Atlantic and several low and high latitude areas at millennial and shorter time scales are reviewed with special emphasis to oceanic teleconnections and feedbacks. Paleo-observations and modeling results suggest that changes in the THC mode exert significant control on the spatial and temporal variability of heat and moisture exchange between the ocean and atmosphere and hence on the global climate at orbital and sub-orbital timescales. The widely debatable "bipolar seesaw" and "eddy" hypotheses were proposed to explain the anti-phase millennial-scale temperature changes in high northern and southern latitudes. The assumed coupling between weakened upwelling in the North Pacific and reduced NADW formation during the stadials of the DO cycles still needs further investigation. The atmospheric teleconnections operate mainly on short-term scales via the major wind systems, ENSO, NAO, and other oscillations.

Key words Teleconnection · Linkage · Antarctica · Low latitudes · Abrupt change · Reorganization of the thermohaline circulation · Greenhouse gases · Bipolar seesaw · Feedback · Decoupling · ENSO · Monsoon · Trade winds

7.1 Introduction

The correlation of multiple short-term events (1,500–1,000 years and less) in distant areas of the world, including low latitudes is undoubtedly among the most outstanding achievements of paleoceanographic and paleogeographic research in the last decade (Figs. 7.1 and 7.2, Color Plate 7.1). Well-documented fluctuations in the oxygen minimum layer at intermediate depths in the Santa Barbara Basin, northeastern Pacific (Hendy and Kennett 2000; Hendy et al. 2004) and Arabian Sea, northwestern Indian Ocean (Schulz et al. 1998; Schulte and Müller 2001; Altabet et al. 2002) provide good examples of short-term variations at low latitudes correlated to DO cycles and H-events in the North Atlantic. As these regional variations cannot be explained

E.V. Ivanova, *The Global Thermohaline Paleocirculation*,
DOI 10.1007/978-90-481-2415-2_7,

by the sole mechanism, by reorganization of surface circulation at high latitudes, or by simple coincidence, they must be linked via the oceanic and/or atmospheric teleconnections (Color Plate 7.2). Therefore, we face the challenge of understanding the mechanisms of teleconnections and the determining role of a particular area or latitudes in the global change. I imply the term teleconnections to denote rapid remote transfer of climatic signals that trigger the paleoenvironmental changes (via the positive and/or negative feedbacks).

As shown in previous chapters and numerous publications, teleconnections are documented at different time scales including Heinrich events (duration ~0.5–2 ka, mean periodicity ~7 ka), Dansgaard–Oeschger cycles (periodicity ~1.5 ka), and shorter decadal–centennial events. The general pattern of these cycles is relatively slow cooling and, as a rule, more rapid or even abrupt warming. The latter is caused by the quick response of the ocean surface layer to the external forcing, such as increase in insolation or freshwater discharge. Heat transfer by surface currents was much faster than the reorganization of the entire system of the global thermohaline circulation, which probably lasted about 1,000 years.

Although climatic teleconnections are known and studied on the synoptic timescale (e.g., IPCC 2007), the modern hydrological observations cannot explain all aspects of the climate system functioning. The incomprehensible origin of El-Niño, which has been discussed at least since 1924 (Walker 1924) and is extensively investigated under the umbrella of the TOGA program during 10 years, represents an example of such still unsolved problems. Thus, paleoceanographic studies are especially promising for investigation of the mechanisms and controlling factors of rapid climate change in a way that cannot be addressed with instrumental and historical records.

Previous chapters offered numerous examples of multi-proxy studies that call on the teleconnections to explain the seemingly unrelated climatic events over great distances. For example, warming phases in the North Atlantic during interglacials and DO cycles are correlated with intervals of enhanced bioproductivity in some regions of the monsoon area in the Indo-Pacific (see Sections Section 5.5, Section 6.2, and Section 6.3; Schulz et al., 1998; Wang et al. 1999a; Altabet et al. 2002; Ivanova et al. 2003b) whereas the increased primary production in the Sulu Sea is associated with the DO stadials (de Garidel-Thoron et al. 2001). Such simultaneous paleoenvironmental changes can be explained not only by the response of different proxies to orbitally driven variations of insolation but also by the teleconnection via the global

Fig. 7.1 Generalized climate features during interstadials (*top*) and stadials (*bottom*) of Dansgaard–Oeschger cycles during MIS 3 (after Voelker et al. 2002 with Elsevier permission). T (↑↓) indicate trends, i.e., warmer (colder), more (less), increased (lowered), etc. *Gray dots* mark the core sites. SST = sea surface temperature, T = temperature, SSS = sea surface salinity, mons. = monsoon, NE = northeast, SW = southwest, prod. = productivity, cond. = conditions, IRD = ice-rafted debris, NADW = North Atlantic Deep Water, DW = deep water, NPIW = North Pacific Intermediate Water, GNAIW = Glacial North Atlantic Intermediate Water, BW = bottom water, AABW = Antarctic Bottom Water, LCDW = Lower Circumpolar Deep Water , OMZ = oxygen minimum zone

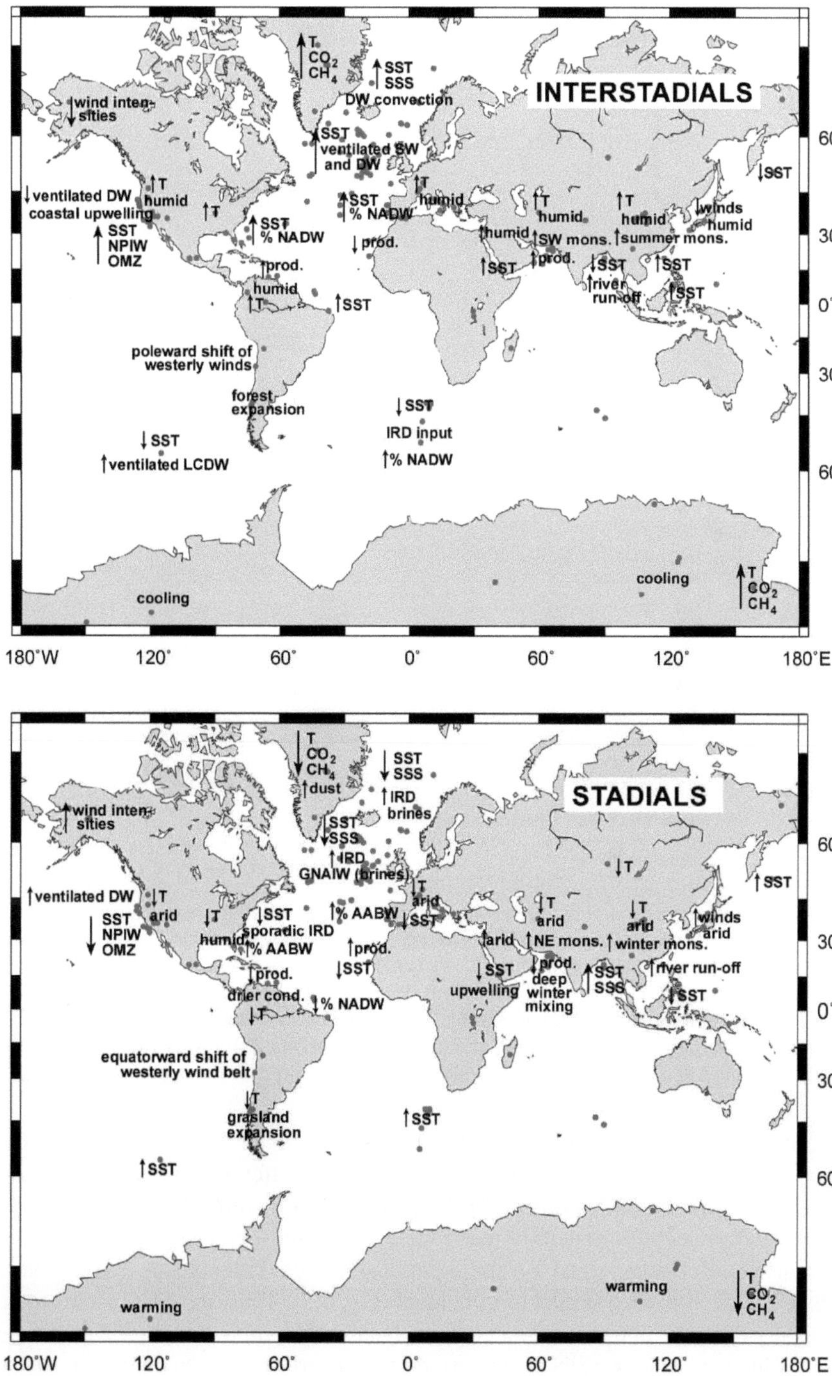

INTERSTADIALS
wind intensities
ventilated DW
coastal upwelling
SST
NPIW
OMZ
T
humid
T
SST
% NADW
prod.
humid
T
poleward shift of westerly winds
forest expansion
SST
ventilated LCDW
cooling
T
CO2
CH4
SST
SSS
DW convection
SST
ventilated SW and DW
SST
% NADW
prod.
SST
SST
IRD input
% NADW
T
humid
T
humid
humid
SW mons.
SST
prod.
T
humid
summer mons.
SST
river run-off
SST
SST
winds
humid
SST
cooling
T
CO2
CH4
60°N
30°
0°
30°
60°S
180°W
120°
60°
0°
60°
120°
180°E
STADIALS
wind intensities
ventilated DW
SST
NPIW
OMZ
T
arid
T
humid
SST
sporadic IRD
% AABW
prod.
drier cond.
T
% NADW
equatorward shift of westerly wind belt
T
grasland expansion
SST
warming
T
CO2
CH4
dust
SST
SSS
IRD
brines
SST
SSS
IRD
GNAIW (brines)
% AABW
SST
prod.
SST
SST
T
arid
arid
SST
upwelling
T
arid
NE mons.
prod.
deep winter mixing
SST
SSS
T
T
arid
winter mons.
river run-off
SST
winds
arid
SST
warming
T
CO2
CH4
60°N
30°
0°
30°
60°S
180°W
120°
60°
0°
60°
120°
180°E

Fig. 7.1

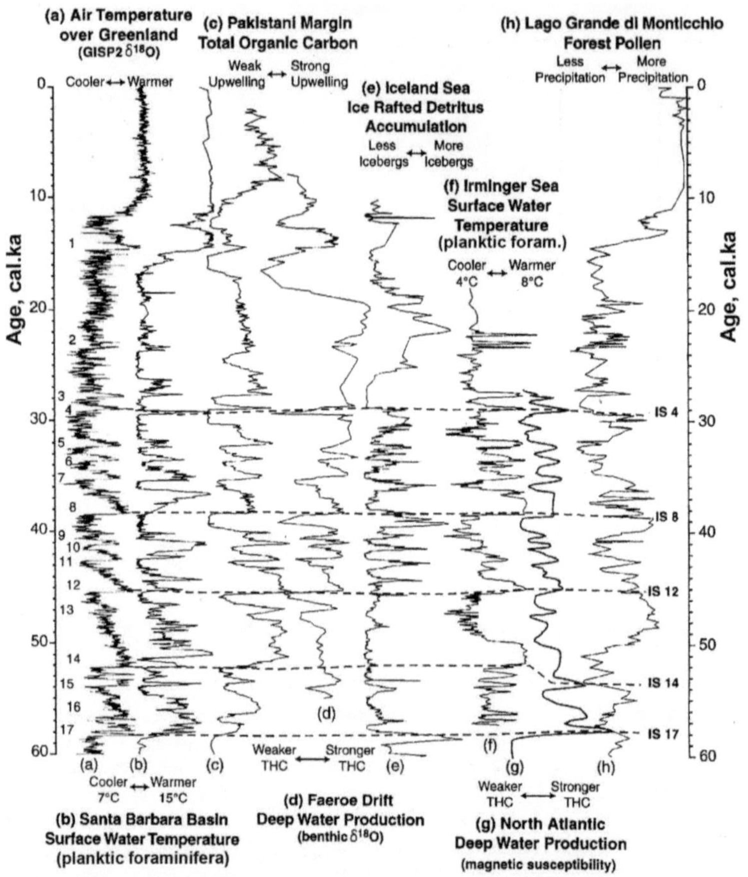

Fig. 7.2 Global comparison of high-resolution rapid change records (Sarnthein et al. 2000a). From left to right: (**a**) τ^{18}O record from the GISP2 ice core; (**b**) *N. pachyderma* dextral to sinistral coiling ratio from ODP Hole 893A, Santa Barbara Basin; (**c**) TOC record from Core SO90 136KL Pakistan Margin; (**d**) benthic τ^{18}O of record Faeroe Drift deposits; (**e**) ice-rafted debris accumulation for Core 2644, Western Iceland Sea; (**f**) sea surface temperature based on faunal assemblages for Core SO82-5, Irminger Sea; (**g**) magnetic susceptibility for Core MD95-2010, Norwegian Sea; and (**h**) woody taxa pollen abundance for Largo Grande di Mondicchio. Printed with permission from AGU

thermohaline circulation and via the atmosphere. The major difference between the oceanic and atmospheric teleconnections lies in the different time scales of their operation (e.g., Liu and Alexander 2007 and references therein). Whereas the atmospheric teleconnections generally reach the equilibrium very fast, within a season, the transfer of climate signal via the thermohaline circulation spans a wide range of time scales from interannual to millennial (Fig. 7.3). Thus, the THC seems to be especially effective in transmitting climate signals at the orbital and millennial time scales which exceed the adjustment period of the overturning, while teleconnection via the atmosphere is typical for interannual, decadal, and longer oscillations (Liu and Alexander 2007).

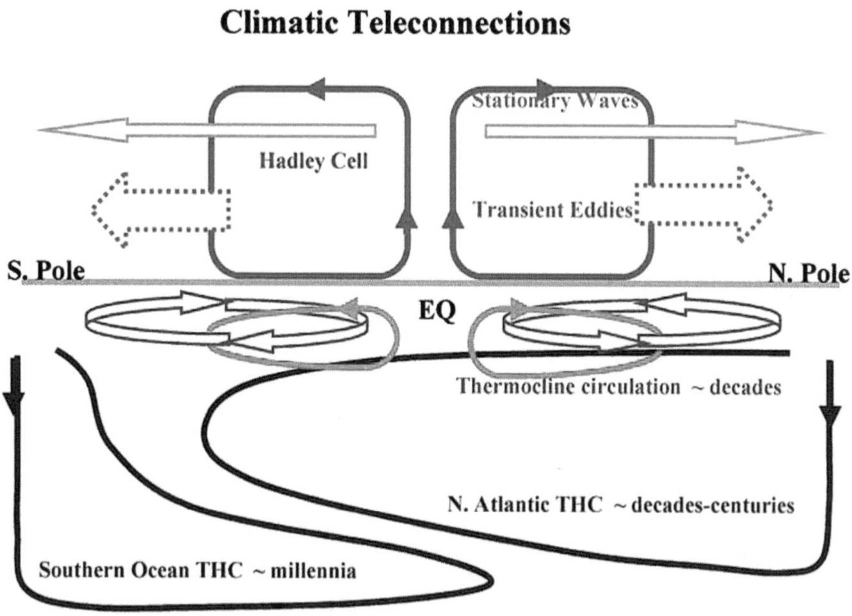

Fig. 7.3 Schematic for the major branches of climatic teleconnections in the atmosphere and ocean (Liu and Alexander 2007). The atmospheric teleconnections occur at fast time scales, usually shorter than a month (not marked). The oceanic teleconnections occur at a wide range of time scales as marked. Printed with permission from AGU

Global signals are characterized by their shape and content, first of all by period, amplitude, and symmetry or asymmetry of variations within the persistent cycles. Periodicity of the signal points to its stability and significance for environmental changes.

The above-mentioned difference in the input and output signals evidences in favor of teleconnections and against simple correlation of events controlled by the same external factors.

Several studies (e.g., Paillard and Labeyrie 1994; Labeyrie and Elliot 1999; Schulz et al. 1999; McManus et al. 1999; Broecker and Hemming 2001) clearly demonstrated that the suborbital DO cycles and Bond cycles are quite persistent regardless of their much shorter duration as compared to the orbital cycles. Orbital factors and changes of greenhouse gases content in the atmosphere cannot explain such simultaneous events as H-1 in the North Atlantic and severe drought in Arabia, California, and Barombi Mbo crater lake in the present-day rainy tropical central Africa. Cold stadials of DO cycles also co-occur with droughts in deserts of Arabia and California (Sirocko et al. 1999).

It is worth mentioning that the precise chronology of each event is crucial for their reliable identification and understanding of the teleconnection mechanisms. Because of the AMS-^{14}C-dating limitations, the short-term events that could be considered synchronous in remote areas are restricted to the last glacial and the Holocene, i.e., the last 25–45 ka. However, even within this time interval there is uncertainty related to the unknown value of the regional reservoir effect, especially during glacials and Heinrich events, when it might be substantially different from the modern value (e.g., Bard et al. 1997, 1998; Sarnthein et al. 1988, 2000b; Mangerud et al. 2006; see also Section 6.2). Reliability of age-to-depth models of proxy time series beyond the limits of the radiocarbon dating can be significantly improved using the dated ash layers of wide geographic distribution. For example, Toba ash corresponding to the boundary between stages MIS 4 and 5, or between DO cycles 19 and 20 is considered to be an excellent time marker for distant

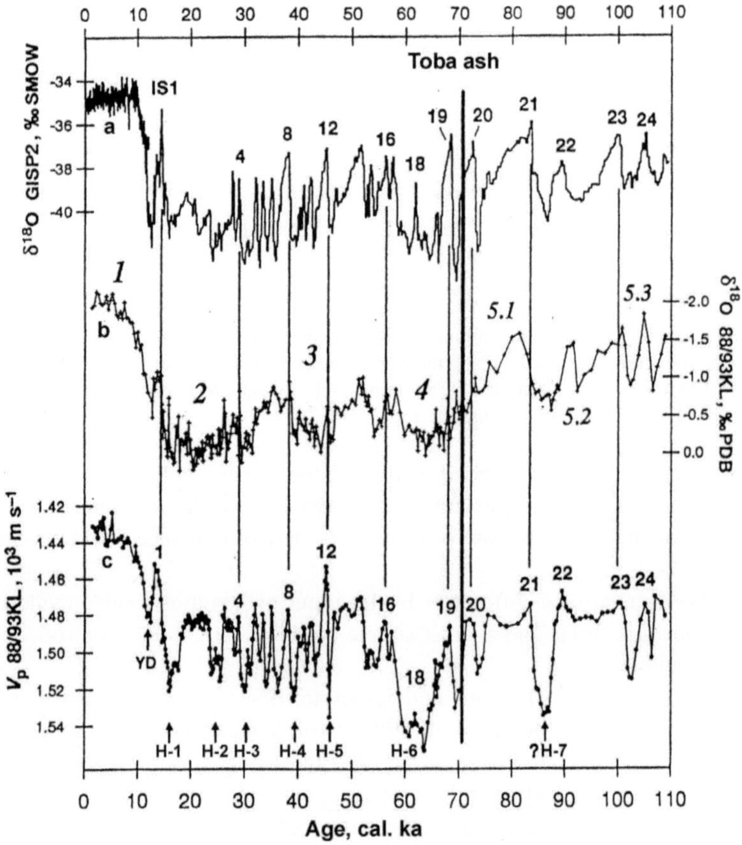

Fig. 7.4 Correlation of high-frequency climate variability for the past 110 ka between ice core and marine sediment core records (Schulz et al. 1998 with Nature permission).(**a**) Greenland GISP2 $\delta^{18}O$ ice record; (**b**) the marine $\delta^{18}O$ record of the planktic foraminifer *Globigerinoides ruber*, (**c**) the record of sonic velocity (*Vp*) of sediment cores SO90-88/93KL. Numbers indicate Greenland interstadials IS1–18, and equivalent Arabian Sea monsoonal events 1–18; H-1–H-6 indicate northern North Atlantic Heinrich meltwater events, coinciding with the deposition of bioturbated intervals in the Arabian Sea record. YD = Younger Dryas, italic numbers 1–5.3 indicate standard SPECMAP oxygen-isotope stages. In cores 88KL and 93KL, shards derived from the Toba mega-eruption occur at 620–624 and 623–62-cm core depth, respectively, near the isotope stage 4/5 boundary, that is, between Greenland interstadials IS19 and IS20 as well as between Arabian Sea equivalents 19 and 20

correlations between high and low latitudes (e.g., Schulz et al., 1998; Sirocko et al. 1999; Fig. 7.4).

7.2 Interhemispheric Teleconnections

As shown in Section 6.4, 24 warm interstadials of DO cycles were identified in the interval between 20 and 105 ka in Greenland ice cores GRIP and GISP2 (Dansgaard et al. 1993; Grootes et al. 1993; Grootes and Stuiver 1997). M. Bender and co-authors (1994) correlated the oxygen isotope records derived from the ice cores GISP2 from Greenland and Vostok from Antarctica with the SPECMAP oxygen isotope stack and identified nine interstadials within the same time interval (20–105 ka) in the Eastern Antarctic, which can be correlated with relatively long (more than 2 ka) interstadials in Greenland. Unlike Greenland, Antarctic cycles are characterized by slow warming and cooling with relatively gradual transitions between stadials and interstadials (Bender et al. 1994). The amplitude of millennial-scale air temperature change in Antarctica was much smaller than in Greenland, 1–3°C and 8–16°C, respectively (Dansgaard et al. 1993; Blunier and Brook 2001; NGRIP members 2004; EPICA community members 2004, 2006). M. Bender and co-authors (1994) suggested that warmings in Antarctica represent the response of the climate system to changes in the Northern Hemisphere. It is generally recognized that warmings and coolings at high latitudes of both hemispheres are correlated within the glacial–interglacial cycles. However, they are not synchronous within 1,500-year cycles (Broecker 2000; Blunier et al. 1998, 1999; Blunier and Brook 2001).

In 1992, T. Crowley suggested that some of the cold stadials of the DO cycles, in particular 8 and 12, and the Younger Dryas in the North Atlantic corresponded to warming intervals in Antarctica. This idea was supported by several studies (e.g., Blunier et al. 1998; 1999; Labeyrie and Elliot 1999). Comparison of SST variations between the North and South Atlantic during the last glacial confirmed the antiphase pattern of short-term changes in both hemispheres (Vidal et al. 1999). Unlike the North Atlantic, episodes of IRD-enriched sediment deposition in the southeastern part of the ocean coincide with warm DO interstadials and intervals of intensified NADW formation (Kanfoush et al. 2000). The abundant IRD discharge might result from sea level rise and/or more intensive NADW formation, which led to destabilization of ice shelves in the Weddell Sea (see Section 6.4).

Independent evidence of the anti-phase pattern of NADW and AABW distribution during the abrupt climate changes comes from the studies in the mid-latitude North Atlantic. For example, the extreme values of benthic oxygen isotope record in core MD95-2042 from 3,500-m water depth off the Portugal coast (Shackleton et al. 2000) correspond to oxygen isotope maxima and minima in the Antarctic ice core Byrd and are in anti-phase with the extreme oxygen isotope values in the Greenland ice core GISP 2 (Fig. 7.5). In other words, benthic oxygen isotope record in core MD95-2042 reflects the Antarctic warmings. This observation confirms the AABW propagation far to the north (Color Plate 3.2) during the Northern Hemisphere coolings and weakening of the convection in the North Atlantic. The

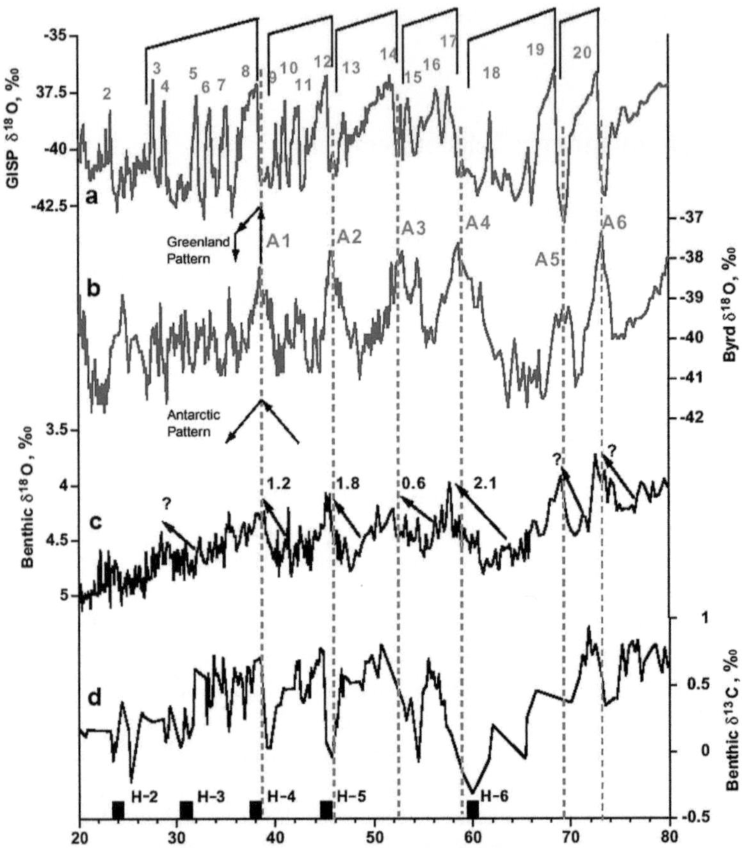

Fig. 7.5 Ice core and deep ocean records of climate change during the last glacial period (Adkins et al. 2005 with Elsevier permission). (**a**) Oxygen isotope variation of the GISP2 ice core on the Blunier and Brook (2001) timescale. $\delta^{18}O$ is a proxy for atmospheric temperature above the ice core (~3,000-m altitude). The interstadial Dansgaard/Oeschger (DO) warm events indicated by numbers are grouped into Bond cycles between Heinrich events. (**b**) Oxygen isotope variation of the Antarctic Byrd ice core on the Blunier and Brook (2001) timescale. Numbers are the Southern Hemisphere events that correspond to the larger DO events in the north. (**c**) Benthic foraminifera record of deep-ocean temperature and ice volume from 3,500 m off of Portugal (core MD95-2042, Shackleton et al. 2000). Lower values represent warming and/or continental ice sheet melting. Black numbers are degrees centigrade of residual $\delta^{18}O$ warming after subtracting the sea level signal from New Guinea coral reef profiles. *Question marks* are possible deep temperature increases that are not covered by the Huon record. The *black arrows* show the sense of temperature rise. (**d**) Benthic $\delta\ ^{13}C$ from the same core MD95-2042. *Black bars* at the bottom of the figure correspond to the ages of Heinrich Events, the age uncertainty increases downcore

recent faithful correlation of the benthic carbon isotope record from the same core MD95-2042 with several proxy time series from the nearby IMAGES core MD01-2444 ascertained that the area was influenced by AABW during the longest DO stadials (Martrat et al. 2007).

The alkenone-based SST records of the IMAGES core MD95-2043 and ODP site 977A, from the Alboran Sea, Mediterranean, well mirror the Greenland DO and H-events pattern during the interstadial MIS 3 (Fig. 7.6; Moreno et al. 2005) and during the last glacial–interglacial cycle (Martrat et al. 2004), respectively. In core MD95-2043, SST and surface water productivity generally demonstrated lower values over H-events and DO stadials as compared to DO interstadials whereas the deep-water ventilation (at water depth 1,841 m) increases over the stadials. The coolings are also correlated with the increase in the aeolian dust transport from Sahara, strengthening of northwesterlies, and concomitant aridization of the Iberian Peninsula. This tight correlation points to the vigorous Western Mediterranean Deep-Water formation over the stadials and to the strong oceanic and atmospheric teleconnections with the North Atlantic at millennial or even centennial scale. The Mediterranean Outflow intensified over the stadials adding the heat and salt into the intermediate depth North Atlantic and thus increasing the subsurface water density and amplifying the further AMOC resume at DO interstadials (Voelker et al. 2006).

On the contrary, the alkenone-derived SST record in the western tropical Atlantic core M35003-4 (12°05'N, 61°15'W; water depth 1,299 m) displays warmings during the abrupt Northern Hemisphere coolings like the YD and H-1 (Fig. 7.7), whereas the eastern tropical part of the ocean cooled in phase with the northern part due to the southward cold water advection by the Canary Current (Zhao et al. 1995; Rühlemann et al. 1999). Thus, the warmings in the western tropical North Atlantic and South Atlantic are coupled with those in the Antarctic ice cores as a result of the slowdown of NADW formation expressed by positive shifts of *Cd/Ca* ratio in benthic foraminifera during the YD and H-1 in core EN120 GGC1 (33°40'N, 57°37'W; water depth 4,450 m) from the Bermuda rise (Fig. 7.7; Rühlemann et al. 1999).

As was mentioned in Section 3.5, Heinrich events were much more pronounced in the South Atlantic than DO cycles. This difference is explained by the stronger influence of Heinrich events on the AMOC triggering its temporal collapse (Ganopolski and Rahmstorf 2001). The cross-equatorial oceanic heat transfer decreased correspondingly leading to the Antarctic warming. According to some authors (Bender et al. 1994; Ganopolski and Rahmstorf 2001), warmings (DO interstadials) in Greenland were accompanied by coolings in Antarctica. Other researchers (Blunier et al. 1998, 1999; Blunier and Brook 2001) argue that the interstadials in Greenland were delayed compared to the initiation of Antarctic warmings but their peaks coincided (Brook et al. 2005; Adkins et al. 2005, Fig. 7.5, Color Plate 7.1). The latter assumption is confirmed by simulations using the general ocean circulation model (Knorr and Lohmann 2003, 2004). Modeling experiments showed that warming with southward sea ice retreat in Antarctica causes the abrupt re-initiation of the AMOC interglacial mode triggered by increased volume transport into the Atlantic Ocean via the warm (through the Indonesian Throughflow) and cold (through the Drake Passage) surface branches of the THC. Along with convective feedback due to increased overturning the advective feedback owing to increased heat transport to the North Atlantic favored the sea ice retreat and the AMOC resumption in the formerly ice-covered North Atlantic (Knorr and Lohmann 2003, 2004).

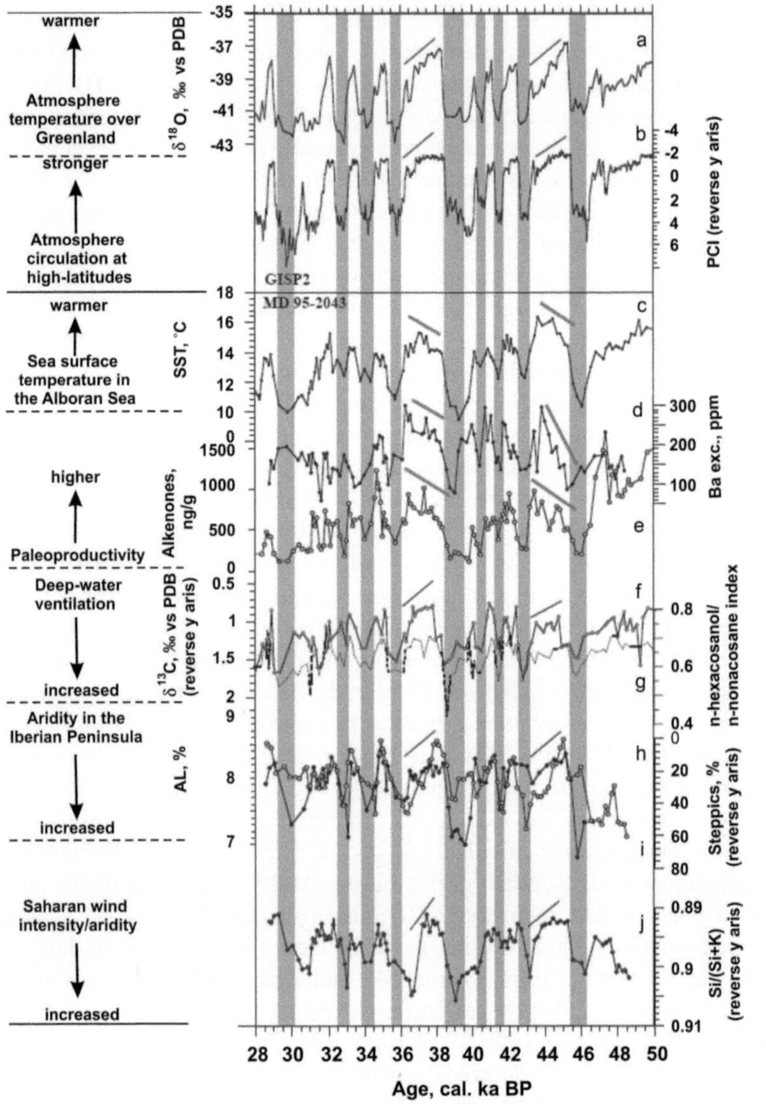

Fig. 7.6 Comparison of different proxies selected from core MD95-2043 (36° 8.6'N, 2°; 37.3'W; 1,841-m water depth) and ice core GISP2 (modified from Moreno et al. 2005). (**a**) $\delta^{18}O$ and (**b**) PCI from GISP2 (Grootes and Stuiver 1997; Mayewski et al. 1997). MD95-2043: (**c**) U^k_{37}-SST; (**d**) barium excess (*black dots*); (**e**) alkenone total concentration (*white dots*); (**f**) epibenthic $\delta^{13}C$ (reverse *y*-axis); (**g**) n-hexacosanol/n-nonacosane index (*dashed line*); (**h**) steppic vegetation (*black dots*, reverse *y*-axis); (**i**) aluminium percentage (*white dots*); (**j**) Si/(Si+K). The DO stadials and H-events are marked by gray bands. Processes represented are indicated at *left*. *Black and gray lines* are plotted to illustrate the differences in interstadial evolution (DO interstadials 8 and 12) between atmospheric and marine systems. PCI = Polar Circulation Index

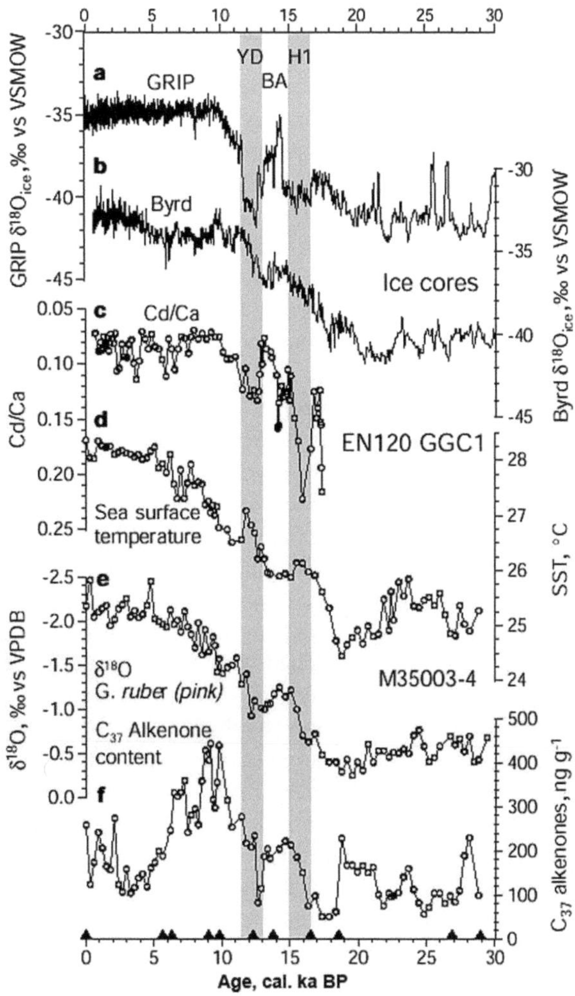

Fig. 7.7 Temperature evolution of the western tropical Atlantic Ocean, Arctic, and Antarctic during the past 30 ka (Rühlemann et al. 1999 with *Nature* permission). Shown is a comparison of alkenone-based sea surface temperatures (SST) (**d**) and oxygen isotope ratios of the planktic foraminifer *G. ruber* (*pink*) (**e**) from sediment core M35003-4 (12° 05' N, 61°15' W; water depth 1,299 m), with oxygen isotope records from Greenland (GRIP) (**a**) and Antarctica (Byrd) (**b**), and with cadmium/calcium ratios of benthic foraminiferal tests in sediment core EN120 GGC1 (33°40'N, 57°37'W; 4,450 m) from the Bermuda rise (**c**; axis inverted); (**f**) C_{37} alkenone content of core M35003-4. AMS-^{14}C control points for M35003-4 are denoted by triangles on lower axis. Age control for the GRIP and Byrd ice cores is after Blunier et al. (1998). H-1, BA, and YD denote Heinrich event H-1 (16,900–15,400 cal. year BP), the Bølling-Allerød period (15,400–12,900 cal. year BP), and the Younger Dryas period (12,900–11,600 cal. year BP), respectively

The assumption about the antiphase changes in high latitudes of both hemispheres was based on the comparison of the variability in methane content in ice cores (Color Plate 7.1) using the Monte Carlo method, as the methane is known to be quickly and evenly (within a few years) spread in the atmosphere on a global scale. T. Blunier and co-authors (1998) argued that the lead of the Antarctic warming by 1-2.5 ka as compared to Greenland during MIS 3 most likely indicates the oceanic teleconnection, as the atmospheric response time is much shorter. If warmings had started in the tropics, as suggested by some authors (e.g., Charles et al. 1996; Lea et al. 2000, 2006), and was transmitted through the atmosphere, the oxygen isotope composition of Antarctic ice would have been changing simultaneously with the methane content at all latitudes. However, the oxygen isotope records shown in Color Plate 7.1 demonstrate the lightening of the oxygen isotope values of the Antarctic ice prior to changes in the methane content and in oxygen isotope values in the Greenland ice, whereas the peaks on the latter two records are in phase. Synchronization of climatic records from Greenland ice cores and Antarctica with the methane content in the atmosphere revealed warming of Antarctica during at least four Heinrich events (H-4, H-5, H-5.2 and H-6) within the interval 65–35 ka BP (Sachs 2005).

Anti-phase climatic trends between the North Atlantic and Antarctic were detected within the last 1,500-year cycle, including the Medieval Warm Period and the Little Ice Age (see Section 4.5.4; Broecker 2001). During the Little Ice Age, the deep-water formation in the Southern Ocean intensified, while in the North Atlantic it weakened (Broecker et al. 1999). This interpretation agrees with the hypothesis of "bipolar seesaw" (Broecker 1997, 1998, 2000; Stocker 1998), which provides the link between intensity of the THC variations and pole-to-pole teleconnections.

In order to explain slow changes in Antarctica during Heinrich events, R. Keeling and M. Visbeck (2005) proposed an alternative "eddy" hypothesis. They postulate the enhanced heat transport via eddies, which split from the ACC as a result of strong mixing and then continued to move southward during intervals of weakened AMOC. It is assumed that intensive formation of eddies was caused by the increase of salinity and density gradient across the ACC due to a southward propagation of salinity anomalies associated with the meltwater injection into the North Atlantic, and that the anomalous density contrast initiated the improvement of intermediate Antarctic water ventilation (Keeling and Visbeck 2005; Sachs 2005). The resulted warming must have occurred to the south of 60ºC.

In the twenty-first century, new oxygen isotope and other proxy records were obtained from the Greenland and Antarctic ice cores (e.g., Grootes et al. 2001; NGRIP members 2004; EPICA community members 2004, 2006). The oxygen isotope record from the relatively small Taylor Dome from the Eastern Antarctica (Grootes et al. 2001) demonstrated DO cycles during MIS 2–MIS 5.1 better than other Antarctic cores, and their amplitude is much higher. P. Grootes and co-authors (2001) consider this record to be the evidence of the global thermohaline circulation influence on the regional heat transfer by the ocean. The high-resolution Antarctic climate record derived from the Dronning Maud Land ice core (EPICA community members 2004, 2006) allowed synchronization with the North Greenland ice core (NGRIP members 2004) using the global atmospheric change in methane concentration. The correlation of oxygen isotope records displays a one-to-one coupling of all Antarctic warmings and stadials of Greenland Dansgaard–Oeschger cycles (EPICA community members 2006). This finding together with the above-mentioned recent data from the Iberian Margin (Martrat et al. 2007) strongly supports the hypothesis of "bipolar seesaw."

As the amplitude of the Antarctic warmings is ascertained to be linearly dependent on the duration of the concomitant DO stadials, it seems that the reduction of oceanic overturning accounts for all these events. The oxygen isotope record from the Dronning Maud Land ice core indicates the similar Antarctic warming rates and perhaps also overturning rates for all events within MIS 3 (EPICA community members 2006). Thus, despite some controversial modeling results (Seidov et al. 2005) most researchers are convinced today that the teleconnection between the North Atlantic and Antarctic is primarily accomplished by the thermohaline circulation (e.g., Ganopolski and Rahmstorf 2001; Adkins et al. 2005; Brook et al. 2005; EPICA community members 2006). Moreover, it seems that millennial-scale climate variability in the Antarctic represents the global pattern, whereas the DO cycles in northern high latitudes are in fact perturbations of this natural scenario.

7.3 Linkages Between the North Atlantic and Eurasian Arctic Seas

As mentioned in Section 3.5, simultaneous coolings, though of different amplitude, in the North Atlantic between 60 and 20°N during several Heinrich events indicate that these events strongly affected the oceanic hydrology far outside the iceberg discharge areas. Meanwhile, H-events might be triggered by iceberg calving from one or more ice sheets around the Nordic seas (van Kreveld 2000). For example, H-1 is supposed to be a result of the intensive iceberg breakouts from the Barents Sea ice sheet (Sarnthein et al. 2000b, see Sections 3.5 and 4.4). This assumption is in line with our data on the pronounced iceberg calving at the beginning of deglaciation in the Barents Sea (Murdmaa et al. 2006; see also Section 4.5.2). In turn, the deglaciation in the Barents Sea was strongly amplified by the Atlantic water input via the northern and western troughs (Section 4.5.2; Lubinski et al. 2001; Ivanova 2003; Murdmaa et al. 2006; Ślubowska-Woldengen et al. 2007). The eastward subsurface Atlantic water penetration along the Eurasian continental slope since ca. 16 cal. ka BP (Taldenkova et al. 2008) likely resulted from the rapid AMOC recovery after the H-1 event.

In the early Holocene, 10–6 cal. ka BP, tropical SSTs rose and then stabilized (Bard et al. 1997; Rühlemann et al. 1999; Pelejero et al. 1999; Kienast et al. 2001). At high and middle latitudes of the Northern Hemisphere, the Holocene thermal optimum is recorded within the same time interval, between 9 and 6 cal. ka BP (Duplessy et al. 1992; Koç et al. 1993; Kerwin et al. 1999; Sarnthein et al. 2003b). In the northern Barents Sea, the optimum corresponds to the interval 7.8–6.9 cal. ka BP and is associated with the culmination of the early Holocene tendency to intensify the Atlantic water inflow (Duplessy et al. 2001, 2005; Section 4.5.3). In the lower Holocene sediments from the southeastern Kara Sea, high percentages of benthic foraminifers and diatoms tolerant to relatively high salinity (Section 4.5.3, Ivanova 2002b, 2003; Polyakova and Stein 2002) at about 9–7 cal. ka BP is likely related to strongly transformed Atlantic water penetration as well. Thus, the above-mentioned early Holocene events in the Barents and Kara seas are roughly correlated with each other, as well as with the warming and intensification of the thermohaline circulation in the Norwegian–Greenland Basin. This caused an enhanced Atlantic water inflow eastward, along the Norwegian coast and continental slope of Eurasia, and its input via the troughs into the Barents, Kara, and Laptev seas since the beginning of the

Holocene (Boucsein et al. 2000; Ivanova 2003; Ivanova et al. 2007a), that is the climatic teleconnection between the North Atlantic and the Arctic. The increase in the Atlantic water input into the Arctic seas is manifested not only by temperature rise but also by an abrupt change in planktic and benthic foraminiferal assemblages (Fig. 4.9; Color Figs. 4.7 and 7.3; Sarnthein et al. 2003b; Ivanova et al. 2007a, Hald et al. 2007; Chistyakova et al. in press), and marine polynomorphs (Polyakova et al. 2005) at the YD–Preboreal transition.

As shown in Section 4.5.4, our oxygen isotope data from the northern Barents Sea reveal two short-term coolings, one in the early Holocene (about 8.2 cal. ka BP, Color Plate 7.2; Fig. 4.10) and the other within the Holocene thermal optimum (about 7.2 cal. ka BP). The second cooling remains poorly investigated and, probably, is restricted to high latitudes (Duplessy et al. 2001), while the first one is of great interest, since it was established in the North Atlantic (Bond et al. 1997; Broecker and Hemming 2001; Risebrobakken et al. 2003; Section 3.6), Subpolar North Atlantic (Hald et al. 2007), Arabian Sea (Sirocko et al. 1993, 1996; Section 5.4), South China Sea (Section 6.2; Wang et al. 1999a, b), and other regions (e.g., Rohling and Pälike 2005). Since cooling at 8.2 cal. ka is suggested to have been triggered by an abrupt short-term perturbation of the AMOC (Barber et al. 1999, see also Section 3.6), our data from the Barents Sea are consistent with the weakening of the Atlantic water input via the Franz Victoria Trough (Duplessy et al. 2001). It is noteworthy, that several abrupt events possibly related to variations in the THC are more simultaneous in the Arctic, Greenland, and North Atlantic (Fig. 7.2) than the longer-lasting events like the so-called Holocene Thermal Maximum, which seems to be diachronous.

Climatic changes reconstructed in the Barents Sea for the last 1,000 years are correlated with the Medieval Warming and cooling of the Little Ice Age (Ivanova et al. 2003a; Murdmaa et al. 2004; Polyak et al. 2004). We found the link between the enhanced sedimentation rates in the Russian Gavan' Fjord controlled by Shokal'sky glacier at the northwestern coast of Novaya Zemlya and the intensification of cyclonic activity at high latitudes of the northern hemisphere during intervals characterized by the positive NAO index (Sections 4.5.4 and 1.2.1; Color Figs. 1.6 and 4.8). Our assumption is in line with other paleoceanographic and paleoclimatological data on atmospheric teleconnections of decadal and centennial scales.

7.4 Teleconnections Between North Atlantic and Monsoon Regions

L. Wang and co-authors (Wang et al. 1999a, Wang and Sarnthein 1999) recognized 10 DO cycles in core S-17940 from the northern South China Sea. Later, all DO cycles were identified in the oxygen isotope record from ODP Site 1144 with the fluctuation amplitude of up to 1.5‰ (Sarnthein et al. 2001). In the Arabian Sea, variations in bioproductivity, SST, and intensity of the oxygen minimum zone were correlated to changes in oxygen isotope composition of the Greenland ice cores over the last 65 ka (Figs. 7.4, 5.8; Schulz et al. 1998; Schulte and Müller 2001; Altabet et al. 2002). In the northeastern part of the sea, the intervals of laminated sediments reflect an increase of primary production caused by the strengthening of the summer monsoon (Schulz et al. 1998). These laminated intervals correspond to DO interstadials in the North Atlantic. On the contrary, intervals of weakened southwestern monsoon and less-developed oxygen minimum zone are represented by bioturbated sediments with low organic matter content, indicating the improved bottom waters ventilation. These bioturbated layers correlate with intervals of cooling and freshwater discharge into the North Atlantic. H. Schulz and co-authors (1998) suggested that the reason for this coincidence is the same controlling factor, such as global warming caused by the increase in water vapor and other greenhouse gases concentration in the atmosphere. It seems, however, that this coupling can be better explained by climatic teleconnections between the North Atlantic and the monsoon area during large-scale reorganization of atmospheric circulation which resulted from weakening of northward heat transport by the thermohaline circulation after massive iceberg melting.

The linkages between the monsoon area and the North Atlantic during the deglaciation and Holocene are also documented. The first abrupt strengthening of the Indian summer monsoon during the postglacial time (13–12 ^{14}C ka BP) coincides with warming and northward migration of the Subpolar Front in the northeastern Atlantic, and with the beginning of deglaciation in Antarctica. The second event of the summer monsoon strengthening (10–9.5 ^{14}C ka BP) corresponds to significant warming and ice sheets melting in Europe and Greenland. The maximum intensity of the summer monsoon in the Indian Ocean between 9.5 and 5.5 ^{14}C ka BP is roughly concurrent with the interval of maximum temperature at mid-latitudes North Atlantic (9.5–4 ^{14}C ka BP), while the subsequent interval of decreasing temperatures is supposed to be linked to the monsoon weakening (Overpeck et al. 1996; Svendsen and Mangerud 1992, 1997; Grootes et al. 1993; Jouzel et al. 1995). F. Sirocko and co-authors (1999) suggested different origin for the specific events, in particular, they concluded that the short-term oxygen isotope spikes during LGM and termination reflect the teleconnection between ENSO and Asian monsoon. The transferring effect was attributed to the water vapor which is known to have a large potential to cause the asymmetric heating in both hemispheres (Sirocko et al. 1996). The importance of the water vapor in interhemispheric teleconnections due to the rapid redistribution of low-latitude moisture by the meridional atmospheric circulation and large-scale eddies was also addressed by E. Bard and co-authors (1997). According to F. Sirocko and co-authors (1996), short-term (decadal to centennial) events in the monsoon area were affected, but not controlled, by changes in high northern latitudes because of the very slow response of ice sheets to external forcing.

Short-term events in the Arabian and South China seas within Termination I were most likely synchronous (Sirocko et al. 1999, Wang et al. 1999b). Alternate intervals of dominating winter and summer monsoons in the Arabian Sea are well correlated with abrupt events (about 100 years) identified in the Santa Barbara Basin, California, and in Barombi Mbo Lake in tropical Africa over the last 25 ka (Sirocko et al. 1999). The coincidence of short-term events in the South China Sea and Greenland is testified by our data (Figs. 6.4 and 6.9; Wang et al. 1999a, b) and confirmed by the study of two sediment cores from the southern part of the sea (Kienast et al. 2001). According to the alternative point of view (e.g., Hong et al. 2005), the Indian Ocean and East Asian monsoons are oscillating in antiphase. There is no unambiguous answer to the question about the leading role of either high latitudes or tropical dynamics in climate changes (e.g., Berger et al. 2006; Ashkenazy and Tziperman 2006) especially keeping in mind the dating uncertainty. However, increasing number of detailed proxy time series from different regions undoubtedly confirm the teleconnections between high and low latitudes at different time scales.

7.5 Teleconnections Among North Pacific, North Atlantic, and Southern Ocean

Data obtained during the last decade from the North Pacific deep-sea cores suggest the antiphase pattern of millennial-scale events like DO cycles in this region and in the North Atlantic, that is at the two "ends" of the global conveyor (de Vernal and Pedersen 1997; Kiefer et al. 2001).

In the Santa Barbara Basin, California, the Dansgaard–Oeschger cycles were recognized from abrupt changes in planktic foraminiferal assemblages, oxygen isotope composition of planktic and benthic foraminifera, and from variations in major and trace element content in laminated sediments recovered by ODP Sites 893 and 1017 within the interval 60–25 cal. ka BP (Fig. 7.8, Behl and Kennett 1996; Hendy and Kennett 1999, 2000, 2003; Hendy et al. 2002, 2004). Higher oxygen levels ascertained off California coast during the DO stadials in the North Atlantic are interpreted in terms of California Current and coastal upwelling weakening (Behl and Kennett 1996). Foraminifera-based SST reconstructed using transfer functions was changing by 3–5°C within about 70 years. I. Hendy and J. Kennett (1999, 2000) interpreted these variations as the response of the California Current and related upwelling to reorganization of atmospheric circulation in the North Pacific, in particular, the intensity of the Aleutian Low, during the abrupt climatic oscillations. This explanation is similar to that mentioned above for the monsoon area. So, remote teleconnection via the oceanic conveyor within 70 years is considered impossible while regional surface water currents might respond much faster. The modeling experiments also demonstrated that stadial coolings in the North Atlantic due to a weakened AMOC could induce the strengthening of the Aleutian Low and consequent large-scale cooling in the central North Pacific (Mikolaewich et al. 1997; Zhang and Delworth 2005).

I. Hendy and J. Kennett (2003) recognized very fast changes in the intermediate water circulation mode off California coast which preceded the atmospheric reorganization at or after the initiation of DO interstadials by 60–200 years. The authors assumed that the interstadial warmings were associated with the northward advection of modified Antarctic Intermediate Water by the California Undercurrent.

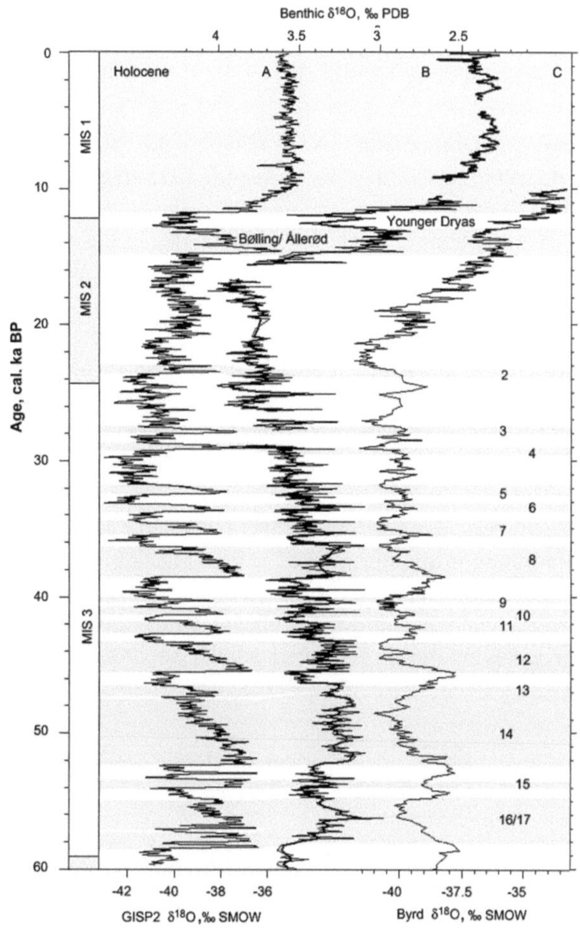

Fig. 7.8 Comparison between (**a**) GISP2 $\delta^{18}O$ (SMOW) isotope time series (Grootes et al. 1993); (**b**) the benthic paleoclimatic record at ODP Site 893A, Santa Barbara Basin, and (**c**) Byrd $\delta^{18}O$ (SMOW) isotope time series (Blunier and Brook 2001) for the last 60 ka (Hendy and Kennett 2003 with Elsevier permission). Average benthic foraminiferal $\delta^{18}O$ values are represented by the *light line*, while smoothed data with a 5-point running average is represented by the *heavy line*. Breaks in these lines represent core intervals with insufficient benthic foraminifera abundance for analysis. Variability in these parameters clearly defines the Dansgaard/Oeschger (DO) climate oscillations (numbers 17–3) during MIS 3 and the Bølling/Allerød. Gray bands represent warm intervals (interstadials and the Holocene). DO interstadials are numbered according to GISP2 scheme. MIS = Marine Isotopic Stages

On the contrary, stadial coolings at intermediate depth likely resulted from the rapid southward transport of the cold North Pacific Intermediate Water within the regional gyre system (Hendy and Kennett 2003). Thus, changes in the oceanic overturning at intermediate depths seem to be involved in the centennial-scale interhemispheric teleconnections in the Pacific. Previously, the evidences of the greater deep-water ventilation in the northeast Pacific at millennial scale within MIS 3 and 2 were found by D. Lund and A. Mix (1998).

Short aridity maximum between about 15 and 14.5 ka BP (17.8–16.5 cal. ka BP) derived from pollen spectra at the same ODP Site 893 in the Santa Barbara Basin is coeval with H-1 event in the North Atlantic (Heusser and Sirocko 1997). Decadal–centennial intervals of reduced intermediate water ventilation during the last glacial and Termination I (Behl and Kennett 1996) correspond to the peaks of maximum abundance of *Pinus* well adapted to droughts in pollen spectra and to intervals of frequent El-Niños (Sirocko et al. 1999).

Very fast intermittent warming between 16 and 11 cal. ka BP was recognized from alkenone-based SST in the vicinity of Vancouver Island (about~50°N), on the continental slope of Canada (Kienast and McKay 2001). It corresponds to the post-glacial warming in the North Atlantic between H-1 and YD events and, according to the rate of temperature rise (1°C within 40–80 years), points to the atmospheric teleconnection, as well as above-mentioned variations in the California Current during the DO cycles.

On the contrary, intervals of SST rise in the northwest Pacific seem to coincide with cold stadials of DO cycles. Unlike the above-mentioned data on the northeastern Pacific, these warmings are considered to be caused by the weakening of the upwelling at the North Pacific end of the global conveyor (Kiefer et al. 2001). If this correlation is correct, it means that the out-of-phase SST changes in the northwest Pacific and North Atlantic reflect the transfer of the at least longest signals within DO cycles via the thermohaline circulation. This seems to be an argument in favor of THC behavior rather as a single whole than as a system of several more or less independent limbs.

Numerical modeling of the Greenland ice sheet response to the short-term warmings in the northwest Pacific (Fig. 7.9, Kiefer et al. 2002) showed possible precipitation increase in the eastern Greenland by 40% compared to its average interannual value during the glacial, which must have compensated melting and iceberg calving during DO interstadials. According to T. Kiefer and co-authors (2002),

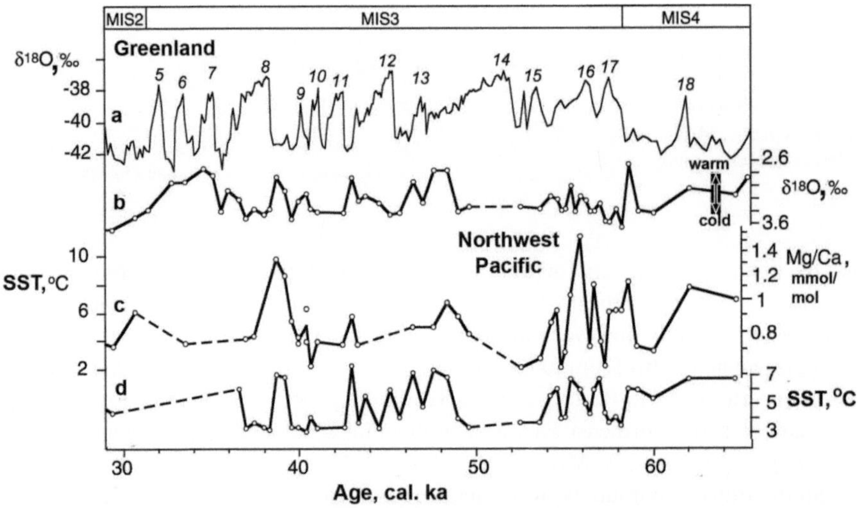

Fig. 7.9 Planktic $\delta^{18}O$ values of *N. pachyderma* (sin.) (**b**) and sea surface temperatures (SSTs) (**c–d**) at Subarctic NW Pacific ODP Site 883 (51°N, 168°E; 2,385-m water depth) between 29 and 65 cal. ka BP compared to the GISP2 $\delta^{18}O$ temperature record (**a**) after (Kiefer et al. 2001, 2002). (**c**) summer SST, estimated from planktic foraminiferal assemblages by means of the SIMMAX transfer function. (**d**) Mg/Ca ratios and deduced SST from *N. pachyderma* (sin.). Gaps in SST records result from lack of foraminifera specimens. MIS = Marine Isotope Stages

this shows the positive feedback between the two regions, promoting the stabilization of the climate system. At the same time, a later detailed comparison of the deglaciation time series from several North Pacific cores revealed spatiotemporal heterogeneity of SST variability in the area (Kiefer and Kienast 2005). This implies more complex ocean–atmosphere interaction when transferring climatic signals, than it was suggested before.

7.6 Pole-to-Equator Teleconnections in the Eastern Equatorial Pacific

The investigation of pole-to-equator teleconnections in the eastern equatorial Pacific (EEP) is of particular importance in the context of the contradictory recent observations concerning the leading of warming either at low latitudes (e.g., Lea et al. 2000, 2006; Koutavas et al. 2002) or at high southern latitudes (Stott et al. 2007) during the last termination. EEP represents one of the major areas of CO_2 release from the ocean into the atmosphere, and hence can exert a significant impact on the global changes. Besides, the region is known to exhibit high seasonal, interannual (ENSO-like), and longer-term variability with possible worldwide consequences (Clement et al. 1999; Mix 2006).

It was suggested that ITCZ shifted northward during the DO interstadials (Bard 2002) and southward during the DO stadials and H-events (Peterson et al. 2000; Stott et al. 2002; Schmidt et al. 2004 Leduc et al. 2007) thus strongly affecting the moisture supply and hydrological balance in the Caribbean and EEP. Millennial-scale variations in sea surface salinity of up to 2–4 psu were recently reconstructed in the EEP north of the equator subtracting the U^k_{37} and Mg/Ca-based SSTs from the ice volume-corrected oxygen isotope values of *Globigerinoides ruber* that vary with both temperature and salinity over the last 30 ka (ODP site 1242, Benway et al. 2006) and 90 ka (IMAGES Core MD02-2529, Leduc et al. 2007; Figs. 7.10 and 7.11). High salinities were found to associate with the southward ITCZ shift in the tropical Atlantic during the H-events, Younger Dryas, and DO stadials, thus reflecting the teleconnections between high northern latitudes and tropics. G. Leduc and co-authors (2007) suggested that the fluctuations of moisture transport have to be considered as an important positive feedback for abrupt climate changes. If the North Atlantic freshwater budget can be modulated by variations in moisture supply from Caribbean to EEP across the Panama Isthmus, corresponding freshwater "storage" in the Atlantic might finally affect the NADW formation (Schmidt et al. 2004; Leduc et al. 2007).

Along with the sea surface salinity, other paleoceanographic parameters seem to experience noticeable changes associated with H-events. The records of planktic foraminiferal SST estimated by MAT and bioproductivity-proxy FP index estimated as the total faunal portion of *G. bulloides* + *G. glutinata* + *N. dutretrei* in foraminiferal assemblage (Ivanova et al. in prep., see also Chapter 2) demonstrate remarkable millennial-scale variability in temperature and productivity over the last

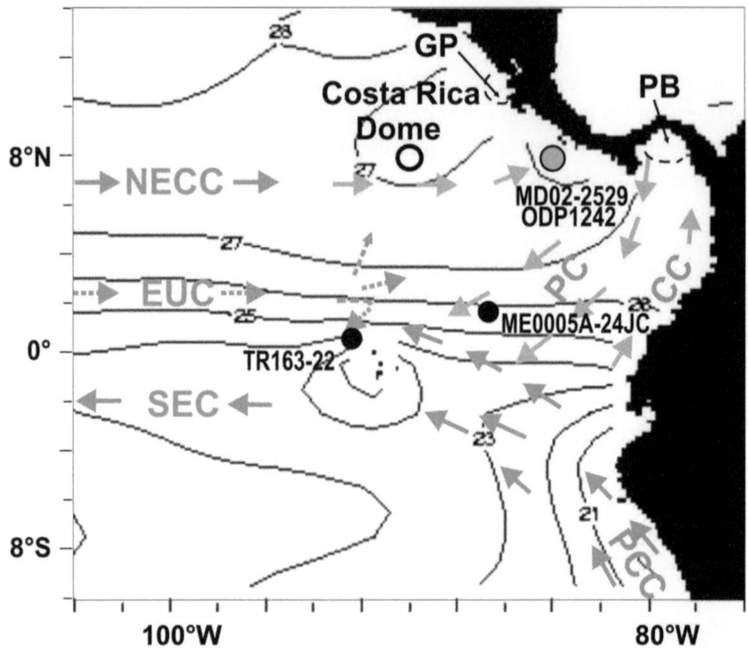

Fig. 7.10 Surface temperature and circulation in the eastern equatorial Pacific and location of the deep-sea cores discussed in Section 7.6. The isotherms represent modern SST at 10 m w.d. (WOA 2005); mean annual position of the Costa Rica Dome and coastal upwellings in Panama Bight (PB), Gulf of Papagayo (GP) are shown after Fiedler (2002). Currents (after Glynn and Wellington 1983; Fiedler and Talley 2006): SEC (NEC) – South (North) Equatorial Current, NECC – North Equatorial Countercurrent, EUC – Equatorial Undercurrent, PCC – Peru Coastal Current, CC – Colombia Current, PC – Panama Current

60 ka in Core MD02-2529 off Costa Rica (Fig. 7.12). FP index increases concurrently with the salinity rise during six H-events (from H-6 to H-1) and during the YD relative to the adjacent values (Fig 7.12b,c). On the contrary, foraminiferal SST record in Core MD02-2529 does not show any consistent positive or negative correlation with H-events (Fig 7.12a,d). This result is consistent with the conclusion by D. Lea and colleagues (2006) about the absence of the North Atlantic climate imprint in the tropical Pacific sea surface temperatures based on the *Mg/Ca* SST record in Core TR163-22 nearby the Galapagos Islands. However, the visual match of the SST record from Core MD02-2529 with the ice δ^{18}O record from Byrd glacier (Blinier and Brook 2001; Fig. 7.12d, f) demonstrate some similarity to the three of four Antarctic warmings (A1, A2, and A4) within MIS 4-2. The SST rise reached ~1–1.5°C within these warming intervals relative to the nearby values (Fig. 7.11d). The resemblance of the Pacific SST record from Core MD02-2529 and Antarctic ice core record may reflect the oceanic teleconnection between the high southern latitudes and tropics previously suggested for the last glacial interval (e.g., Feldberg and Mix 2003; Martinez et al. 2003). However, the core location a few degrees

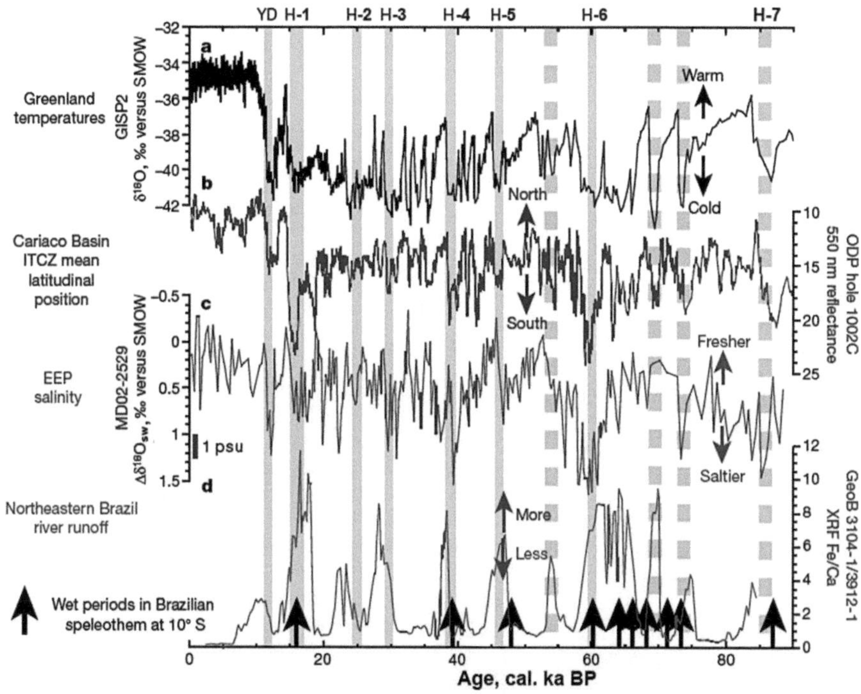

Fig. 7.11 Temporal variations of the calculated $\Delta\delta^{18}O$ sw of MD02-2529 compared to other paleoclimatic records. The data are presented on their published original time scales without further tuning (Leduc et al. 2007 with *Nature* permission). (**a**) Greenland GISP2 paleotemperature record; (**b**) Cariaco basin sediment reflectance, monitoring the latitudinal mean position of the ITCZ in the northwestern equatorial Atlantic; (**c**) $\Delta\delta^{18}O$ sw of core MD02-2529; (**d**) Fe/Ca measurements, a proxy for riverine input, performed on sediments retrieved off Brazil; *arrows* indicate time intervals of the ITCZ southward expansion over Brazil based on well-dated speleotherm growth intervals. *Gray vertical bars* mark the North Atlantic H-events. YD = Younger Dryas, H1–H6 = Heinrich events

north of the equatorial cold tongue which serves a strong hydrographic barrier for the southern surface water advection along the American coast makes this assumption rather unlikely. As was mentioned by several authors (e.g., Stott et al. 2002; Lea et al. 2006), the driving force of the millennial-scale SST changes in the tropics remains the key challenge which need further investigation.

The increase in bioproductivity during several H-events may be interpreted in terms of intensified nutrients supply at the core site. At present, bioproductivity at site MD02-2529 is influenced by the seasonal cycle and associated shifts of the Costa Rica Dome, and by propagation of eddies and filaments from the nearby coastal upwellings in the Panama Bight and Gulf of Papagayo with the Ekman drift. As today, these seasonal upwellings are caused by coastal wind jets associated with the northeast trade winds (Chelton et al. 2000). They might also be intensified during the H-events. The strengthening of the northeast trade winds and consequently Walker circulation during the LGM was suggested by D. Andreasen and

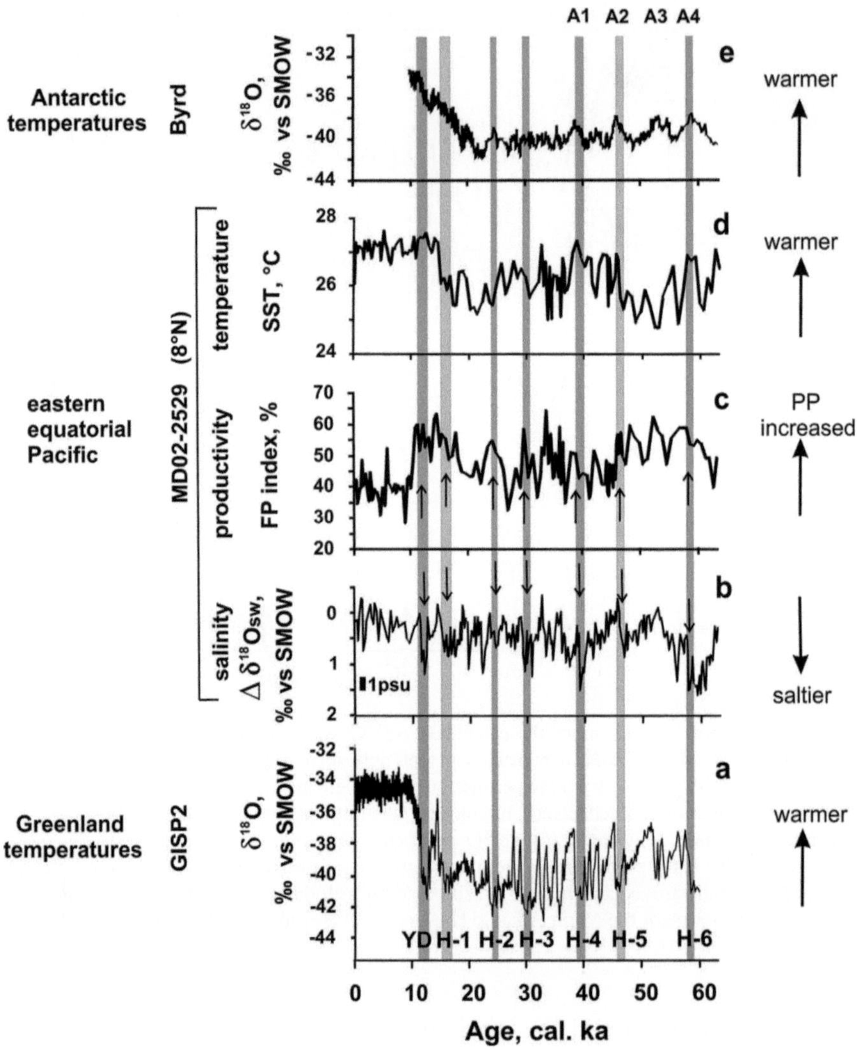

Fig. 7.12 Correlation of proxy time series in the EEP core MD02-2529 (8°12.33′N, 84°07.32′W, water depth 1,619 m) with Greenland and Antarctic paleotemperature records. (**a**) Oxygen isotope variation of the Greenland GISP2 ice core, a proxy of air temperature above the ice sheet, on Stuiver and Grootes (2000) timescale; (**b**) $\Delta\delta^{18}O$ sw of core MD02-2529, a proxy of sea surface salinity (Leduc et al. 2007); (**c**) foraminiferal productivity (FP) index, a qualitative proxy of primary production (author's data); (**d**) foraminiferal MAT-derived SST (author's data); (**e**) oxygen isotope variation of the Antarctic Byrd ice core, a proxy of air temperature above the glacier, on the Blunier and Brook (2001) timescale. All data for core MD02-2529 are presented on the published time scales (Leduc et al. 2007). *Arrows* indicate time intervals of the ITCZ southward expansion over Brazil (see Fig. 7.10). Gray vertical bars mark the correlation between the North Atlantic H-events and peaks of Antarctic warmings. YD = Younger Dryas, H1–H6 = Heinrich events, A1-A4 = Antarctic warmings

C. Ravelo (1997). Besides, M. Kienast and co-authors (2006) argued about intensified equatorial upwelling at the EEP site ME0005A-24JC (Fig. 7.10) due to stronger northeast trade winds during the AMOC slowdown at H-1. On the other hand, the migration of the Costa Rica Dome is also known to be subject of changes in the wind stress curl and surface water currents (Fiedler 2002). Thus, changes in surface water salinity, bioproductivity (and perhaps, temperature in the Equatorial cold tongue) in the EEP seem to reflect mainly the atmospheric teleconnections with the high northern latitudes during several intervals of AMOC slowdown.

The high values of FP index, and especially spikes in percentage of *G. glutinata* over terminations I and II (Color Plate 7.4), coincide with the maximum primary production estimates based on the relative abundance of *Florisphaera profunda* in the coccolithophore assemblages (using the equation by L. Beaufort et al. 1997, 2004) for the same core MD02-2529. Otherwise, planktic foraminiferal assemblages comprise the significant portion of warm-water oligotrophic species (*G. ruber* and *Globigerinoides sacculifer*) during the interglacials and higher percentage of cold-water species (*Globigerina bulloides* and *Neogloboquadrina pachyderma* dex.) during the glacials on the background of the general dominance of relatively deep-water dissolution resistant species (*Neogloboquadrina dutertrei* and *Globorotalia menardii*) throughout the record (Ivanova et al. in prep.). An increase in primary production at Termination I corresponds to carbon isotope minimum of subsurface planktic foraminifer as it was previously established in the EEP and interpreted in terms of an increased nutrients supply from the deep ocean to the surface in the Southern Ocean, and their further propagation with the Subantarctic Mode Water to the Equatorial Undercurrent and tropical thermocline (Toggweiler et al. 1991; Spero and Lea 2002; Spero et al. 2003; Pena et al. 2008; Loubere and Bennett 2008). This transfer of nutrient-rich water masses ~20–14 cal. ka BP. concomitant with the CO_2 uplift from the deep ocean and release to the atmosphere is thought to represent the postglacial re-initiation of the global oceanic overturning (Spero and Lea 2002; Marchitto et al. 2005, 2007).

Meanwhile, the enhanced nutrients advection from the south might cause unfavorable conditions for the deep-dwelling minor foraminiferal species *Globorotalia inflata* which is found to disappear in Core MD02-2529 prior to ~16 cal. ka BP (Ivanova et al. in prep.). The species does not occur in the foraminiferal assemblages since 16.6 cal. ka BP. likely due to limiting nutrients or oxygen content.

Today, the phosphorus content is relatively high at site MD02-2529, whereas the oxygen is low within and below the thermocline layer (Schlitzer 2000), where *G. inflata* is supposed to thrive in the tropics and subtropics (Patrick and Thunell 1997; Wilke et al. 2006). As this nutrient and oxygen distribution pattern seems to have been rather persistent since the $\delta\ ^{13}C$ excursion at Termination I, it might hamper the species proliferation.

On the other hand, benthic foraminiferal assemblages monitor the enhanced intermediate water ventilation over the termination I as manifested by the maximum abundance and percentage of epifaunal species within the interval of the last 50 cal. ka in Core MD02-2529 (Ovsepyan and Ivanova 2009, Fig 7.13b). The timing of this maximum coincides with two events of the enhanced Southern Ocean overturning and coeval reduced NADW export (Marchitto et al. 2007, Fig. 7.13c,d)

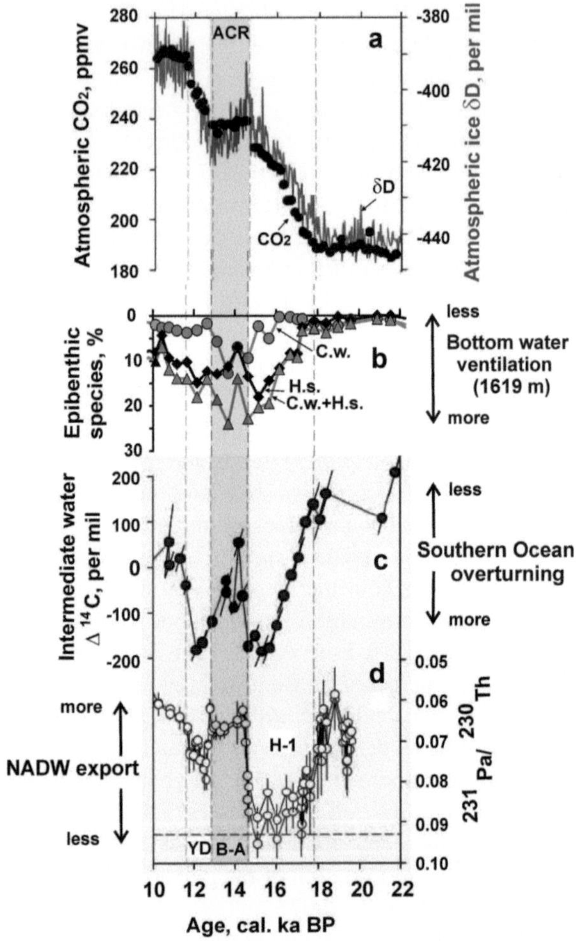

Fig. 7.13 Southern and northern ocean-atmosphere changes during the last deglaciation, compared to intermediate water $\delta^{14}C$ and epibenthic foraminiferal abundance. (**a**) Atmospheric CO_2 and ice δD temperature proxy from Antarctica Dome C placed on GISP2 timescale. (**b**) Percent of epibenthic foraminiferal species *Cibicidoides wuellerstorfi* (C.w.) and *Heterolepa subhaidingeri (H.s.)*. (**c**) Baja California intermediate water $\Delta^{14}C$. (**d**) Inverted decay-corrected record of $^{231}Pa/^{230}Th$ in marine sediments from Bermuda Rise, western North Atlantic from (McManus et al. 2004). Horizontal dashed line shows water column production ratio for these isotopes (0.093); lower values are primarily due to Pa export by vigorous NADW. *Vertical dashed lines* show ages of climatic boundaries. ACR = Antarctic Cold Reversal. (**a, c, d**) from (Marchitto et al. 2007), (**b**) data from Ovsepyan and Ivanova (2009)

just prior to and at the end of the Bølling–Allerød warming and its counterpart Antarctic Cold Reversal. These events, in turn, are roughly concomitant with the timing of radiocarbon depletion in the intermediate water of the northeastern Pacific and atmospheric CO_2 rise due to the carbon redistribution from the deep ocean to the atmosphere during Termination I (Fig. 7.13a, c). Thus, our benthic foraminiferal data supports the pole-to-equator teleconnection associated with the re-arrangement of the thermohaline circulation over the Termination I.

7.7 Mechanisms of Teleconnection

In recent years, several plausible mechanisms of teleconnections have been suggested; however, they are still enigmatic. Because of the coupling between heat and moisture exchange in the global thermohaline circulation, the oceanic overturning seems to be the best candidate for the transfer of climatic signal at glacial–interglacial and millennial time scales. This is confirmed by the data discussed in this and previous chapters that directly indicate changes of paleoceanographical proxies as a result of variations in Atlantic water inflow into the Arctic, waters of the return surface branch of the thermohaline circulation into the western Indian Ocean and South China Sea, intermediate water propagation to the tropical and mid-latitude North Pacific.

Modeling simulations demonstrated the global impact of Heinrich events originating in the North Atlantic. The imprints of these events have been registered in the Indian Ocean and southeastern Pacific (Seidov and Haupt 1999). Different experiments showed that subarctic (Seidov and Haupt 1997) and tropical (Kiefer e t al. 2002) areas of the Pacific were less affected by these events. Although the magnitude of the LGM cooling in the tropics used in the modeling simulations exceeds that reconstructed by CLIMAP (1981), it resulted only in slight changes in the thermohaline circulation (Seidov and Haupt 1999). Most likely, amplitude of the cooling was less important for changing the circulation mode than fluctuations in salinity (see Sections 3.3, 3.4, and 3.5). Very recent data on the salinity oscillations in the low latitudes demonstrate their coupling with the changes in the NADW subduction during the glacial–interglacial and DO cycles. This suggests that variations in advection of saline tropical waters into the North Atlantic might affect the AMOC intensity and contribute to the high-latitude climate change (Schmidt et al. 2004; Benway and Mix 2006; Leduc et al. 2007). On the other hand, several authors (e.g., Liu et al. 2007) argue that the tropical Indo-Pacific is subject to the North Atlantic impact via oceanic Rossby and Kelvin waves at centennial time scales.

Some authors believe that the influence of the thermohaline circulation on the European and northern Asian climates is well understood, but it is still questionable for the tropics largely controlled by the monsoon and/or ENSO dynamics (Sirocko et al. 1999). According to several studies (e.g., Wang et al. 1999b; Oppo et al. 2003) the East Asian monsoon are tightly linked to the North Atlantic and global changes via atmospheric and oceanic teleconnection. Data on the Arabian Sea and western fringe of the subtropical gyre in the Indian Ocean (Sections 5.7 and 5.5; Ivanova et al. 1995, 2003b; Levi et al. 2007) evidence the influence of the global conveyor on the tropical hydrography. The shallowing and weakening of the Indonesian Throughflow during the glacials triggered the surface salinity rise in the tropical Indian Ocean (Section 5.7; Ivanova et al. 1995) and further export of salt to the tropical Atlantic. The latter in turn favored the AMOC resumption after the cold phases (Levi et al. 2007).

Besides the prevalent concept of heat transport from high to low latitudes exerting the control on global changes, there is an alternative point of view that the monsoon response to the insolation changes precedes the more gradual response of ice sheets. According to this hypothesis, significant increase in the land-to-sea contrast in the monsoon area promotes heat transfer to high and mid-latitudes and eventually to melting of ice sheets that is to the deglaciation (Wang et al. 1999b). It is also supposed that changes in the tropics controlled the hydrological cycle and hence, the high latitudinal climate of the Northern Hemisphere (Clement and Cane 1999; Clement et al. 1999; de Garidel-Thoron et al. 2005). Further investigations are necessary to prove these hypotheses (Wang et al. 2005), but it is already well ascertained that excessive heat is effectively transported from the tropics to the subtropics by Hadley cells and then poleward by synoptic storms in the atmosphere (e.g., Liu et al. 2007). Teleconnections between the two hemispheres could be brought about by the intensification of the cross-equatorial heat exchange caused by the intensification of the trade winds-driven surface currents in the subtropical gyres (Little et al. 1997).

Another assumption postulates the significant impact of the ENSO on the global high-frequency climatic variations within the last 70 ka, since low latitudes and especially the equatorial Pacific accumulate the most part of solar insolation and then export it to high latitudes. F. Sirocko and co-authors (1999) consider changes associated with the ENSO in the tropics to cause significant variations in the heat balance at high latitudes. Paleogeographic studies revealed opposite changes during simultaneous Holocene events in different regions of tropical Africa, India, California, northern Australia, and South America, which are typical for the ENSO (Sirocko et al. 1999).

The ENSO phenomenon is known to be closely linked to subtropical jet stream in the upper troposphere and low stratosphere which penetrates far into the tropics during the boreal winter, but is mainly restricted to middle and subtropical latitudes in summer. According to the concept suggested by J. Bjerknes (1969) and further developed by several scientists (e.g., Sirocko et al. 1999), intense El-Niño events affect the location, duration, and intensity of the subtropical jet stream owing to SST increase in the eastern tropical Pacific. Hence, ENSO can be considered as a mechanism of teleconnection. This hypothesis is rather difficult to test as the sediment cores from the eastern Pacific are generally characterized by low accumulation rates and pronounced dissolution of calcareous microfossils. However, several authors (e.g., Stott et al. 2002; Beaufort et al. 2004) suggest that the data from the Santa Barbara Basin and millennial–centennial scale variations in the eastern tropical Pacific, mentioned earlier, can be interpreted in terms of ENSO variability, implying the existence of short intervals with abnormally strong and/or frequent El-Niño within the last climatic cycle.

L. Stott and colleagues (2002) argued that Indian and Asian monsoons behaved in phase with each other and with ENSO during the DO cycles, and that their concert action can explain many paleo-observations. In particular, it concerns the reduced productivity and denitrification in the Santa Barbara Basin in the northeastern Pacific, Oman upwelling in the northwestern Indian Ocean, and Cariaco Basin in the eastern tropical Atlantic which imply a less intense oxygen minimum zone during the stadials. On the contrary to the "Super-El Nino" mode during the stadials, the DO interstadials were characterized by stronger summer monsoons and more persistent La Niñas.

The associated normal or higher rainfall resulted in productivity increase and corresponding accumulation of laminated sediments in all three basins. Moreover, L. Stott and co-authors (2002) suggest that the ENSO variability has to be considered a plausible mechanism to affect the centennial- to millennial-scale variations in atmospheric concentration of greenhouse gases. At present, El Niño events associate with the sharp weakening of the equatorial upwelling in the EEP which represents the major source of the CO_2 and N_2O flux to the atmosphere. Thus, the authors submit that ENSO could provide a strong impact on the extratropical climate and ocean variability in the past.

In the modern ocean, ENSO probably controls the intensity of Pacific water inflow into the Indonesian Straits and therefore indirectly influences the SST in the central Indian Ocean. During El Niño events, the Pacific water transport via the Indonesian Gateway weakens (Potemra et al. 1997; Murtugudde et al. 1998), whereas the surface water salinity increases in the Mozambique Channel, western Indian Ocean (Levi et al. 2007). This mechanism of teleconnections might have been active during short-term and even millennial-scale events. C. Levi and colleagues (2007) addressed the possible link between the AMOC resumption and ENSO-like variability during the last glacial and deglaciation. They assumed that the pattern of millennial climate variability during the last glacial and deglaciation observed in two deep-sea cores from the tropical Indian Ocean originated from the interaction between the relatively slow grows and melting of the ice sheet in the Northern Hemisphere and short-term changes in ENSO with associated latitudinal shifts of the ITCZ. The generated ENSO-like oscillator operated with millennial time response when the ice sheets were sufficiently extended, as at H-1, YD and DO stadials.

Several recent studies (e.g., Oppo et al. 2003; Dürkop et al. 2008) call on a tight coupling of the upper limb of thermohaline circulation, monsoon, ENSO, and ITCZ dynamics in the Indonesian Archipelago region. This implies a strong involvement of both oceanic and atmospheric teleconnections in the volume and moisture transport between this area and other parts of the globe, notably the North Atlantic, tropical Indo-Pacific, South China Sea, and Southern Ocean.

The modeling experiments by A. Schmittner and A. Clement (2002) demonstrated that if the steady state overturning is significantly weaker than today, the tropical ENSO forcing can cause the THC collapse via the perturbation of the freshwater balance in the low-latitude Atlantic. Thus, the effect of the tropical forcing may be of the same order of magnitude as the middle- and high-latitude freshwater forcing associated with the ice sheets growth and melting during the last ice age except for the termination. However, the mechanism of teleconnections between the North Atlantic and remote tropical–subtropical ocean basins with oxygen minimum zones remains elusive (Schmittner et al. 2007).

The recent modeling simulations by A. Schmittner and colleagues (2007) show that the millennial-scale variability of the subsurface oxygen concentration in the Indo-Pacific oxygen minimum zones was largely affected by changes in the AMOC. In fact, the stadial weakening of the NADW transport brought about the decrease in oxygen content in the deep Atlantic and in the Southern Ocean below 1-km depth, while the upper ocean oxygen concentration increased in the northern Pacific and Indian oceans. Additional impacts of changes in the regional wind stress and in the North Pacific Intermediate Water formation were found significant but smaller. These promising experiments again raise a question about the THC as one of the best candidates for being considered one of the major mechanisms of teleconnection.

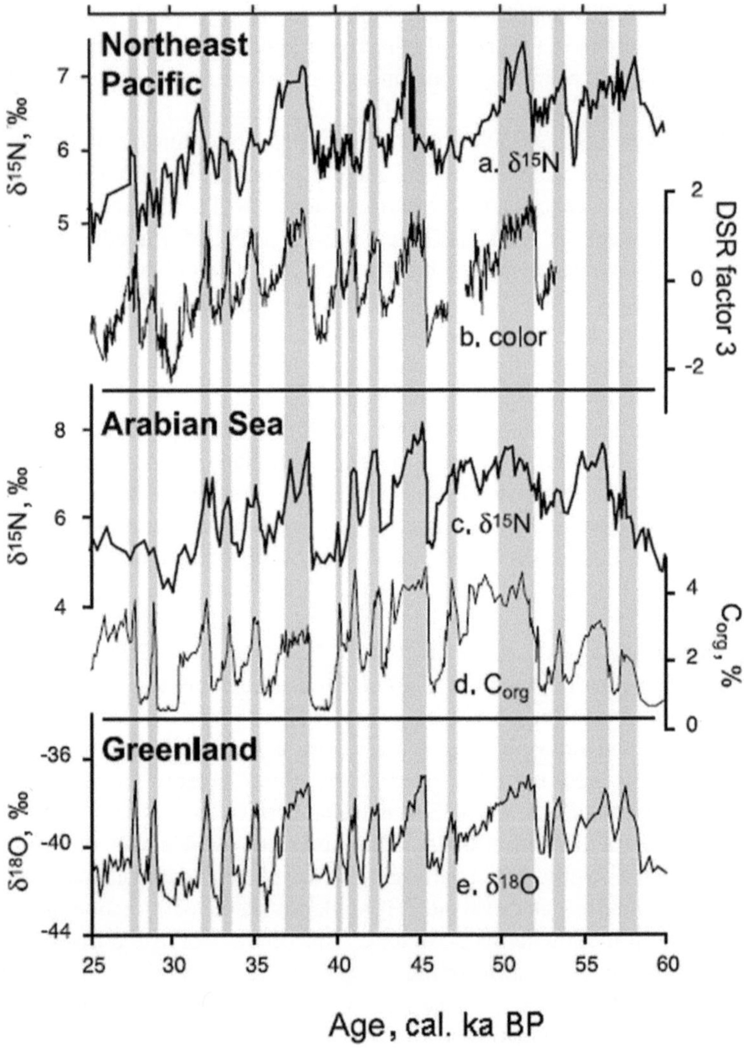

Fig. 7.14 High-resolution records from the last glacial period (60–25 ka B.P.) from the Pacific (**a and b**) and Indian (**c and d**) oceans indicating apparently synchronous fluctuations of sub-surface oxygen levels and biological productivity with (**e**) Greenland climate (Schmittner et al. 2007). Nitrogen isotopes from ODP site 1017 (water depth 955 m) from the northeastern Pacific (**a**) and NIOP site 905 (water depth 1,586 m) from the Arabian Sea (**c**) record changes in water column denitrification, which is controlled by the intensity of suboxia (<5 mM oxygen) such that high values of $\delta^{15}N$ indicate more denitrification and, by inference, lower oxygen concentrations. Sediment color (core MV99-PC08, **b**) and organic carbon content (core SO90-136KL, **d**) are interpreted to record increased productivity during warm periods in Greenland, which are shown by higher oxygen isotopes values (**e**). All records are plotted on their original, published age scales. Printed with permission from AGU

On the other hand, the productivity increase during the H-events 1–6 in the eastern equatorial Pacific (EEP) at 8°N, as derived from the foraminiferal record of Core MD02-2529 (Fig. 7.12c), seems to be in conflict with low-productivity values in the Santa Barbara Basin off California (34°N) during the DO stadials (Fig. 7.14). The decoupling might result from the local differences in the wind stress (Schmittner et al. 2007) or to the time-transgressive fashion of the intensification of upwelling-favorable winds along the northwest American margin, as was proposed by S. Kienast et al. (2002) for the postglacial time. It is noteworthy that the events of enhanced PP inferred from the coccolithophore study in the IMAGES core MD97-2141 from the Sulu Sea are roughly coeval with several DO stadials in GRIP ice core and with four H-events (de Garidel-Thoron et al. 2001). The authors assume that this correlation implies yet obscure dynamics common for both high and low latitudes. The possible correlation of the enhanced productivity intervals in the EEP (core MD02-2529) and Sulu Sea (core MD97-2141) during three H-events presents an interesting puzzle to provide its coherent explanation.

Several modeling experiments ascertained that global teleconnections occur via the atmosphere (e.g., Mikolajewich et al. 1997). Not only Hadley cells, monsoons, and zonal wind flows but also ENSO, NAO, AO, and other oscillations account for this process. For instance, our data from the Russian Gavan' Fjord of the Novaya Zemlya (see Section 4.5.4; Ivanova et al. 2003a) demonstrate the role of NAO, which changed the frequency of cyclones at high latitudes, in the teleconnection between the North Atlantic and Arctic. However, even in this case the thermohaline circulation contributes to the teleconnection by controlling the location of sea ice margin and Atlantic water penetration into the Barents Sea and further into the fjord.

Variations in greenhouse gases concentration in the atmosphere, except for the water vapor, can hardly be considered as a major mechanism of teleconnections on millennial and centennial time scales. Atmospheric greenhouse forcing should cause roughly synchronous temperature changes on a global scale. This suggestion is not consistent with the data discussed above. Apparently, variations in greenhouse gases concentration are more essential for long-term climatic oscillations such as glacial–interglacial cycles (e.g., Lea at al. 2000, 2006). However, the strength of the greenhouse gases effect could be amplified by the ocean. The thermohaline circulation can modulate the atmospheric water vapor content and thus trigger rapid interhemispheric teleconnection via the redistribution of low-latitude moisture (Broecker 1996; Bard et al. 1997).

7.8 Summary

The global thermohaline circulation should be considered as one of the most important agents of climate teleconnections at glacial–interglacial, millennial, and perhaps even centennial scales. This statement is confirmed by the data presented above from several remote areas, for example, the early Holocene warming in the Barents

Sea strongly amplified by the strengthening of Atlantic water inflow; surface water salinity increase on the western fringe of the subtropical gyre in the Indian Ocean associated with the weakening of the return branch of the global conveyor during the LGM; variations of hydrological parameters in the South China Sea triggered by the changing inflow of Pacific waters from the northeast and south. The millennial-scale variability in bioproductivity, denitrification, and intermediate water ventilation in several tropical and mid-latitude areas seem to be also driven by the changes in the intensity of NADW formation and northward propagation of waters formed in the south. Especially conspicuous changes in several paleo proxies are found to associate with the reorganization of the THC during the terminations.

Along with the oceanic "tunnel", the atmospheric "bridge" seems to be particularly important at centennial and shorter time scales because of the faster response of the atmosphere to perturbations in the climate system. The atmospheric teleconnections operate not only via the major wind systems like monsoons, trades, and westerlies but also via ENSO, NAO, and other oscillations. An example of predominantly atmospheric teleconnection in northern high latitudes during the Last Millennium comes from the Russian Gavan' Fjord.

Modern observations along with paleoceanographic data suggest that the ENSO phenomenon might be one of the main mechanisms of the global teleconnections. Although the origin of this phenomenon still has to be explained, the ENSO teleconnection apparently involves both the ocean and the atmosphere.

Conclusion

The most intriguing question I ask myself is whether the past global thermohaline circulation behaved as the united system or as a combination of more or less independent limbs over the last climatic cycle. To address this challenge properly, first of all we have to be certain about the precise chronology of the proxy time series collected worldwide. Hence, there is a strong need to develop new techniques of dating and to eliminate existing uncertainties in the local radiocarbon reservoir effects.

The data presented in this book generally confirm existing opinions on the coherent behavior of the global thermohaline circulation at glacial–interglacial scale and conceivably at millennial scale. It seems that the reorganization of the deep-water convection at high latitudes over the terminations was accompanied by the strengthening of the warm return route which weakened during the glacials and recovered by their end or during the glacial–interglacial transitions. This assumption is consistent with the data on the variability in the Agulhas Linkage and in the surface water salinity in the western Indian Ocean discussed in Chapter 5. The propagation of the Southern Ocean signal to the low latitudes (pole-to-equator teleconnection) via the intermediate water and tropical thermocline shown in Chapter 7 also points to the general reorganization of the global overturning during terminations.

Paleoceanographic data from different parts of the World Ocean demonstrate that the oceanic conveyor should be considered as the most important channel for transfer of climatic signals, especially for the events longer than the time of the overall oceanic overturning. For the shorter, decadal-to-centennial-scale events, the climatic signal is mainly transmitted via the atmosphere. However, in this case the global thermohaline circulation is indirectly handing the signal transfer, since atmospheric parameters are predominantly controlled by the heat and moisture exchange with the ocean.

The convincing evidence of the above statement comes from the Barents Sea where the stable isotope and microfossil time series clearly demonstrate the influence of Atlantic water penetration via the deep-sea troughs on the local paleoenvironments at the orbital and suborbital time scales, whereas the atmospheric teleconnections are documented in the marine and terrestrial archives of the Last Millennial. The latter in turn is defined by the NAO index, which is strongly linked to the Atlantic meridional overturning circulation.

E.V. Ivanova, *The Global Thermohaline Paleocirculation*,
DOI 10.1007/978-90-481-2415-2_BM2,

Variations in temperature, salinity, surface water bioproductivity, and river runoff within another marginal part of the ocean, the South China Sea, were controlled not only by oscillations in summer and winter monsoons but also by the intensity of water exchange with the western Pacific. Whereas the inflow of warm waters of the return branch of the global conveyor into the sea via the Borneo Strait intensified during the warmings (that is sea level high stands), the input of colder waters from the subtropical Pacific through the Luzon Strait increased during the coolings (that is sea level low stands). Shorter oscillations within the Dansgaard–Oeschger cycles established in the region likely operated via the atmospheric teleconnection.

Growing number of evidence has been obtained from marine and terrestrial paleoarchives over the last few years in favor of the strong linkage between the Asian (Indian and East Asian) monsoon and the global thermohaline circulation at glacial–interglacial and shorter time scales. The interaction of these systems is manifested in the variability of the temperature, salinity, upwellings, mixed-layer and thermocline depths, surface and intermediate circulation in the northern Indian Ocean and Indonesian Straits, oxygen minimum zone in the Arabian Sea. This interaction is affected by several external factors including climate changes in high northern and southern latitudes, El Niño–Southern Oscillation, intensity of the equatorial westerlies, and shifts in the intertropical convergence zone operating at different timescales.

Further development and combining of the existing and new paleoceanographic and terrestrial proxies will be crucial to monitor the entire range of hydrological and climatic parameters. The paleoceanographic potential of different traditional and new promising tracers has to be used to decipher the individual impact of each of the above-mentioned phenomena in the worldwide teleconnections. Together with the modeling experiments, modern observations, and historical records, high-resolution paleoarchives provide a new insight into our understanding of the thermohaline circulation functioning and enable a reliable forecast of the global change.

Appendix

Table 1 Benthic foraminifera distribution in the Holocene sediments of Core ASV-880 from the Barents Sea (%)

Species/**Depth in core** (cm)	0	5	10	15	20	25	30	35	40
E. clavatum	46	49	36	41	–	57	36	32	44
Islandiella spp.	13	14	12	12	–	11	9	7	14
Buccella spp.	10	8	17	11	–	9	15	21	11
C. reniforme	15	4	15	10	–	10	10	13	9
C. teretis	0	3	0	0	–	0	5	0	0
M. barleeanus	0	1	0	0	–	0	0	1	0
N. labradoricum	5	4	3	4	–	1	5	10	9
Other	11	17	16	22	–	12	20	16	14

Species/**Depth in core** (cm)	45	50	55	60	65	70	75	80	85
E. clavatum	–	–	38	–	39	51	46	42	57
Islandiella spp.	–	–	15	–	8	6	5	2	2
Buccella spp.	–	–	5	–	7	10	15	17	17
C. reniforme	–	–	15	–	5	4	4	10	9
C. teretis	–	–	0	–	3	1	4	1	0
M. barleeanus	–	–	0	–	0	0	0	1	0
N. labradoricum	–	–	9	–	10	9	13	14	8
Other	–	–	18	–	27	17	14	14	6

Species/**Depth in core** (cm)	90	95	100	105	110	115	120	125	130
E. clavatum	32	38	35	37	31	29	31	40	37
Islandiella spp.	2	1	5	5	5	4	2	4	4
Buccella spp.	31	25	23	20	22	23	21	20	24
C. reniforme	6	5	13	22	16	24	30	18	23
C. teretis	1	1	2	3	2	1	0	3	1
M. barleeanus	3	0	1	1	0	1	0	0	2
N. labradoricum	17	6	1	5	13	14	9	8	4
Other	9	24	20	7	11	5	5	6	6

Note: "–" Percentages are not calculated due to insufficient total foraminiferal abundance in the sample.

Table 1 (continued)

Species/**Depth in core** (cm)	135	140	145	150	155	160	165	170	175
E. clavatum	24	22	37	34	22	17	15	8	13
Islandiella spp.	4	4	5	6	21	17	32	15	9
Buccella spp.	24	16	17	16	13	24	2	3	6
C. reniforme	32	29	19	30	22	26	46	70	54
C. teretis	1	2	2	2	0	0	1	0	3
M. barleeanus	0	2	4	1	4	6	0	0	0
N. labradoricum	7	14	8	6	14	7	1	1	5
Other	7	9	8	6	4	4	2	2	0

Species/**Depth in core** (cm)	180	185	190	195	200	205	210	215	220
E. clavatum	9	10	10	19	15	25	7	17	14
Islandiella spp.	34	19	11	6	7	15	23	19	15
Buccella spp.	2	5	2	3	1	3	6	5	3
C. reniforme	52	64	73	70	73	56	52	38	59
C. teretis	0	0	0	0	0	0	0	1	1
M. barleeanus	0	0	1	0	0	0	0	1	1
N. labradoricum	1	1	0	0	0	0	0	4	0
Other	1	1	3	0	4	1	12	17	7

Species/**Depth in core** (cm)	225	230	235	240	245	250	255	260	265
E. clavatum	9	16	10	7	–	15	19	17	–
Islandiella spp.	20	11	9	8	–	19	4	6	–
Buccella spp.	5	6	7	7	–	4	5	2	–
C. reniforme	54	50	69	71	–	38	54	51	–
C. teretis	1	0	0	0	–	0	0	1	–
M. barleeanus	1	4	1	2	–	1	0	1	–
N. labradoricum	0	2	3	1	–	3	2	0	–
Other	9	11	2	2	–	19	15	21	–

Species/**Depth in core** (cm)	270	275	280	285	290	295	300	305
E. clavatum	–	8	15	20	21	22	–	14
Islandiella spp.	–	1	12	5	7	19	–	5
Buccella spp.	–	2	7	7	6	5	–	3
C. reniforme	–	63	9	50	45	41	–	64
C. teretis	–	0	1	0	1	0	–	1
M. barleeanus	–	0	0	0	1	0	–	0
N. labradoricum	–	0	3	1	5	1	–	4
Other	–	25	54	17	14	11	–	9

Note: "–" Percentages are not calculated due to insufficient total foraminiferal abundance in the sample.

Table 2 Benthic foraminifera distribution in the Holocene sediments of Core ASV-858 from the Barents Sea (%)

Species/**Depth in core** (cm)	0–3	5–6	10–11	15–16	20–21	25–26	30–31	35–36	40–41	45–46	50–51	55–56	60–61
Elphidium sp.	–	1.03	3.02	3.60	5.15	1.61	6.15	0.91	2.61	5.61	–	–	–
E. clavatum	–	22.34	23.12	14.86	20.59	32.53	20.90	25.45	18.70	20.13	–	–	–
E. subarcticum	–	14.09	17.59	12.61	5.15	9.24	15.16	17.27	16.96	15.51	–	–	–
E. incertum	–	0.34	0.00	0.00	1.47	0.00	0.41	0.91	0.43	1.65	–	–	–
I. helenae	–	0.00	0.50	1.35	1.47	0.00	0.00	1.36	0.00	0.66	–	–	–
I. norcrossi	–	13.75	3.02	24.32	19.12	8.84	9.43	17.73	4.35	1.98	–	–	–
C. teretis	–	0.69	0.50	0.00	0.00	0.00	0.00	0.00	0.43	0.33	–	–	–
C. reniforme	–	19.24	18.09	11.71	7.35	26.91	16.39	4.55	19.13	18.15	–	–	–
M. barleeanus	–	0.34	2.01	2.70	0.00	1.61	2.05	2.27	1.30	1.65	–	–	–
N. labradoricum	–	0.69	1.51	0.45	1.47	0.80	2.87	0.00	1.74	1.98	–	–	–
A. gallowayi	–	1.03	4.02	0.90	0.74	1.20	3.28	0.00	4.35	3.63	–	–	–
C. lobatulus	–	3.44	2.01	4.95	13.24	4.42	7.38	10.45	3.91	4.95	–	–	–
V. loeblichi	–	0.34	1.01	0.00	0.00	1.20	0.82	0.00	1.30	0.33	–	–	–
Buccella sp.	–	1.37	0.00	6.76	9.56	2.01	5.33	4.55	6.96	0.99	–	–	–
B. frigida	–	0.00	6.53	2.25	4.41	2.41	0.82	7.73	0.43	2.31	–	–	–
B. hannai arctica	–	1.37	6.03	9.01	4.41	0.40	0.82	0.91	1.74	2.31	–	–	–
B. tenerrima	–	0.00	1.01	0.45	0.74	0.00	0.00	0.00	0.00	0.00	–	–	–
Fissurina/Parafissurina	–	0.69	0.00	0.00	0.00	0.40	0.82	0.00	0.00	0.00	–	–	–
Trifarina sp.	–	0.34	0.50	0.00	0.00	0.40	0.82	0.45	0.00	0.33	–	–	–
Oolina sp.	–	0.00	0.00	0.00	0.00	0.00	0.00	0.00	0.00	0.00	–	–	–
A. glomerata	–	4.81	0.00	0.90	0.00	1.20	0.82	1.82	3.04	2.31	–	–	–
Trochammina sp.	–	2.41	5.53	1.80	1.47	3.21	1.23	0.91	1.74	0.33	–	–	–
Reophax sp.	–	7.22	2.01	0.45	2.94	0.00	0.41	0.45	2.17	2.97	–	–	–
Cyclammina sp.	–	1.72	0.50	0.45	0.00	0.00	0.41	0.00	0.87	1.98	–	–	–
Spiroplectammina sp.	–	0.69	0.00	0.00	0.00	0.00	0.00	0.00	0.00	0.00	–	–	–
other	–	2.06	1.51	0.45	0.74	1.61	3.69	2.27	7.83	9.90	–	–	–

Note: "–" Percentages are not calculated due to insufficient total foraminiferal abundance in the sample.

Table 2 (continued)

Species/**Depth in core** (cm)	65–66	70–71	75–76	80–81	85–86	90–91	95–96	99–100	105–106	110–111	115–116	120–121	125–126
Elphidium sp.	1.75	0.98	1.55	1.46	0.65	3.47	2.73	1.43	–	1.03	0.90	5.00	0.00
E. clavatum	21.75	24.51	22.68	24.27	17.74	17.33	27.73	20.48	–	18.56	11.66	11.50	0.00
E. subarcticum	13.33	16.18	13.40	5.34	11.61	6.44	6.36	7.14	–	12.37	9.87	4.00	0.00
E. incertum	0.70	1.96	1.03	0.97	0.00	0.99	0.00	0.00	–	0.00	1.35	0.00	0.00
I. helenae	0.00	0.98	2.06	0.00	1.29	0.00	0.00	0.00	–	0.00	0.45	0.00	0.00
I. norcrossi	21.05	3.43	6.70	4.85	6.45	4.46	3.18	1.43	–	3.09	3.14	2.50	0.00
C. teretis	0.35	0.00	0.52	2.91	0.00	0.00	0.45	0.48	–	0.00	0.00	0.00	0.00
C. reniforme	10.53	16.18	10.31	1.94	14.84	15.35	40.45	47.62	–	46.39	38.12	35.50	0.00
M. barleeanus	1.40	3.92	6.19	3.40	6.45	3.47	1.82	0.48	–	2.06	3.59	2.50	0.00
N. labradoricum	0.00	0.00	0.52	0.00	0.32	0.00	1.36	1.43	–	4.12	0.90	3.00	0.00
A. gallowayi	0.70	2.94	2.06	0.49	1.61	3.96	0.45	0.48	–	0.00	2.24	1.50	0.00
C. lobatulus	5.96	5.39	8.76	8.74	10.97	8.42	3.64	4.76	–	0.52	5.83	3.50	0.00
V. loeblichi	0.35	1.47	0.00	0.00	0.00	0.50	0.00	0.48	–	0.00	0.00	1.00	0.00
Buccella sp.	4.56	4.41	0.00	1.46	0.97	5.94	1.36	1.43	–	3.61	6.28	5.50	0.00
B. frigida	2.81	3.92	3.61	1.46	2.90	2.97	0.91	0.95	–	1.03	5.38	7.50	0.00
B. hannai arctica	5.96	2.45	5.67	1.46	0.00	0.50	0.45	1.43	–	3.61	0.45	5.00	0.00
B. tenerrima	0.70	0.98	0.52	0.49	0.00	0.00	0.45	0.00	–	0.00	0.00	0.00	0.00
Fissurina/Parafissurina	0.00	0.49	1.03	0.00	0.00	0.00	0.00	0.00	–	0.00	0.00	0.00	0.00
Trifarina sp.	0.00	0.98	0.52	0.49	0.32	0.99	0.91	0.00	–	0.00	0.00	0.00	0.00
Oolina sp.	0.00	1.47	0.00	0.97	0.00	0.00	0.00	0.00	–	0.00	0.00	0.00	0.00
A. glomerata	0.35	0.49	3.61	13.11	1.94	5.45	0.45	0.00	–	0.52	1.35	0.50	0.00
Trochammina sp.	2.81	1.96	1.03	4.85	8.71	6.44	1.82	3.81	–	0.00	3.59	3.00	0.00
Reophax sp.	0.70	0.98	3.61	16.02	0.97	5.94	1.36	0.48	–	0.00	0.00	4.50	0.00
Cyclammina sp.	0.00	0.00	0.00	0.49	0.00	0.00	0.00	0.00	–	0.00	0.00	0.00	0.00
Spiroplectammina sp.	0.00	0.00	0.00	0.97	0.97	1.49	0.91	0.00	–	0.00	0.00	0.50	0.00
other	4.21	3.92	4.64	3.88	11.29	5.94	3.18	5.71	–	3.09	4.93	3.50	0.00

Note: "–" Percentages are not calculated due to insufficient total foraminiferal abundance in the sample.

Table 2 (continued)

Species/**Depth in core** (cm)	130–131	135–136	140–141	145–146	150–151	155–156	160–161	165–166	170–171	175–176	180–181	185–186	190–191
Elphidium sp.	2.34	0.46	1.57	0.34	0.38	0.88	0.87	0.00	0.45	0.00	0.00	0.00	0.96
E. clavatum	22.66	17.43	17.65	18.79	31.54	36.84	41.05	57.83	37.50	41.23	61.19	52.50	29.19
E. subarcticum	8.59	4.59	7.84	3.69	12.69	3.51	3.93	0.43	0.89	0.88	0.91	0.00	1.91
E. incertum	1.56	0.46	0.00	0.34	0.00	0.00	0.44	0.00	0.00	0.00	0.00	0.00	0.00
I. helenae	1.56	0.46	0.39	1.01	1.15	0.00	0.00	0.00	0.89	0.00	1.37	0.50	0.96
I. norcrossi	3.91	3.67	4.31	2.35	1.92	7.89	0.44	3.91	8.48	11.84	2.28	7.00	3.35
C. teretis	0.00	0.00	0.00	0.00	0.00	0.00	0.00	0.00	0.45	0.00	0.00	0.00	1.44
C. reniforme	7.03	12.84	23.53	43.29	25.38	10.53	20.52	14.78	24.11	19.74	21.00	22.00	57.42
M. barleeanus	3.13	2.29	2.75	3.02	0.00	7.89	7.86	3.48	5.36	3.07	1.37	0.50	0.48
N. labradoricum	7.03	21.56	10.59	2.68	1.92	4.39	1.31	1.30	4.02	9.65	1.37	8.50	0.48
A. gallowayi	3.91	0.00	0.78	1.34	0.38	2.63	0.44	0.87	0.00	0.44	0.00	0.00	0.00
C. lobatulus	1.56	3.67	3.14	3.36	8.85	7.89	0.44	0.87	0.45	0.88	0.91	1.00	0.00
V. loeblichi	0.00	0.46	0.00	0.00	0.00	0.88	0.00	0.00	1.34	1.75	0.46	0.50	0.48
Buccella sp.	7.03	5.96	5.10	2.68	4.23	13.16	8.30	13.48	4.02	0.00	4.11	1.50	0.48
B. frigida	4.69	6.42	3.53	1.68	1.92	1.75	4.37	0.00	4.02	0.00	0.00	0.00	0.00
B. hannai arctica	7.03	4.59	2.75	1.01	0.38	0.00	6.11	0.00	4.02	3.07	4.57	4.00	0.48
B. tenerrima	0.00	0.00	0.00	0.67	0.00	0.00	0.00	0.00	0.00	0.00	0.00	0.00	0.48
Fissurina/Parafissurina	0.00	0.00	0.00	0.00	0.00	0.00	0.00	0.43	0.45	0.44	0.00	0.00	0.00
Trifarina sp.	0.00	0.00	0.00	0.00	0.00	0.00	0.44	0.43	0.00	0.00	0.00	0.00	0.00
Oolina sp.	0.00	0.00	0.00	0.00	0.00	0.00	0.00	0.00	0.00	0.00	0.00	0.00	0.00
A. glomerata	2.34	0.92	3.14	2.01	0.77	0.00	0.00	0.00	0.89	0.00	0.46	1.00	0.00
Trochammina sp.	4.69	6.42	3.14	7.72	1.54	0.00	0.44	0.00	1.34	0.00	0.00	0.00	0.00
Reophax sp.	6.25	4.59	4.31	1.01	0.00	0.00	0.00	0.00	0.45	0.00	0.00	0.50	0.00
Cyclammina sp.	0.00	0.92	0.00	0.00	0.00	0.00	0.00	0.00	0.00	0.00	0.00	0.00	0.48
Spiroplectammina sp.	0.78	1.38	0.39	0.00	0.00	0.00	0.00	0.00	0.45	0.00	0.00	0.00	0.00
other	3.91	0.92	5.10	3.02	6.92	1.75	1.75	0.43	0.00	0.44	0.00	0.50	1.44

Note: "–" Percentages are not calculated due to insufficient total foraminiferal abundance in the sample.

Table 3 Benthic foraminifera distribution in the Holocene sediments of Core ASV-1200 from the Barents Sea, (%)

Species/**Depth in core** (cm)	0	5	10	15	20	25	30	35	40	45	50	55	60
Elphidium sp.	0.97	1.05	2.29	–	–	4.98	7.41	–	8.11	1.98	2.80	1.46	3.50
E. clavatum	11.59	10.99	17.14	–	–	16.74	28.89	–	15.68	24.75	8.88	32.04	13.23
E. subarcticum	2.90	11.52	9.71	–	–	16.74	4.44	–	30.81	18.81	6.54	3.88	13.62
I. helenae	1 .93	0.00	0.42	–	–	0.45	0.00	–	0.54	2.97	0.47	0.49	0.78
I. norcrossi	3.86	13.61	5.91	–	–	4.52	13.33	–	6.49	5.94	2.34	5.34	0.78
C. teretis	0.00	0.00	0.00	–	–	0.45	0.00	–	1.08	0.00	0.00	0.00	0.00
C. reniforme	16.43	13.09	27.43	–	–	17.19	8.15	–	10.81	10.40	34.11	20.39	31.52
M. barleeanus	0.48	3.66	2.29	–	–	0.45	2.00	–	1.08	1.49	0.00	3.40	0.39
N. labradoricum	0.97	1.57	2.29	–	–	0.45	5.93	–	1.08	0.99	0.93	1.46	0.00
C. lobatulus	3.86	9.95	2.95	–	–	12.67	10.37	–	9.73	11.00	1.87	8.25	2.33
Buccella sp.	0.97	5.24	0.00	–	–	0.45	0.00	–	0.54	0.00	0.93	4.85	0.78
B. frigida	1.93	4.71	4.22	–	–	3.62	5.19	–	3.24	2.97	3.27	5.34	2.72
B. hannai arctica	1.93	0.00	0.00	–	–	0.00	0.00	–	0.00	2.97	0.47	0.00	0.39
B. tenerrima	0.97	1.05	0.00	–	–	0.45	0.00	–	0.54	0.50	0.47	0.49	0.00
Agglutinated	44.93	16.23	25.16	–	–	3.62	3.70	–	0.00	0.00	24.30	0.00	17.90
Other	6.28	7.33	0.20	–	–	17.19	10.59	–	10.27	15.24	12.62	12.62	12.06

Note: "–" Percentages are not calculated due to insufficient total foraminiferal abundance in the sample.

Table 3 (continued)

Species/**Depth in core** (cm)	65	70	75	80	85	90	95	100	105	110	115	120	125
Elphidium sp.	4.63	3.51	–	3.24	6.78	0.91	4.32	4.55	1.19	1.15	10.34	43.18	1.79
E. clavatum	15.74	23.39	–	13.77	18.64	13.24	25.93	16.82	26.98	10.77	20.20	3.25	17.49
E. subarcticum	5.09	3.51	–	7.29	3.39	0.91	3.70	5.45	0.79	2.69	14.78	1.30	6.73
I. helenae	0.93	0.00	–	0.81	1.02	0.46	0.62	1.82	0.40	0.00	0.00	8.12	0.45
I. norcrossi	9.72	5.85	–	4.45	7.12	0.46	8.33	6.82	3.97	1.92	8.87	1.30	8.07
C. teretis	0.00	0.00	–	0.81	0.34	0.00	0.00	0.00	0.00	0.00	0.00	8.12	0.00
C. reniforme	16.67	15.79	–	44.53	21.02	56.62	22.84	37.27	34.52	48.46	10.34	10.39	22.42
M. barleeanus	2.78	2.92	–	4.86	4.07	0.46	4.32	0.00	3.17	0.00	8.37	1.95	4.93
N. labradoricum	0.93	0.00	–	0.40	0.68	0.46	0.31	0.00	2.38	2.69	1.97	5.52	1.79
C. lobatulus	4.17	1.75	–	6.48	12.54	0.91	7.72	15.00	6.35	4.62	10.84	2.27	8.07
Buccella sp.	1.85	2.92	–	0.40	6.10	0.46	3.40	3.18	2.38	9.23	3.45	3.90	7.17
B. frigida	3.24	4.68	–	2.43	2.37	2.74	0.62	2.73	6.35	6.15	1.48	5.84	3.59
B. hannai arctica	0.93	0.00	–	1.21	0.00	0.46	0.00	1.36	1.98	2.31	0.49	0.00	0.00
B. tenerrima	0.46	0.00	–	0.40	0.00	0.00	0.00	0.00	0.00	0.00	0.00	0.32	0.45
Agglutinated	8.80	24.56	–	1.62	3.39	13.70	2.16	0.45	0.00	4.62	1.48	1.30	4.04
Other	24.07	11.11	–	7.29	12.54	8.22	15.74	4.55	11.51	5.38	7.39	3.90	13.00

Note: "–" Percentages are not calculated due to insufficient total foraminiferal abundance in the sample.

Table 3 (continued)

Species/**Depth in core** (cm)	130	135	140	145	150	155	160	165	170	175	180	185	190
Elphidium sp.	3.40	3.16	0.00	0.69	0.44	0.76	1.18	1.00	0.44	0.00	0.45	0.00	0.64
E. clavatum	11.65	24.68	36.04	36.55	14.85	24.24	18.90	33.00	36.68	40.43	35.91	48.68	57.32
E. subarcticum	5.34	1.27	0.00	0.00	1.31	1.52	5.12	3.00	0.00	0.00	0.00	0.00	1.91
I. helenae	0.97	0.63	1.35	0.00	0.00	0.00	0.79	0.50	2.18	0.00	0.91	0.38	0.64
I. norcrossi	5.34	5.70	5.41	5.52	5.24	9.85	3.15	6.50	8.30	13.91	5.00	6.04	4.46
C. teretis	0.00	0.00	0.00	0.00	0.00	0.00	0.00	0.00	0.00	0.00	0.45	0.00	0.64
C. reniforme	25.24	15.82	18.47	8.97	39.74	12.12	24.41	16.50	23.14	13.91	38.64	30.19	12.10
M. barleeanus	2.91	7.59	1.80	4.14	2.18	19.70	11.02	7.00	2.62	5.22	3.64	4.53	1.27
N. labradoricum	6.31	1.27	4.50	0.69	3.49	2.27	2.36	0.00	1.75	3.91	1.82	4.53	5.73
C. lobatulus	7.28	6.33	9.46	15.86	2.62	2.27	6.69	1.00	2.62	4.78	1.82	1.13	1.27
Buccella sp.	3.40	8.23	0.90	0.69	3.93	1.52	0.39	6.50	2.18	6.52	3.64	1.89	7.64
B. frigida	1.94	5.06	0.00	2.07	12.66	6.06	12.60	10.00	10.92	1.30	1.82	0.00	1.27
B. hannai arctica	0.00	1.27	3.60	0.00	0.87	1.52	4.33	6.00	2.62	3.04	2.73	0.75	1.27
B. tenerrima	0.49	0.00	0.90	0.69	0.00	6.06	0.00	0.00	0.00	0.00	0.00	0.00	0.64
Agglutinated	15.53	10.13	11.71	1.38	8.30	1.52	0.00	0.00	0.44	0.00	1.36	0.00	0.00
Other	10.19	8.86	5.86	22.76	4.37	16.67	9.06	9.00	6.11	6.96	1.82	1.89	3.18

Note: "–" Percentages are not calculated due to insufficient total foraminiferal abundance in the sample.

Table 3 (continued)

Species/**Depth in core** (cm)	195	200	205	210	215	220	225	230	235
Elphidium sp.	1.13	0.00	0.00	1.94	0.58	–	–	–	0.00
E. clavatum	55.37	41.15	64.19	29.13	16.28	–	–	–	38.52
E. subarcticum	1.69	0.44	0.44	1.46	4.07	–	–	–	1.64
I. helenae	0.56	0.00	0.00	0.00	0.58	–	–	–	0.00
I. norcrossi	7.91	0.44	2.62	1.94	6.40	–	–	–	1.64
C. teretis	0.00	0.44	0.44	0.97	0.58	–	–	–	0.00
C. reniforme	21.47	37.61	16.59	44.66	40.70	–	–	–	29.51
M. barleeanus	2.82	0.88	0.44	1.46	1.74	–	–	–	0.80
N. labradoricum	1.69	6.64	0.00	0.00	1.74	–	–	–	4.92
C. lobatulus	0.56	2.65	1.31	0.49	1.74	–	–	–	0.00
Buccella sp.	4.52	0.00	5.24	2.43	4.65	–	–	–	5.00
B. frigida	0.00	0.00	0.00	0.49	1.16	–	–	–	0.00
B. hannai arctica	0.00	2.21	2.18	1.94	2.91	–	–	–	5.74
B. tenerrima	0.56	0.00	0.00	0.49	0.58	–	–	–	0.00
Agglutinated	0.00	0.44	0.00	7.77	0.00	–	–	–	4.00
Other	1.69	7.08	6.55	4.85	16.28	–	–	–	8.23

Note: "–" percentages are not calculated due to insufficient total foraminiferal abundance in the sample.

Table 4 Benthic foraminifera distribution in Core BP00-07/5 from the Kara Sea(%)

Species/**Depth in core** (cm)	0	5	15	25	35	45	55	65
E. clavatum	10.9	12.9	10.1	13.4	8.5	10.0	4.8	7.9
E. subarcticum	0.0	0.0	0.3	0.3	0.3	0.0	0.0	0.0
E. bartletti	0.7	2.3	1.3	0.0	0.8	0.7	1.3	0.7
E. groenlandicum	0.2	0.2	1.3	0.3	0.0	0.7	0.4	0.7
E. asklundi	2.7	3.0	9.1	1.7	1.4	3.0	0.9	1.8
Elphidium spp.	0.5	1.6	2.7	1.0	0.8	3.3	0.9	0.4
Triloculina spp.	0.0	0.0	0.0	0.0	0.3	0.0	0.0	0.0
H. orbiculare	3.0	3.2	6.1	10.7	4.0	5.6	5.7	2.9
Haynesina sp.	0.5	0.9	0.0	0.3	0.3	0.0	0.0	0.0
N. labradoricum	0.0	0.0	1.7	0.3	0.3	0.7	0.4	0.7
C. reniforme	22.3	19.4	19.5	12.4	14.2	19.3	14.8	15.1
I. norcrossi	0.2	0.7	4.7	0.7	0.3	1.3	3.5	0.7
I. helenae	0.0	0.0	0.0	0.0	0.3	0.3	0.0	0.4
P. williamsoni	9.7	1.8	2.0	0.3	1.1	1.3	1.3	0.7
Polymorphinidae	0.7	2.3	1.7	1.3	0.6	0.7	1.3	0.4
F. fusiformis	0.7	0.0	0.0	0.0	0.0	0.0	0.0	0.4
Stainfortia spp.	0.0	0.2	1.3	0.3	0.3	0.0	0.0	0.0
Cornuspira spp.	2.2	0.7	0.3	0.0	0.0	0.7	0.0	0.0
Buccella spp.	2.5	2.8	4.4	12.4	4.0	3.0	3.9	3.2
Lagena spp.	0.5	0.0	0.7	0.0	0.3	0.3	0.0	0.0
Quinqueloculina spp.	0.7	2.8	0.0	0.0	0.0	0.3	0.0	0.0
Miliolinella spp.	0.0	0.9	0.7	0.3	0.0	0.3	0.0	0.0
Other calcareous	3.0	3.7	2.4	0.7	2.8	2.0	2.6	2.2
P. karika	9.9	9.7	13.8	17.4	25.5	15.6	10.0	14.0
L.crassimargo	2.0	0.9	0.0	1.0	1.1	1.7	0.0	0.4
A.cassis	1.2	8.3	3.7	5.7	3.1	4.7	2.6	7.9
Textularia spp.	6.2	1.4	0.0	1.7	2.3	1.0	0.9	0.7
R.turbinatus	14.6	14.3	5.1	13.1	19.0	14.6	27.9	24.8
S.biformis	2.5	0.9	0.0	0.0	0.6	0.0	0.0	0.4
Reophax spp.	0.0	0.9	5.4	3.0	1.7	1.7	3.5	4.7
Other agglutinated	2.5	4.1	1.7	1.3	6.2	7.3	13.1	9.0

Table 4 (continued)

Species/**Depth in core** (cm)	75	85	95	105	115	125	135	145
E. clavatum	9.2	5.8	5.2	1.2	2.4	1.2	2.7	9.1
E. subarcticum	0.0	0.0	0.0	0.0	0.0	0.0	0.0	0.0
E. bartletti	1.1	0.7	3.5	5.4	2.4	1.8	1.7	0.3
E. groenlandicum	0.8	0.7	1.7	3.1	1.2	0.6	0.7	0.3
E. asklundi	0.0	3.1	5.2	10.8	0.0	0.6	3.7	8.1
Elphidium spp.	0.0	2.7	2.9	3.1	0.0	0.6	1.0	1.3
Triloculina spp.	0.0	0.0	0.0	0.0	0.0	0.0	0.0	0.0
H. orbiculare	13.4	9.2	14.0	22.8	7.1	5.8	12.0	4.4
Haynesina sp.	0.0	0.0	0.0	0.0	0.0	0.0	0.0	0.0
N. labradoricum	1.9	0.3	1.2	0.8	0.0	1.2	0.0	1.0
C. reniforme	12.2	16.3	14.5	7.7	1.2	1.2	3.7	15.4
I. norcrossi	4.2	0.3	1.2	1.2	0.0	0.0	2.0	0.7
I. helenae	1.9	0.3	0.0	0.4	0.0	0.0	0.0	0.0
P. williamsoni	0.8	0.3	0.0	1.2	0.0	0.0	0.3	1.3
Polymorphinidae	1.1	1.0	1.7	3.1	2.4	0.0	1.7	1.0
F. fusiformis	0.0	0.0	0.0	0.0	0.0	0.0	0.0	0.0
Stainfortia spp.	0.0	1.4	0.6	1.2	0.0	0.0	0.3	0.0
Cornuspira spp.	0.4	1.4	0.0	0.4	0.0	0.0	0.0	0.3
Buccella spp.	0.8	5.1	10.5	0.0	0.0	0.6	1.7	3.4
Lagena spp.	0.4	0.3	0.0	1.2	0.0	0.6	0.0	0.0
Quinqueloculina spp.	0.0	0.0	0.0	0.8	1.2	0.0	0.0	0.0
Miliolinella spp.	0.4	0.7	0.0	0.4	0.0	0.0	0.0	0.0
Other calcareous	4.6	1.0	6.4	4.6	4.7	1.8	3.0	1.0
P. karika	9.5	7.1	9.3	6.9	51.8	36.3	40.1	26.8
L. crassimargo	1.5	0.7	1.2	0.8	1.2	2.9	1.0	0.3
A. cassis	7.3	8.2	2.3	6.6	9.4	12.3	3.3	4.7
Textularia spp.	0.4	0.7	0.0	0.0	2.4	0.6	1.0	0.3
R. turbinatus	14.9	19.4	3.5	7.7	9.4	15.8	11.0	7.7
S. biformis	0.8	0.0	0.0	0.0	0.0	0.6	2.0	0.0
Reophax spp.	2.7	1.7	5.2	3.1	2.4	2.3	2.0	5.7
Other agglutinated	9.9	11.6	9.9	5.8	1.2	13.5	5.0	6.7

Species/**Depth in core** (cm)	155	165	175	185	195	205	215	225
E. clavatum	0.9	11.4	11.0	17.8	9.0	5.0	10.7	12.0
E. subarcticum	0.3	0.0	0.0	0.3	0.0	0.0	0.0	0.3
E. bartletti	3.3	2.6	5.4	1.3	4.3	13.6	3.0	10.1
E. groenlandicum	0.3	0.3	1.0	0.6	0.7	1.0	0.0	0.0
E. asklundi	1.2	7.5	10.4	6.5	13.0	7.6	5.0	12.7
Elphidium spp.	0.6	2.9	4.0	3.9	4.0	2.6	1.3	1.3

Table 4 (continued)

Triloculina spp.	0.0	0.0	0.0	0.3	0.0	0.0	0.0	0.0
H. orbiculare	2.4	19.9	20.4	12.0	40.0	39.7	39.6	16.9
Haynesina sp.	0.0	0.0	0.0	0.0	0.0	0.0	0.0	0.0
N. labradoricum	0.6	2.3	1.0	0.6	0.0	0.0	0.0	0.3
C. reniforme	3.3	26.1	25.4	26.2	12.0	10.9	14.8	31.8
I. norcrossi	0.0	0.7	2.0	4.2	0.3	0.7	0.0	1.0
I. helenae	0.9	0.3	0.7	0.0	0.0	0.0	0.0	0.0
P. williamsoni	0.0	3.3	1.3	1.6	4.0	0.7	1.0	1.9
Polymorphinidae	0.3	0.7	1.3	3.6	2.3	2.3	0.7	1.9
F. fusiformis	0.0	0.3	0.3	0.0	0.0	0.0	0.0	0.0
Stainfortia spp.	0.6	1.0	1.7	1.0	0.3	0.7	0.3	3.6
Cornuspira spp.	0.0	0.7	0.0	0.0	0.3	0.0	0.0	0.3
Buccella spp.	0.9	4.2	3.7	9.1	0.7	6.6	2.7	3.2
Lagena spp.	0.3	0.7	0.0	0.6	0.0	0.0	0.7	0.0
Quinqueloculina spp.	0.0	0.3	0.0	0.0	0.7	0.3	0.3	0.3
Miliolinella spp.	0.0	1.0	0.0	0.3	0.7	0.3	0.3	0.3
Other calcareous	0.9	2.9	6.0	4.9	2.0	1.0	5.4	1.3
P. karika	68.4	6.8	1.3	2.6	3.3	3.0	9.7	0.0
L. crassimargo	0.0	0.0	0.0	0.0	0.0	1.0	0.0	0.6
A. cassis	2.1	1.6	0.0	0.0	0.0	1.0	0.7	0.0
Textularia spp.	1.2	0.0	0.0	0.0	0.0	0.0	0.0	0.0
R. turbinatus	0.0	0.3	0.3	1.6	0.7	0.7	1.3	0.0
S. biformis	0.0	0.0	0.0	0.0	0.0	0.0	0.0	0.0
Reophax spp.	3.6	0.3	0.7	0.3	0.7	0.3	0.0	0.0
Other agglutinated	8.1	2.0	2.0	0.6	1.0	1.0	2.3	0.0

Species/**Depth in core** (cm)	235	245	255	265	275	285	295	305
E. clavatum	13.8	11.2	5.0	5.9	12.2	4.0	11.4	4.4
E. subarcticum	0.0	0.7	0.0	0.3	0.0	0.3	0.0	0.3
E. bartletti	8.9	6.6	26.4	12.2	10.8	18.3	2.0	10.8
E. groenlandicum	0.3	0.3	0.0	0.0	0.7	2.0	0.0	0.9
E. asklundi	12.5	8.2	15.0	8.9	10.1	15.0	5.5	23.7
Elphidium spp.	3.6	3.3	5.7	3.8	1.7	1.7	5.5	3.2
Triloculina spp.	0.0	0.0	0.0	0.0	0.3	0.0	0.0	0.0
H. orbiculare	14.8	15.1	27.1	45.3	28.8	37.3	18.1	30.1
Haynesina sp.	0.0	0.0	0.0	0.0	0.0	0.0	0.0	0.0
N. labradoricum	1.0	0.0	0.0	0.0	0.3	0.0	0.0	0.0
C. reniforme	27.6	16.8	3.6	10.9	18.1	4.3	35.8	12.3
I. norcrossi	0.3	0.3	0.0	0.0	0.0	0.0	0.0	0.0
I. helenae	0.0	0.0	0.0	0.0	0.0	0.0	0.0	0.0
P. williamsoni	3.3	2.0	2.9	1.5	1.4	3.0	2.8	2.3
Polymorphinidae	2.6	1.0	0.7	1.5	1.7	5.0	3.5	2.6
F. fusiformis	0.0	0.0	0.0	0.0	0.3	0.0	0.0	0.0
Stainfortia spp.	0.7	0.7	0.7	0.0	0.7	0.3	0.0	0.9
Cornuspira spp.	0.3	0.0	0.0	0.0	0.0	0.0	0.0	0.0

Table 4 (continued)

Buccella spp.	5.3	2.3	2.9	4.1	3.5	3.7	6.7	0.9
Lagena spp.	0.3	0.3	0.0	0.5	0.0	0.0	0.0	0.3
Quinqueloculina spp.	0.3	1.0	0.7	0.0	0.0	0.0	0.0	0.0
Miliolinella spp.	0.3	0.0	0.0	0.0	0.3	0.0	0.8	0.0
Other calcareous	2.3	3.0	3.6	4.8	2.8	2.0	3.1	5.3
P. karika	1.3	11.8	2.1	0.0	4.9	2.7	1.6	0.6
L. crassimargo	0.0	2.0	0.7	0.0	0.0	0.0	0.0	0.0
A. cassis	0.0	1.6	0.0	0.0	0.0	0.0	0.0	0.0
Textularia spp.	0.0	0.0	0.0	0.0	0.3	0.0	0.0	0.0
R. turbinatus	0.0	4.9	0.0	0.0	0.0	0.0	1.2	0.6
S. biformis	0.0	0.0	0.0	0.0	0.0	0.0	0.0	0.0
Reophax spp.	0.0	1.3	2.1	0.0	0.0	0.0	0.0	0.0
Other agglutinated	0.3	5.6	0.7	0.3	1.0	0.3	2.0	0.9

Species/**Depth in core** (cm)	315	325	335	345	355	365	370	380
E. clavatum	1.1	9.5	10.5	8.4	3.9	9.2	6.4	20.2
E. subarcticum	0.0	0.0	0.0	0.0	0.0	0.0	0.0	0.0
E. bartletti	19.8	32.9	6.8	18.3	7.3	9.8	7.8	1.9
E. groenlandicum	4.9	1.9	1.4	1.0	1.7	0.6	3.2	0.0
E. asklundi	11.0	8.9	10.6	17.8	20.6	16.5	18.4	20.5
Elphidium spp.	8.2	4.7	8.6	8.9	11.8	3.1	3.2	4.0
Triloculina spp.	0.0	0.0	0.0	0.0	0.0	0.0	0.0	0.0
H. orbiculare	42.3	14.2	14.7	23.0	38.3	20.2	33.7	12.1
Haynesina sp.	0.0	0.0	0.0	0.0	0.0	0.6	0.4	1.9
N. labradoricum	0.0	0.0	0.0	0.0	0.0	0.0	0.0	0.0
C. reniforme	1.6	12.0	17.8	6.8	4.2	16.8	9.2	13.4
I. norcrossi	0.0	0.0	0.0	0.0	0.0	0.0	0.0	0.0
I. helenae	0.0	0.0	0.0	0.0	0.0	0.0	0.0	0.0
P. williamsoni	0.0	4.1	1.4	1.6	1.7	8.9	1.8	3.4
Polymorphinidae	3.3	3.5	6.2	4.2	2.5	3.4	2.1	3.4
F. fusiformis	0.0	0.0	0.0	0.0	0.0	0.0	0.0	0.0
Stainfortia spp.	0.0	0.0	0.3	0.5	0.0	0.0	0.0	0.0
Cornuspira spp.	0.0	0.3	0.0	0.0	0.0	0.3	0.0	1.6
Buccella spp.	0.0	2.5	5.5	2.6	4.8	7.3	5.3	8.7
Lagena spp.	0.0	0.0	0.3	0.0	0.0	0.3	0.0	0.0
Quinqueloculina spp.	0.0	1.6	0.0	0.5	0.0	0.6	1.1	2.8
Miliolinella spp.	0.0	0.0	0.0	0.0	0.0	0.3	0.0	0.3
Other calcareous	3.3	2.5	2.4	4.7	2.8	1.8	5.3	2.8
P. karika	0.0	0.0	0.0	0.0	0.0	0.0	0.0	0.6
L. crassimargo	0.5	0.6	0.3	0.5	0.0	0.0	0.0	0.3
A. cassis	0.5	0.0	1.4	0.0	0.3	0.0	0.4	0.0
Textularia spp.	0.0	0.0	0.0	0.0	0.0	0.0	0.0	0.0
R. turbinatus	3.3	0.3	9.6	0.5	0.0	0.0	1.1	1.6
S. biformis	0.0	0.0	0.0	0.0	0.0	0.3	0.4	0.0
Reophax spp.	0.0	0.0	0.0	0.0	0.0	0.0	0.0	0.0
Other agglutinated	0.0	0.3	2.1	0.5	0.0	0.3	0.4	0.6

Table 4 (continued)

Species/**Depth in core** (cm)	**385**	**390**	**395**	**405**	**415**	**425**	**435**	**445**
E. clavatum	22.3	15.1	8.1	9.1	16.8	10.7	10.0	9.8
E. subarcticum	0.0	0.0	0.0	0.0	0.0	0.0	0.0	0.0
E. bartletti	1.5	2.2	6.6	3.3	5.4	2.4	0.5	6.2
E. groenlandicum	1.0	1.8	1.4	0.5	1.6	0.9	1.0	0.7
E. asklundi	15.4	16.9	22.3	21.1	18.5	10.4	14.5	12.1
Elphidium spp.	2.7	3.4	6.6	10.0	6.5	0.3	2.0	2.3
Triloculina spp.	0.0	0.0	0.0	0.0	0.0	0.0	0.0	0.0
H. orbiculare	12.9	11.4	27.5	8.1	10.9	13.5	17.5	24.4
Haynesina sp.	0.0	1.8	0.0	0.5	0.5	1.8	3.5	0.0
N. labradoricum	0.0	0.0	0.0	0.0	0.0	0.0	0.0	0.0
C. reniforme	27.5	23.4	2.4	26.8	25.0	37.9	37.0	25.7
I. norcrossi	0.0	0.0	0.0	0.0	0.0	0.0	0.0	0.0
I. helenae	0.0	0.0	0.0	0.0	0.0	0.0	0.0	0.0
P. williamsoni	2.5	6.2	1.9	4.8	2.2	5.2	1.5	2.6
Polymorphinidae	3.0	3.7	3.8	0.5	1.6	4.0	2.5	2.3
F.fusiformis	0.0	0.0	0.0	0.0	0.0	0.0	0.0	0.0
Stainfortia spp.	1.0	0.0	0.0	0.0	0.0	0.0	0.0	0.0
Cornuspira spp.	1.2	0.9	0.5	0.0	1.1	1.2	0.0	0.7
Buccella spp.	3.2	3.4	5.2	6.7	5.4	5.2	7.5	3.6
Lagena spp.	0.0	0.0	0.0	1.0	0.5	0.0	0.0	0.0
Quinqueloculina spp.	0.7	1.5	1.4	1.9	0.5	1.5	1.0	3.3
Miliolinella spp.	0.5	0.6	0.0	0.0	1.1	1.5	0.0	0.3
Other calcareous	4.0	4.0	7.1	2.4	1.1	1.5	1.0	5.5
P. karika	0.5	0.6	0.5	0.0	0.0	0.9	0.0	0.0
L. crassimargo	0.0	0.0	2.8	0.0	0.0	0.3	0.0	0.0
A. cassis	0.0	0.0	0.5	1.9	0.0	0.3	0.0	0.0
Textularia spp.	0.0	0.0	0.0	0.0	0.0	0.0	0.0	0.0
R. turbinatus	0.0	0.6	0.0	0.0	0.0	0.3	0.5	0.0
S. biformis	0.0	0.0	0.0	0.0	0.0	0.0	0.0	0.0
Reophax spp.	0.0	0.0	0.0	0.0	0.0	0.0	0.0	0.0
Other agglutinated	0.0	2.5	1.4	1.4	1.1	0.0	0.0	0.7

Species/**Depth in core** (cm)	**455**	**465**	**475**	**485**	**495**	**505**	**515**	**525**
E. clavatum	8.9	10.7	7.7	2.9	47.6	11.3	14.1	12.1
E. subarcticum	0.0	0.0	0.3	0.0	0.0	0.0	0.0	0.0
E. bartletti	3.9	1.5	1.3	5.9	1.4	3.1	2.3	0.3
E. groenlandicum	0.0	0.0	0.0	0.0	0.0	0.2	0.0	0.0
E. asklundi	20.4	11.7	32.7	23.8	4.1	5.3	4.4	5.1
Elphidium spp.	2.6	4.6	3.3	0.3	1.0	2.6	6.7	4.0
Triloculina spp.	0.0	0.0	0.0	0.0	0.0	0.0	0.0	0.0
H. orbiculare	29.3	14.3	18.3	24.1	5.1	9.4	7.0	12.5
Haynesina sp.	2.0	1.5	4.0	0.0	5.4	6.7	5.4	8.1
N. labradoricum	0.3	0.0	0.0	0.0	0.0	0.0	0.0	0.0
C. reniforme	14.1	26.0	14.7	19.2	22.6	44.7	46.3	37.4
I. norcrossi	0.0	0.0	0.0	0.3	0.0	0.0	0.0	0.0
I. helenae	0.0	0.0	0.0	0.0	0.0	0.0	0.0	0.0

Table 4 (continued)

P. williamsoni	6.3	7.1	6.3	5.5	0.7	2.2	1.0	4.4
Polymorphinidae	2.6	2.0	3.7	1.6	1.0	1.0	2.0	0.3
F. fusiformis	0.0	0.0	0.0	0.0	0.0	0.0	0.3	0.0
Stainfortia spp.	0.7	0.5	0.0	0.0	0.0	0.0	0.0	0.0
Cornuspira spp.	2.0	0.0	1.0	2.0	0.0	1.9	1.7	4.4
Buccella spp.	2.6	5.6	3.7	9.4	5.7	5.3	4.0	2.7
Lagena spp.	0.0	0.0	0.0	0.0	0.0	0.0	0.0	0.0
Quinqueloculina spp.	2.0	2.6	0.7	1.0	2.0	2.6	1.0	3.7
Miliolinella spp.	1.0	2.0	0.0	1.0	1.0	1.0	0.0	1.0
Other calcareous	1.0	9.2	1.7	1.3	2.4	2.6	1.7	3.7
P. karika	0.0	0.0	0.0	0.0	0.0	0.0	0.0	0.0
L. crassimargo	0.0	0.5	0.3	0.3	0.0	0.0	0.3	0.0
A. cassis	0.3	0.0	0.0	0.0	0.0	0.0	0.3	0.0
Textularia spp.	0.0	0.0	0.0	0.0	0.0	0.0	0.0	0.0
R. turbinatus	0.0	0.0	0.0	0.0	0.0	0.0	0.0	0.0
S. biformis	0.0	0.0	0.0	0.0	0.0	0.0	0.0	0.0
Reophax spp.	0.0	0.0	0.0	0.0	0.0	0.0	0.3	0.3
Other agglutinated	0.0	0.0	0.3	1.3	0.0	0.0	1.0	0.0

Species/**Depth in core** (cm)	**535**	**545**	**555**	**565**	**575**	**585**	**595**	**605**
E. clavatum	20.8	29.4	14.0	15.8	9.6	8.8	10.3	11.8
E. subarcticum	0.0	0.0	0.0	0.0	0.0	0.0	0.0	0.0
E. bartletti	0.3	0.3	0.0	2.4	1.6	1.6	1.0	1.6
E. groenlandicum	0.0	0.0	0.0	0.0	0.0	0.3	0.0	0.0
E. asklundi	0.8	1.0	1.9	3.1	1.0	3.9	0.3	7.2
Elphidium spp.	4.7	1.0	2.8	0.3	0.3	2.3	1.0	2.3
Triloculina spp.	0.0	0.7	0.0	0.0	0.0	0.0	0.0	0.0
H. orbiculare	5.2	4.9	5.6	13.7	7.0	14.1	3.1	9.9
Haynesina sp.	4.2	0.0	9.0	5.1	3.8	1.3	5.8	13.2
N. labradoricum	0.0	0.0	0.0	0.0	0.0	0.0	0.0	0.0
C. reniforme	50.4	32.4	41.3	41.1	47.9	42.5	19.6	29.9
I. norcrossi	0.0	0.0	0.0	0.0	0.3	0.0	0.0	0.0
I. helenae	0.0	0.0	0.0	0.0	0.0	0.0	0.0	0.0
P. williamsoni	1.8	2.9	4.0	4.5	0.3	3.6	11.7	3.9
Polymorphinidae	0.3	0.3	1.2	1.0	0.3	1.6	0.7	0.7
F. fusiformis	0.0	0.3	0.0	0.0	0.0	0.0	1.4	0.0
Stainfortia spp.	0.0	0.0	0.0	0.0	0.0	0.0	0.0	0.0
Cornuspira spp.	1.6	12.4	2.8	2.1	7.7	0.3	18.9	3.3
Buccella spp.	5.7	1.3	4.0	2.7	4.8	2.3	1.0	1.0
Lagena spp.	0.0	0.3	0.9	0.3	0.3	0.0	1.4	0.7
Quinqueloculina spp.	2.6	8.5	7.1	5.8	11.8	14.4	16.8	11.2
Miliolinella spp.	0.0	1.0	0.0	0.0	0.3	1.0	3.1	1.0
Other calcareous	1.6	1.0	2.2	1.4	2.9	0.7	3.4	0.3
P. karika	0.0	0.0	0.0	0.0	0.0	0.0	0.3	0.0
L. crassimargo	0.0	0.0	0.3	0.0	0.0	0.0	0.0	0.3
A. cassis	0.0	0.3	0.0	0.3	0.0	0.3	0.0	0.3

Table 4 (continued)

Textularia spp.	0.0	0.0	0.0	0.0	0.0	0.0	0.0	0.0
R. turbinatus	0.3	0.0	0.0	0.0	0.0	0.0	0.0	0.0
S. biformis	0.0	0.0	0.0	0.0	0.0	0.0	0.0	0.0
Reophax spp.	0.0	0.0	0.0	0.3	0.0	0.0	0.0	1.0
Other agglutinated	0.0	2.0	2.8	0.0	0.0	1.0	0.0	0.3

Species/**Depth in core** (cm)	**615**	**625**	**630**
E. clavatum	6.3	14.9	8.8
E. subarcticum	0.0	0.0	0.0
E. bartletti	1.3	0.0	2.9
E. groenlandicum	0.0	0.0	0.7
E. asklundi	0.6	0.0	0.7
Elphidium spp.	1.9	1.0	2.9
Triloculina spp.	0.0	0.0	1.5
H. orbiculare	12.7	18.2	17.6
Haynesina sp.	4.4	6.6	0.0
N. labradoricum	0.0	0.0	0.0
C. reniforme	27.2	28.7	26.5
I. norcrossi	0.0	0.0	0.0
I. helenae	0.0	0.0	0.0
P. williamsoni	6.3	5.0	6.6
Polymorphinidae	0.0	1.3	0.7
F. fusiformis	0.0	0.0	0.0
Stainfortia spp.	0.0	0.0	0.0
Cornuspira spp.	6.3	0.7	0.0
Buccella spp.	1.3	2.6	0.0
Lagena spp.	3.2	0.3	0.0
Quinqueloculina spp.	17.1	12.5	14.7
Miliolinella spp.	7.6	4.6	11.0
Other calcareous	2.5	2.0	0.0
P. karika	0.0	0.0	0.7
L. crassimargo	0.6	0.0	0.7
A. cassis	0.0	0.0	0.0
Textularia spp.	0.0	0.0	0.0
R. turbinatus	0.0	0.0	0.0
S. biformis	0.0	0.0	0.0
Reophax spp.	0.0	0.0	0.0
Other agglutinated	0.6	1.7	3.7

Table 5 Benthic foraminifera distribution in Core BP01-62/6 from the Kara Sea (%)

Species/**Depth in core** (cm)	0	20	40	60	80	100	120	140	160	180	200
Buccella spp.	0	7	1	9	8	5	6	8	7	4	7
C. reniforme	2	19	6	21	26	32	41	36	39	31	52
E. asklundi/incertum	0	2	1	1	0	1	2	1	3	8	0
E. bartletti	0	0	0	0	2	0	0	3	2	2	1
E. clavatum	3	16	9	11	14	18	9	12	11	14	4
H. orbiculare	2	5	6	6	9	6	3	6	5	4	6
I. helenae	0	2	0	1	0	1	1	0	2	0	0
I. norcrossi	1	10	11	5	19	12	12	4	12	7	5
N. labradoricum	1	2	5	1	0	2	4	5	3	1	0
P. williamsoni	0	0	0	0	0	0	0	0	1	1	0
E. subarcticum	2	4	1	8	0	1	1	0	2	4	4
Quinqueloculina spp.	0	0	0	0	0	0	0	0	0	0	0
Miliolinella spp.	0	2	1	2	1	0	1	1	1	0	0
Agglutinated	86	29	49	12	8	12	7	16	4	11	5
Other calcareous	4	4	7	22	12	7	11	8	8	14	11

Species/**Depth in core** (cm)	220	240	260	280	290	300	320	340	350	360	370
Buccella spp.	3	4	5	1	6	5	12	6	2	1	3
C. reniforme	51	59	40	54	62	63	40	39	53	53	58
E. asklundi/incertum	4	3	0	0	0	1	0	3	1	1	0
E. bartletti	1	0	5	5	0	1	3	6	0	1	0
E. clavatum	3	5	6	9	3	8	2	6	28	7	7
H. orbiculare	7	5	9	7	7	6	12	10	6	3	3
I. helenae	1	0	3	1	0	2	3	1	0	2	0
I. norcrossi	5	6	19	13	7	6	3	5	2	2	4
N. labradoricum	1	1	2	0	0	0	2	10	0	0	0
P. williamsoni	0	0	2	0	3	0	9	5	0	3	4
E. subarcticum	3	0	0	3	0	0	0	0	0	0	0
Quinqueloculina spp.	0	1	0	1	0	1	2	2	0	9	5
Miliolinella spp.	0	0	1	0	1	0	1	1	0	0	0
Agglutinated	7	8	0	0	2	2	0	1	3	11	7
Other calcareous	12	8	6	5	7	5	10	7	4	5	6

Table 5 (continued)

Species/**Depth in core** (cm)	380	390	400	420	440	460	480	500	520	540	550
Buccella spp.	3	3	0	7	5	4	1	2	3	2	6
C. reniforme	58	56	50	48	44	38	22	39	31	27	29
E. asklundi/incertum	0	1	1	0	5	4	12	6	9	6	4
E. bartletti	0	0	1	1	1	3	4	1	9	0	0
E. clavatum	12	19	23	18	9	15	5	21	5	14	34
H. orbiculare	7	7	8	12	11	10	23	11	12	6	6
I. helenae	2	2	1	1	1	4	7	3	1	1	1
I. norcrossi	5	1	0	1	4	3	6	5	3	5	4
N. labradoricum	0	2	0	1	0	0	0	2	0	1	1
P. williamsoni	6	5	12	5	10	5	6	3	6	21	3
E. subarcticum	0	0	0	0	0	0	0	0	0	0	3
Quinqueloculina spp.	2	0	1	0	1	5	2	1	5	0	0
Miliolinella spp.	0	0	0	1	0	0	1	2	1	0	0
Agglutinated	0	0	1	1	1	1	0	0	1	4	0
Other calcareous	4	5	2	4	5	8	11	5	15	12	9

Species/**Depth in core** (cm)	560	580
Buccella spp.	6	8
C. reniforme	35	12
E. asklundi/incertum	2	11
E. bartletti	0	0
E. clavatum	19	28
H. orbiculare	4	15
I. helenae	2	0
I. norcrossi	7	0
N. labradoricum	1	0
P. williamsoni	14	8
E. subarcticum	2	0
Quinqueloculina spp.	2	0
Miliolinella spp.	0	0
Agglutinated	2	2
Other calcareous	5	7

Table 6 Benthic foraminifera distribution in Core PS 2718/6 from the Kara Sea(%)

Species/**Depth in core** (cm)	340	350	360	370	380	390	400	410	420
Buccella spp.	6.2	–	–	–	1.9	–	3.2	–	–
C. reniforme	35.1	–	–	–	20.4	–	46.1	–	–
E. bartletti	0.0	–	–	–	5.1	–	0.4	–	–
E. clavatum	29.0	–	–	–	20.4	–	16.7	–	–
H. orbiculare	6.5	–	–	–	7.0	–	5.3	–	–
Islandiella spp.	7.2	–	–	–	8.3	–	5.7	–	–
P. williamsoni	0.7	–	–	–	0.0	–	0.4	–	–
V. loeblichi	5.1	–	–	–	5.1	–	6.4	–	–
Agglutinated	0.7	–	–	–	16.6	–	1.1	–	–
Other calcareous	9.4	–	–	–	15.3	–	14.9	–	–

Note: "–" Percentages are not calculated due to insufficient total foraminiferal abundance in the sample.

Table 6 (continued)

Species/**Depth in core** (cm)	430	440	450	460	470	480	490	500	510
Buccella spp.	11.1	–	–	–	–	0.6	1.1	0.9	–
C. reniforme	33.3	–	–	–	–	55.8	53.7	51.9	–
E. bartletti	0.0	–	–	–	–	0.0	1.7	5.1	–
E. clavatum	20.2	–	–	–	–	10.3	11.4	6.0	–
H. orbiculare	6.1	–	–	–	–	1.3	4.0	6.0	–
Islandiella spp.	10.6	–	–	–	–	23.1	8.6	9.3	–
P. williamsoni	0.5	–	–	–	–	0.0	2.3	0.0	–
V. loeblichi	3.5	–	–	–	–	1.9	2.3	6.9	–
Agglutinated	4.0	–	–	–	–	2.6	0.0	0.0	–
Other calcareous	10.6	–	–	–	–	4.5	14.9	13.9	–

Species/**Depth in core** (cm)	520	530	540	550	560	570	580	590	600
Buccella spp.	1.1	0.8	–	1.7	0.0	1.3	0.7	4.0	2.6
C. reniforme	57.3	45.9	–	40.3	67.5	59.1	61.5	44.4	54.7
E. bartletti	2.7	5.7	–	5.0	2.5	0.6	0.4	5.6	0.0
E. clavatum	6.5	16.4	–	10.1	5.0	10.1	19.6	25.4	23.1
H. orbiculare	7.6	6.6	–	3.4	4.0	6.9	3.0	4.8	1.7
Islandiella spp.	8.6	15.6	–	14.3	11.5	9.4	0.0	5.6	10.3
P. williamsoni	0.5	0.0	–	1.7	1.0	1.9	0.7	0.8	0.9
V. loeblichi	0.5	2.5	–	1.7	2.5	3.8	5.9	2.4	2.6
Agglutinated	0.0	0.0	–	1.7	0.0	0.0	0.4	0.0	0.0
Other calcareous	15.1	6.6	–	20.2	6.0	6.9	7.8	7.1	4.3

Species/**Depth in core** (cm)	610	620	630	640	650	660	670	680	690
Buccella spp.	2.9	2.0	–	8.7	–	–	1.96	6.80	7.03
C. reniforme	59.6	55.4	–	53.1	–	–	66.67	61.22	44.32
E. bartletti	0.0	4.0	–	3.6	–	–	0.00	0.00	6.49
E. clavatum	16.4	3.0	–	4.4	–	–	19.61	18.37	3.24
H. orbiculare	4.1	16.8	–	11.6	–	–	3.92	2.04	21.08
Islandiella spp.	1.8	7.9	–	1.8	–	–	1.96	2.72	2.70
P. williamsoni	0.6	3.0	–	0.7	–	–	0.98	2.72	5.41
V. loeblichi	3.5	4.0	–	2.9	–	–	2.94	2.72	1.08
Agglutinated	2.3	1.0	–	0.0	–	–	0.98	0.00	0.00
Other calcareous	8.8	3.0	–	13.1	–	–	0.98	3.40	8.65

Note: "–" Percentages are not calculated due to insufficient total foraminiferal abundance in the sample.

Table 6 (continued)

Species/**Depth in core** (cm)	700	710	720	730	740	750	760	770	780
Buccella spp.	5.19	0.68	0.46	4.15	2.94	2.67	0.84	1.02	0.00
C. reniforme	34.91	57.53	49.77	22.58	27.06	34.76	21.01	8.16	6.11
E. bartletti	3.30	1.37	0.46	4.15	0.59	0.00	0.00	0.00	0.00
E. clavatum	6.13	12.33	20.09	15.67	27.65	41.18	56.30	73.47	77.10
H. orbiculare	21.23	5.48	5.48	34.56	14.71	5.35	1.68	2.04	1.53
Islandiella spp.	4.72	8.90	3.65	1.38	4.71	3.21	0.84	3.06	0.00
P. williamsoni	13.21	7.53	11.42	3.23	14.12	7.49	10.08	6.12	0.00
V. loeblichi	3.77	1.37	1.37	0.92	0.00	0.00	0.00	0.00	0.00
Agglutinated	0.47	0.00	0.00	0.00	0.00	0.00	0.00	0.00	0.00
Other calcareous	7.08	4.79	7.31	13.36	8.24	5.35	9.24	6.12	15.27

Note: "–" Percentages are not calculated due to insufficient total foraminiferal abundance in the sample.

Table 7 Benthic foraminifera distribution in Core ACB-987 from the Russian Gavan' Fjord (%)

Species/**Depth in core** (cm)	0	5	10	20	30	40	45	50	55	60	70
E. clavatum	40.60	20.47	50.00	36.68	14.86	42.02	35.95	23.15	22.36	26.88	57.63
C. reniforme	10.90	15.35	6.25	49.78	78.83	45.71	60.46	59.61	62.11	51.16	20.34
Buccella spp.	1.50	0.00	0.89	1.31	0.00	0.61	0.00	0.49	0.62	0.00	0.42
N. labradoricum	0.00	0.00	0.00	0.87	0.45	0.00	0.00	0.00	0.00	0.00	1.27
E. bartletti	23.68	50.23	17.86	3.49	1.35	2.15	0.65	1.97	3.11	10.12	11.02
C. lobatulus	4.14	3.26	15.18	1.31	1.35	0.92	2.94	2.96	11.18	5.20	5.93
Other	19.17	10.70	9.82	6.55	3.15	8.59	0.00	11.82	0.62	6.65	3.39

Species/**Depth in core** (cm)	75	80	85	90	95	100	105	110	115	120	125
E. clavatum	37.09	9.33	20.28	10.85	10.00	72.16	19.46	–	25.71	33.78	–
C. reniforme	38.97	59.00	50.47	84.88	85.33	11.34	69.73	–	49.29	26.35	–
Buccella spp.	1.41	1.33	0.94	0.39	0.33	1.03	1.08	–	25.00	29.05	–
N. labradoricum	0.00	0.00	0.00	0.00	0.00	0.00	0.00	–	0.00	0.68	–
E. bartletti	4.69	0.00	1.89	0.00	0.00	0.34	5.95	–	0.00	0.00	–
C. lobatulus	17.37	3.33	26.42	1.94	4.00	9.62	3.24	–	0.00	5.41	–
Other	0.47	27.00	0.00	1.94	0.33	5.50	0.54	–	0.00	4.73	–

Species/**Depth in core** (cm)	130	135	140	145	150	155	160	165	170	180	185
E. clavatum	–	58.62	55.35	52.73	55.38	60.33	48.52	34.14	22.57	24.71	22.44
C. reniforme	–	36.40	35.77	35.69	39.84	37.67	46.20	58.60	69.47	67.24	69.27
Buccella spp.	–	1.15	1.31	4.50	1.20	0.67	0.63	2.42	0.44	0.29	0.98
N. labradoricum	–	0.00	0.78	0.00	0.40	0.00	1.27	0.00	3.98	4.89	1.46
E. bartletti	–	0.77	0.00	0.32	0.00	0.33	0.00	0.00	0.00	0.00	0.00
C. lobatulus	–	2.30	2.09	1.93	1.99	0.67	1.27	0.48	0.44	1.15	0.49
Other	–	0.77	4.70	4.82	1.20	0.33	2.11	4.36	3.10	1.72	5.37

Note: "–" Percentages are not calculated due to insufficient total foraminiferal abundance in the sample.

Table 7 (continued)

Species/**Depth in core** (cm)	190	200	210	220	230	240	250	260	270	280	290
E. clavatum	28.10	42.92	52.99	34.37	67.20	43.14	48.58	67.22	65.29	72.77	52.42
C. reniforme	64.29	39.48	38.92	58.20	27.65	36.27	41.01	23.41	27.39	15.96	40.00
Buccella spp.	0.95	0.43	1.20	1.24	1.61	1.31	1.26	0.67	0.64	0.00	0.61
N. labradoricum	0.48	5.15	0.90	1.24	0.64	13.73	2.52	4.68	1.27	3.29	0.30
E. bartletti	0.00	0.00	0.00	0.00	0.00	0.00	0.32	0.00	0.00	0.00	0.00
C. lobatulus	1.43	0.86	0.60	0.31	0.00	0.98	0.95	0.00	1.59	2.35	1.21
Other	4.76	11.16	5.39	4.64	2.89	4.58	5.36	4.01	3.82	5.63	5.45

Species/**Depth in core** (cm)	300	310	320	330	340	350	360	370	380	390	400
E. clavatum	67.68	39.92	22.67	49.01	44.86	46.86	62.42	34.96	40.60	61.90	66.52
C. reniforme	21.21	46.37	55.78	33.55	44.55	42.26	20.50	45.13	53.85	26.79	24.78
Buccella spp.	2.53	2.42	9.11	5.59	3.43	3.35	1.55	4.87	0.43	4.76	2.61
N. labradoricum	1.01	3.63	4.67	0.33	0.31	2.09	0.31	0.88	0.85	0.60	1.74
E. bartletti	0.00	0.40	0.00	0.33	0.00	0.00	0.00	0.00	0.00	0.30	0.00
C. lobatulus	1.52	1.21	1.56	0.33	0.00	0.00	0.00	2.65	0.85	0.60	0.87
Other	6.06	6.05	6.22	10.86	6.85	5.44	15.22	11.50	3.42	5.06	3.48

Species/**Depth in core** (cm)	410	420	430	440	450	460	470	475	480	485	490
E. clavatum	67.18	62.78	67.74	64.62	65.71	50.00	70.09	56.29	51.17	77.96	–
C. reniforme	24.15	29.77	24.52	11.14	20.51	33.33	18.69	39.74	27.23	2.30	–
Buccella spp.	1.86	0.32	1.29	3.06	4.49	1.28	2.80	0.66	2.35	3.62	–
N. labradoricum	0.31	1.29	0.65	14.21	5.77	33.33	1.25	0.00	5.63	0.00	–
E. bartletti	0.00	0.00	0.32	0.56	0.00	1.28	1.87	0.00	6.57	13.82	–
C. lobatulus	1.86	1.62	0.65	1.67	0.64	2.56	1.56	0.66	4.69	1.64	–
Other	4.64	4.21	4.84	4.74	2.88	2.99	3.74	2.65	2.35	0.66	–

Species/**Depth in core** (cm)	495	500	505	510	515	520	525	530	535	540	545
E. clavatum	41.42	76.55	50.27	46.92	27.74	39.68	76.71	65.80	66.45	57.46	65.64
C. reniforme	42.26	19.54	40.98	38.86	41.77	23.02	4.66	22.80	23.59	28.07	23.93
Buccella spp.	5.44	0.00	2.73	0.47	3.05	0.00	0.55	3.26	4.32	1.32	5.83
N. labradoricum	0.00	0.65	0.55	11.37	3.96	30.16	0.00	3.91	1.00	10.96	0.61
E. bartletti	1.26	0.33	0.00	0.47	0.61	0.79	0.00	0.00	0.00	0.00	0.31
C. lobatulus	2.09	0.65	0.55	0.00	3.96	0.79	0.00	0.98	1.00	0.00	0.31
Other	7.53	2.28	4.92	1.90	0.61	5.56	18.08	3.26	3.65	2.19	3.37

Species/**Depth in core** (cm)	550	560	570	580	590	595
E. clavatum	75.15	29.56	60.20	50.69	80.33	55.65
C. reniforme	15.95	57.23	25.99	36.11	13.67	26.78
Buccella spp.	1.84	1.57	0.33	4.17	0.67	1.67
N. labradoricum	3.07	3.77	9.87	2.08	0.33	9.21
E. bartletti	0.61	0.63	0.66	2.78	0.67	0.00
C. lobatulus	0.31	1.26	0.66	2.08	0.33	2.93
Other	3.07	5.97	2.30	2.08	4.00	3.77

Note: "–" Percentages are not calculated due to insufficient total foraminiferal abundance in the sample.

Table 8 Planktic foraminifera distribution in Core GeoB-3011-1 from the Arabian Sea (%)

Species/**Depth in core** (cm)	3	8	13	18	23	28	33	38	43	48	53	63
G. adamsi	0.0	0.0	0.0	0.0	0.3	0.0	0.0	0.0	0.0	0.0	0.2	0.0
G. anfracta	0.8	0.3	0.0	0.0	0.0	0.0	0.0	0.0	0.0	0.0	0.0	0.0
B. digitata	0.8	0.6	0.0	0.7	0.3	0.6	0.5	0.0	0.7	0.3	0.2	0.0
G. bermudezi	0.0	0.3	0.0	1.0	0.0	0.9	0.5	0.0	0.7	1.8	0.0	0.3
G. bulloides	35.8	34.0	29.8	31.1	24.1	24.1	17.6	24.9	14.8	24.7	17.1	12.6
G. calida	2.5	1.2	0.7	1.6	1.8	1.3	3.9	1.3	2.3	1.5	1.0	1.2
G. conglobatus	0.4	0.3	0.0	1.0	0.0	1.3	0.0	0.3	0.0	0.0	0.2	0.3
G. crassaformis	0.0	0.0	0.0	0.0	0.0	0.0	0.0	0.0	0.0	0.0	0.2	0.0
N. dutertrei	4.6	5.0	4.7	8.5	5.9	6.3	5.2	9.8	6.0	9.8	7.3	4.7
G. falconensis	0.8	0.6	3.7	0.0	2.6	2.8	3.4	1.6	2.0	2.4	1.0	1.8
G. glutinata	19.6	32.7	18.7	22.3	21.5	21.6	17.1	22.4	22.8	24.7	29.3	40.8
G. hexagona	0.4	0.6	0.3	0.0	0.3	0.0	0.5	0.6	0.7	0.0	0.0	1.2
G. iota	0.0	0.3	1.0	0.0	0.0	0.6	0.3	0.6	0.3	0.0	0.7	0.9
G. menardii	3.3	2.8	6.7	5.6	5.4	2.8	1.0	3.2	3.7	4.0	6.6	7.0
G. uvula	0.4	0.0	0.7	0.0	1.0	0.0	2.1	0.3	1.0	0.0	2.0	2.1
P. obliquiloculata	1.7	0.6	1.7	1.6	2.3	1.9	2.3	0.9	2.7	0.6	1.5	1.2
N. pachyderma dex.	0.8	0.9	0.7	0.0	2.3	0.0	1.3	0.3	4.7	4.6	6.1	5.9
N. pachyderma sin.	0.0	0.0	0.0	0.0	0.0	0.3	0.0	0.0	0.0	0.6	0.2	0.6
G. ruber pink	0.0	0.0	0.0	0.0	0.0	0.0	0.0	0.0	0.0	0.0	0.2	0.0
G. ruber	7.9	9.7	13.4	10.8	15.1	16.6	18.1	15.8	14.4	11.3	9.5	5.6
G. rubescens	6.3	4.0	3.7	3.3	6.4	6.6	6.5	5.7	6.0	3.7	3.2	6.2
G. sacculifer	4.2	1.9	6.0	4.6	2.1	4.1	5.7	4.4	2.7	2.4	3.7	0.6
G. scitula	0.4	0.0	0.7	0.0	1.5	1.6	2.3	0.3	4.0	0.0	1.7	0.6
G. aequilateralis	0.4	0.6	2.0	4.9	2.1	3.4	3.9	2.5	1.7	1.5	1.0	0.3
G. tenellus	0.8	0.9	1.0	0.0	1.0	1.3	2.3	1.6	1.3	1.2	1.0	0.9
G. theyeri	0.8	0.6	1.0	1.6	0.5	0.0	0.5	0.3	1.0	1.8	2.7	1.0
G. truncatulinoides sin.	0.0	0.0	0.0	0.0	0.5	0.0	1.0	0.0	1.7	0.0	0.2	0.0
G. truncatulinoides dex.	0.0	0.0	0.0	0.3	0.0	0.6	0.0	0.0	0.0	0.0	0.0	0.0
O.universa	0.4	0.6	1.7	0.3	1.3	0.0	1.3	2.5	1.7	1.2	0.7	0.3
Other	6.7	1.2	2.0	0.7	1.8	1.3	2.8	0.6	3.0	1.8	2.4	2.6

Table 8 (continued)

Species/**Depth in core** (cm)	73	83	93	103	113	123	133	143	153	163	173	183
G. adamsi	0.0	0.0	0.0	0.0	0.0	0.0	0.0	0.3	0.3	0.0	0.0	0.0
G. anfracta	0.0	0.0	0.0	0.3	0.0	0.0	0.0	0.0	0.0	0.0	0.3	0.2
B. digitata	0.0	0.3	0.0	0.0	0.7	1.2	0.3	0.7	0.7	0.8	0.5	1.5
G. bermudezi	0.0	0.0	0.0	0.0	0.0	0.0	0.0	0.0	0.0	0.0	0.0	0.0
G. bulloides	14.5	14.7	7.7	16.0	21.8	28.8	21.4	18.8	22.1	21.1	25.5	25.6
G. calida	0.5	1.6	1.7	2.2	1.7	1.2	1.9	0.7	2.7	1.1	1.1	1.3
G. conglobatus	0.0	0.0	0.0	0.0	0.0	0.0	0.0	0.0	0.0	0.0	0.0	0.0
G. crassaformis	0.2	0.8	0.0	0.0	0.4	0.9	0.6	0.7	0.7	1.7	0.8	1.1
N. dutertrei	6.0	5.0	5.4	8.8	11.1	11.8	7.3	13.9	9.2	6.1	6.8	6.3
G. falconensis	2.1	0.5	2.3	2.5	1.1	0.3	0.6	0.7	0.3	2.5	1.4	1.5
G. glutinata	39.3	36.1	33.6	25.1	22.5	14.2	21.4	23.3	22.1	25.8	27.9	24.4
G. hexagona	0.9	1.0	1.0	0.3	1.1	1.2	1.3	0.3	0.7	1.1	2.4	0.2
G. iota	0.5	0.0	0.3	0.3	0.2	0.0	0.3	0.7	1.0	0.3	0.0	0.6
G. menardii	5.3	5.2	2.3	7.5	8.7	9.1	4.2	8.3	4.1	6.6	4.9	3.2
G. uvula	0.9	0.5	1.0	0.0	0.7	0.6	2.2	0.3	2.0	1.4	3.3	1.9
P. obliquiloculata	2.3	0.8	3.0	1.6	0.9	5.2	0.3	2.4	2.4	3.6	1.9	1.1
N. pachyderma dex.	7.6	8.9	5.4	10.0	8.1	6.7	8.9	6.6	3.1	1.7	2.2	4.4
N. pachyderma sin.	0.5	0.0	0.3	0.9	0.2	0.0	0.0	0.0	0.0	0.3	0.0	1.7
G. ruber pink	0.0	0.0	0.0	0.0	0.0	0.0	0.0	0.0	0.0	0.0	0.3	0.0
G. ruber	7.9	9.7	16.8	11.6	8.5	10.9	19.2	11.1	8.5	10.0	6.5	10.5
G. rubescens	4.6	5.8	7.4	4.7	6.1	2.4	4.5	4.2	9.9	8.3	5.4	5.3
G. sacculifer	1.6	1.0	4.0	0.6	0.9	1.2	1.6	1.4	1.4	1.1	1.1	1.5
G. scitula	0.0	0.5	0.7	0.3	0.0	0.0	0.3	0.3	0.0	0.0	1.1	0.0
G. aequilateralis	0.9	0.8	2.3	0.9	1.1	0.6	1.3	2.1	4.4	3.3	3.0	2.5
G. tenellus	0.5	0.8	0.7	0.3	0.2	0.0	0.3	0.7	0.3	0.0	0.0	0.4
G. theyeri	1.4	1.0	1.3	1.6	0.7	0.6	0.0	1.0	1.4	0.3	0.3	1.1
G. truncatulinoides sin.	0.0	0.0	0.0	0.0	0.0	0.0	0.0	0.0	0.0	0.0	0.0	0.2
G. truncatulinoides dex.	0.0	0.0	0.0	0.0	0.0	0.0	0.0	0.0	0.0	0.0	0.0	0.0
O.universa	0.2	1.0	0.3	0.9	0.7	0.6	0.0	0.3	1.0	0.6	0.5	1.5
Other	2.3	3.9	2.3	3.4	2.6	2.4	1.9	1.0	1.7	2.5	3.0	2.3

Table 8 (continued)

Species/**Depth in core** (cm)	193	203	213	223	233	243	253	263	273	283	293	303
G. adamsi	0.0	0.0	0.0	0.3	0.0	0.0	0.0	0.0	0.0	0.0	0.0	0.0
G. anfracta	0.0	0.0	0.0	0.0	0.0	0.0	0.0	0.0	0.0	0.0	0.0	0.0
B. digitata	0.3	0.3	1.4	0.7	0.3	0.4	0.6	0.7	0.3	0.3	0.3	0.3
G. bermudezi	0.0	0.0	0.0	0.0	0.0	0.0	1.3	1.0	0.7	0.0	0.3	0.3
G. bulloides	22.6	33.0	30.2	30.7	29.4	29.1	26.9	18.7	22.5	15.6	25.6	14.3
G. calida	0.7	2.1	1.1	1.0	2.9	3.4	1.9	2.1	2.0	1.9	1.3	2.6
G. conglobatus	0.0	0.0	0.0	0.0	0.3	0.0	0.2	0.0	0.0	0.3	0.6	0.7
G. crassaformis	0.7	0.6	0.0	0.0	0.0	0.0	0.7	0.0	0.0	0.0	0.0	0.3
N. dutertrei	7.5	6.1	3.6	3.0	2.9	6.0	3.3	6.2	5.6	9.3	6.5	8.8
G. falconensis	2.4	0.3	0.7	1.3	0.3	0.9	1.7	2.4	1.0	2.8	4.2	3.9
G. glutinata	25.0	26.4	29.3	32.7	25.5	30.6	30.1	33.6	31.8	32.1	17.5	19.2
G. hexagona	0.3	0.3	0.5	0.0	0.5	0.9	0.2	0.7	0.3	0.0	0.0	0.0
G. iota	0.7	0.0	0.0	0.7	0.0	0.0	0.2	0.3	0.0	0.0	0.0	0.3
G. menardii	12.7	13.0	4.7	5.0	4.8	5.5	4.1	3.8	4.6	2.2	1.6	0.7
G. uvula	1.4	1.2	1.4	2.0	1.1	1.1	1.7	1.4	2.3	1.6	1.0	2.9
P. obliquiloculata	1.4	1.5	0.5	1.7	1.3	0.4	0.7	1.7	1.7	1.6	1.6	0.7
N. pachyderma dex.	6.2	4.2	4.3	5.6	5.8	0.4	2.2	10.0	8.9	6.2	9.7	14.0
N. pachyderma sin.	1.0	0.6	1.1	1.0	1.3	0.4	0.2	1.4	0.0	0.3	0.3	0.3
G. ruber pink	0.0	0.0	0.0	0.0	0.0	0.0	0.0	0.0	0.0	0.0	0.0	0.0
G. ruber	8.2	3.3	10.4	6.3	10.1	10.2	11.1	5.9	8.3	9.7	13.9	13.7
G. rubescens	4.1	1.5	2.7	2.6	4.0	5.3	6.5	2.8	3.0	5.6	5.8	4.6
G. sacculifer	0.7	1.5	1.1	2.0	1.3	1.3	0.6	1.0	1.3	1.6	1.9	1.3
G. scitula	0.3	0.0	0.0	0.0	0.3	0.4	0.7	0.7	0.0	0.3	0.6	1.3
G. aequilateralis	0.3	0.6	1.4	1.0	2.7	0.6	2.6	2.4	1.7	2.8	2.9	5.2
G. tenellus	0.0	0.0	0.5	0.0	0.3	0.4	0.0	0.0	0.3	0.6	1.0	1.0
G. theyeri	0.3	0.3	1.6	0.3	0.8	0.0	0.4	0.0	0.7	0.0	0.0	0.3
G. truncatulinoides sin.	0.0	0.0	0.0	0.0	0.0	0.0	0.0	0.0	0.0	0.6	1.0	0.3
G. truncatulinoides dex.	0.0	0.0	0.0	0.0	0.0	0.0	0.0	0.0	0.0	0.0	0.0	0.0
O.universa	0.7	0.6	1.6	0.7	0.8	0.2	0.2	0.0	0.3	0.9	0.6	0.7
Other	2.4	2.4	2.3	1.7	3.4	2.3	2.0	3.1	2.6	3.7	1.6	2.3

Table 8 (continued)

Species/**Depth in core** (cm)	313	323	333	343	353	363	373	383	393	403	413	423
G. adamsi	0.0	0.0	0.0	0.0	0.3	0.0	0.0	0.0	0.0	0.0	0.0	0.0
G. anfracta	0.0	0.0	0.0	0.0	0.0	0.0	0.0	0.0	0.0	0.0	0.0	0.0
B. digitata	0.5	0.5	0.5	0.0	0.7	1.0	0.6	0.0	0.6	0.0	0.3	0.0
G. bermudezi	2.1	1.1	0.5	0.3	0.0	1.0	0.3	0.6	1.5	1.9	0.7	0.0
G. bulloides	15.2	12.1	14.3	13.4	17.9	17.5	30.4	32.1	20.8	15.1	22.0	35.2
G. calida	3.0	3.8	1.9	1.7	1.3	2.2	2.0	1.2	1.5	1.6	1.4	1.3
G. conglobatus	0.2	0.3	0.2	0.0	0.0	0.3	0.3	0.0	0.0	0.5	0.0	0.0
G. crassaformis	0.5	0.3	0.2	0.3	0.0	1.0	2.3	0.3	1.2	1.4	2.1	3.5
N. dutertrei	8.9	6.2	7.1	8.9	7.6	4.1	4.6	2.4	7.1	8.0	4.1	3.5
G. falconensis	3.7	1.3	0.7	1.0	1.7	2.2	1.4	1.5	0.9	1.1	0.3	0.3
G. glutinata	16.4	24.1	20.7	22.9	20.2	19.1	19.5	31.2	19.3	18.4	32.6	26.8
G. hexagona	0.2	1.1	1.2	1.0	0.7	0.0	0.3	0.6	0.0	0.8	0.7	0.3
G. iota	0.0	0.0	0.0	0.0	0.0	0.0	0.0	0.0	0.0	0.3	0.0	0.0
G. menardii	3.0	4.6	8.3	12.7	7.9	6.4	7.7	9.0	4.2	3.0	9.3	7.4
G. uvula	1.2	1.1	0.7	1.0	2.0	1.0	0.3	0.6	0.6	0.0	0.0	0.0
P. obliquiloculata	0.5	0.5	1.0	2.1	1.3	1.9	0.9	0.3	2.7	1.6	0.7	0.6
N. pachyderma dex.	10.8	11.3	10.9	10.6	7.3	1.3	1.4	0.0	9.8	12.9	8.6	1.6
N. pachyderma sin.	0.2	0.0	1.0	0.3	0.7	0.3	0.0	0.0	0.3	0.3	0.3	0.0
G. ruber pink	0.0	0.0	0.0	0.0	0.0	0.0	0.0	0.0	0.0	0.0	0.0	0.0
G. ruber	12.2	9.7	8.6	5.8	12.9	19.7	15.2	12.0	12.5	15.7	6.2	9.4
G. rubescens	8.7	5.6	8.1	6.5	8.6	6.1	3.2	3.6	6.5	7.7	6.5	5.2
G. sacculifer	0.7	2.4	2.1	1.4	0.7	4.8	2.9	1.8	1.8	2.7	0.7	1.9
G. scitula	0.9	2.7	1.7	0.7	0.0	0.3	1.1	0.6	1.2	0.8	0.0	0.0
G. aequilateralis	5.4	5.9	5.2	4.1	5.0	5.1	1.7	0.9	2.7	2.5	1.0	0.3
G. tenellus	0.5	0.5	0.7	0.0	0.3	1.0	1.4	0.3	1.2	0.0	0.3	1.0
G. theyeri	0.5	1.1	1.0	2.1	0.3	0.0	1.1	0.6	0.3	0.3	0.3	0.0
G. truncatulinoides sin.	0.0	0.0	0.2	0.0	0.0	0.0	0.0	0.0	0.0	0.3	0.0	0.0
G. truncatulinoides dex.	0.0	0.0	0.0	0.0	0.0	0.0	0.0	0.0	0.0	0.0	0.0	0.0
O.universa	2.1	1.6	1.2	1.4	1.0	2.5	0.6	0.3	2.4	1.9	0.0	0.0
Other	2.6	2.4	2.1	1.7	1.7	1.3	0.9	0.0	0.9	1.1	1.7	1.6

Table 8 (continued)

Species/**Depth in core** (cm)	433	443	453	463	473	483	493	503	513
G. adamsi	0.0	0.0	0.0	0.0	0.0	0.3	0.0	0.0	0.0
G. anfracta	0.0	0.0	0.0	0.0	0.0	0.0	0.0	0.0	0.3
B. digitata	0.3	0.5	0.0	1.0	0.0	0.0	0.5	0.3	0.8
G. bermudezi	0.6	0.5	0.3	0.0	0.0	0.9	0.0	0.6	0.3
G. bulloides	27.5	20.4	17.0	38.1	25.8	15.1	15.4	18.5	18.9
G. calida	1.7	0.7	1.5	1.3	0.3	1.5	2.1	0.9	2.0
G. conglobatus	0.0	0.9	0.6	0.0	0.0	1.5	1.6	0.3	0.0
G. crassaformis	6.4	3.6	5.2	1.9	2.3	1.8	2.5	2.8	2.5
N. dutertrei	7.5	4.7	7.0	1.9	4.8	10.0	9.6	11.4	4.2
G. falconensis	0.0	0.0	1.2	1.6	3.5	3.3	3.4	3.1	2.3
G. glutinata	22.5	28.9	25.2	28.5	25.5	16.3	16.5	17.7	26.2
G. hexagona	0.6	0.0	0.3	0.3	0.0	0.0	0.0	0.3	0.8
G. iota	0.3	0.2	0.3	0.0	0.0	0.0	0.2	0.9	0.3
G. menardii	5.6	7.1	9.4	8.3	4.5	1.2	2.5	3.1	2.0
G. uvula	0.3	0.2	0.0	0.3	0.3	0.6	0.2	0.0	0.8
P. obliquiloculata	0.6	1.4	0.9	0.3	0.3	0.9	2.1	2.3	1.1
N. pachyderma dex.	2.2	1.7	1.2	0.3	1.3	2.4	1.8	2.3	4.8
N. pachyderma sin.	0.3	0.0	0.0	0.0	0.0	0.0	0.2	0.0	0.3
G. ruber pink	0.0	0.0	0.0	0.0	0.6	2.4	3.0	0.9	0.3
G. ruber	13.6	12.3	12.7	7.4	17.1	15.4	16.7	12.3	14.6
G. rubescens	3.1	5.0	2.7	3.5	6.5	4.2	3.0	6.3	4.2
G. sacculifer	2.2	3.8	5.2	1.9	2.6	6.9	10.1	5.7	5.9
G. scitula	0.3	0.7	0.3	0.0	0.0	0.0	0.2	0.9	0.3
G. aequilateralis	2.5	2.4	3.3	0.6	1.9	6.6	3.9	4.3	2.8
G. tenellus	0.3	0.2	0.3	1.0	0.6	0.3	0.9	1.1	2.0
G. theyeri	0.3	0.5	0.0	0.3	0.0	0.0	0.2	0.0	0.3
G. truncatulinoides sin.	0.0	0.0	0.0	0.0	0.0	1.2	1.6	0.9	0.0
G. truncatulinoides dex.	0.0	0.0	0.0	0.0	0.0	0.0	0.0	0.0	0.0
O.universa	0.6	2.6	3.0	0.0	0.3	3.9	1.1	1.1	0.6
Other	1.1	1.7	2.4	1.3	1.6	3.0	0.5	2.3	1.4

Table 9 Planktic foraminifera distribution in Core NAST from the Arabian Sea (%)

Species/**Depth in core** (cm)	5	10	15	20	25	30	35	40	45	50	55	60
O.universa	0.3	0.8	3.0	0.2	0.8	0.7	0.0	1.8	0.8	0.8	2.3	2.8
G. conglobatus	0.3	0.6	0.0	0.7	0.8	0.9	1.2	1.0	1.6	2.0	2.1	0.9
G. ruber pink	0.0	0.0	0.0	0.0	0.0	0.0	0.0	0.0	0.0	0.0	0.0	0.0
G. ruber white	12.1	11.4	12.6	11.9	22.8	22.8	20.1	26.3	33.1	37.0	30.1	19.7
G. tenellus	3.1	3.2	0.5	4.4	2.8	4.5	2.5	3.3	2.3	3.9	1.8	2.9
G.sacculifer	7.9	12.2	22.0	12.6	14.6	13.3	12.1	15.7	14.6	5.2	8.9	6.0
G. aequilateralis	3.1	4.4	6.9	4.0	2.2	3.1	4.0	2.8	4.8	3.2	4.4	7.6
G. calida	1.5	0.9	0.0	0.2	0.4	2.1	1.4	1.3	1.3	0.4	3.0	4.8
G. bulloides	7.7	9.4	4.2	7.5	5.2	5.7	5.8	3.8	4.3	5.2	7.0	14.0
G. falconensis	6.4	2.7	2.8	9.4	6.4	3.5	4.2	6.1	4.9	6.3	5.6	8.4
G. rubescens	2.6	2.5	0.9	4.0	3.6	2.4	4.4	3.0	3.9	3.3	2.5	2.8
G. quinqueloba	3.8	1.4	1.5	0.9	0.2	0.0	0.5	1.3	0.1	0.4	0.8	0.2
N. pachyderma sin.	0.0	0.3	0.4	0.2	0.0	0.2	0.0	0.3	0.0	0.4	0.7	0.0
N. pachyderma dex.	1.0	0.6	2.1	1.2	0.4	0.5	1.1	4.1	2.4	1.9	2.3	2.2
N. dutertrei	3.6	5.1	8.6	5.6	6.0	8.1	6.0	5.6	1.9	1.9	1.8	1.2
P. obliquiloculata	4.9	5.1	8.1	3.7	3.8	4.7	4.8	4.6	3.5	3.4	2.8	2.1
G. truncatulinoides sin.	0.0	0.0	0.0	0.0	0.0	0.0	0.2	0.3	0.0	0.0	0.2	0.0
G. truncatulinoides dex.	0.0	0.0	0.0	0.0	0.2	0.2	0.5	1.5	0.9	1.5	0.9	0.0
G. menardii	9.2	4.0	8.2	4.0	3.0	2.9	2.3	1.3	1.1	0.6	1.9	0.3
G. glutinata	25.6	28.7	17.1	26.5	23.2	20.9	22.2	13.4	15.2	15.9	15.9	17.1
Other	6.9	6.8	1.3	2.8	3.8	3.6	6.7	2.8	3.5	6.7	5.3	7.1

Species/**Depth in core** (cm)	65	70	75	80	85	90	95	100	110	120	130	140
O.universa	4.6	1.1	2.9	3.1	5.2	4.3	4.3	4.3	1.3	6.9	2.8	3.9
G. conglobatus	3.1	2.8	0.8	2.1	0.6	0.0	1.4	1.1	2.2	1.7	1.8	1.9
G. ruber pink	0.0	0.0	0.0	0.0	0.0	0.0	0.0	0.0	0.0	0.0	0.0	0.0
G. ruber white	24.0	30.1	22.1	15.3	23.1	6.7	27.6	19.9	18.5	24.5	16.9	27.2
G. tenellus	2.6	2.5	1.2	0.5	0.8	1.6	0.5	3.0	3.8	1.0	0.7	1.6
G.sacculifer	6.4	5.5	11.0	9.0	12.7	10.8	6.0	10.1	5.0	8.1	6.9	3.0
G. aequilateralis	6.0	4.3	5.1	9.2	9.0	8.8	5.3	2.3	5.8	2.9	4.9	4.4
G. calida	5.9	2.9	3.1	1.5	5.4	6.1	5.3	2.3	3.2	1.9	3.1	3.9
G. bulloides	10.2	6.8	3.5	4.6	6.3	7.3	4.8	2.5	1.8	7.4	8.2	6.5
G. falconensis	11.1	7.7	12.0	7.4	9.8	8.6	11.9	5.0	7.3	12.8	10.2	6.0
G. rubescens	2.2	1.7	12.7	10.7	3.8	4.5	1.9	0.2	2.1	2.4	3.6	3.0
G. quinqueloba	0.7	5.2	3.4	1.1	0.6	6.3	2.6	1.8	0.7	6.4	1.6	0.9
N. pachyderma sin.	1.1	0.6	0.6	0.0	0.0	0.4	0.0	0.7	2.5	1.4	1.4	0.7
N. pachyderma dex.	1.2	1.1	1.4	2.0	2.1	2.0	1.5	3.4	5.3	1.9	2.5	5.3
N. dutertrei	1.4	2.1	2.5	1.0	0.8	2.2	0.5	4.3	0.8	0.7	1.6	3.5
P. obliquiloculata	1.2	3.0	1.9	2.8	3.8	3.7	1.9	7.1	3.2	1.9	2.4	6.5
G. truncatulinoides sin.	0.0	0.0	0.0	0.0	0.0	0.0	0.0	0.0	0.0	0.0	0.0	0.0
G. truncatulinoides dex.	0.1	0.0	0.0	0.1	0.0	0.2	0.5	0.0	0.1	0.0	0.0	0.0
G. menardii	0.8	2.3	1.2	0.0	0.4	1.2	1.0	2.1	2.3	1.0	0.5	1.8
G. glutinata	13.0	15.3	9.9	18.4	11.7	15.3	15.9	19.7	28.3	10.9	21.8	13.1
Other	4.4	5.0	4.8	11.2	4.0	10.0	7.2	10.1	5.8	6.4	8.9	6.9

Table 9 (continued)

Species/**Depth in core** (cm)	150	160	170	180	190	200	210	220	230	240	250	260
O.universa	1.9	2.1	0.7	3.0	5.2	2.8	0.7	2.1	2.6	1.2	1.1	1.9
G. conglobatus	1.8	1.5	1.6	0.3	0.0	1.0	0.9	0.3	0.4	1.1	0.3	0.9
G. ruber pink	0.0	0.0	0.0	0.0	0.0	0.0	0.0	0.0	0.0	0.0	0.0	0.0
G. ruber white	19.4	12.7	11.0	14.6	12.5	11.5	11.8	8.4	18.2	15.5	12.7	7.7
G. tenellus	2.2	2.1	5.6	1.3	2.0	1.5	4.9	1.0	0.6	1.8	0.6	1.4
G.sacculifer	9.3	7.1	5.0	4.6	8.0	4.8	3.7	6.0	3.5	2.0	3.3	5.5
G. aequilateralis	5.8	4.4	5.6	6.9	10.7	5.3	2.6	4.5	5.4	2.8	1.7	4.4
G. calida	5.1	5.3	5.0	6.6	9.5	9.9	8.8	4.5	2.6	2.2	3.8	1.5
G. bulloides	5.3	5.3	6.0	9.4	8.5	20.6	31.1	11.3	5.8	5.3	12.2	4.4
G. falconensis	9.5	10.9	11.4	6.6	9.8	13.0	6.5	2.4	4.5	3.0	7.1	3.9
G. rubescens	1.6	5.0	7.0	4.3	3.7	5.9	1.6	2.6	2.6	4.0	3.5	5.7
G. quinqueloba	1.6	1.5	3.7	2.5	1.7	2.0	2.6	10.3	22.8	4.9	10.1	3.9
N. pachyderma sin.	0.5	0.0	1.8	1.5	2.0	0.5	1.2	3.1	2.2	4.6	4.1	1.5
N. pachyderma dex.	1.3	1.5	3.2	1.9	0.7	1.3	1.2	0.8	1.1	4.6	4.9	0.6
N. dutertrei	1.1	2.1	0.6	1.6	2.2	1.8	3.5	2.9	2.2	4.3	0.9	0.6
P. obliquiloculata	4.5	4.7	2.6	4.6	1.7	2.0	2.3	3.1	3.2	2.3	2.7	2.8
G. truncatulinoides sin.	0.0	0.0	0.0	0.0	0.0	0.0	0.0	0.0	0.0	0.0	0.0	0.0
G. truncatulinoides dex.	0.0	0.0	0.0	0.1	0.0	0.0	0.0	0.0	0.0	0.0	0.0	0.0
G. menardii	2.6	1.5	1.0	1.2	0.7	0.8	0.9	2.1	0.9	1.3	1.7	0.4
G. glutinata	18.9	26.3	18.2	21.5	17.7	9.7	10.9	26.0	16.4	26.4	24.8	42.7
Other	7.5	6.2	10.0	7.2	3.5	5.6	4.9	8.7	5.0	12.7	4.5	10.2

Species/**Depth in core** (cm)	270	280	290	300	310	320	330	340	350	360	370	380
O.universa	1.9	1.1	7.1	3.1	1.9	2.8	4.8	2.6	0.8	3.7	0.2	0.4
G. conglobatus	0.9	1.4	0.3	1.6	2.5	1.1	0.8	0.5	0.4	1.7	0.7	0.0
G. ruber pink	0.0	0.0	0.0	0.0	0.0	0.0	0.0	0.0	0.0	0.0	0.0	0.0
G. ruber white	11.2	7.7	9.7	16.8	15.5	14.6	10.8	15.8	5.9	15.0	3.4	4.6
G. tenellus	2.4	0.0	1.0	2.7	2.2	2.5	1.1	0.3	2.8	0.4	0.9	4.2
G.sacculifer	5.6	4.3	9.2	4.7	7.3	5.3	7.7	5.7	1.9	7.2	3.4	3.6
G. aequilateralis	9.8	6.6	9.7	2.7	4.6	4.2	5.0	4.0	1.1	3.3	0.9	4.2
G. calida	9.5	3.3	2.1	4.3	1.8	1.7	1.1	1.4	4.5	0.9	0.4	1.5
G. bulloides	12.8	7.6	4.2	6.3	4.8	2.0	1.3	5.2	14.6	3.5	5.2	9.0
G. falconensis	6.4	7.7	5.0	9.6	4.0	4.2	1.3	3.5	11.5	1.9	4.3	7.3
G. rubescens	3.6	3.8	2.4	2.2	4.8	1.1	7.4	3.2	3.4	0.7	4.6	5.4
G. quinqueloba	2.5	5.7	6.3	2.0	4.7	0.3	2.1	0.8	4.5	0.7	5.9	2.1
N. pachyderma sin.	0.9	0.0	0.5	0.4	2.1	0.6	1.6	0.0	1.1	0.0	3.2	0.0
N. pachyderma dex.	0.8	1.3	2.4	4.9	3.3	3.7	3.4	1.9	0.6	0.4	2.1	2.5
N. dutertrei	2.2	4.3	8.6	9.8	5.8	7.9	8.5	5.3	6.0	7.8	3.7	6.9
P. obliquiloculata	2.4	2.1	4.7	5.7	6.5	5.9	8.7	4.8	2.6	6.9	6.2	4.4
G. truncatulinoides sin.	0.0	0.0	0.0	0.0	0.0	0.0	0.0	0.0	0.0	0.0	0.0	0.0
G. truncatulinoides dex.	0.0	0.0	0.0	2.5	1.2	0.3	0.0	0.0	0.0	0.0	0.0	0.0
G. menardii	0.4	0.2	1.0	0.4	1.2	3.7	4.2	3.5	5.9	1.5	2.3	4.6
G. glutinata	21.9	35.8	18.1	17.0	17.6	31.2	21.2	35.4	25.7	40.2	38.8	34.7
Other	4.8	7.3	7.9	3.3	7.9	7.0	9.0	6.1	6.6	4.3	13.9	4.6

Table 9 (continued)

Species/**Depth in core** (cm)	390	400	410	420	430	440	450	460	470	480	490	500
O.universa	0.2	3.5	1.6	4.0	0.5	0.5	1.3	0.3	1.6	1.1	0.9	1.4
G. conglobatus	0.0	0.9	2.4	0.0	0.4	0.4	0.8	0.1	2.3	2.2	1.0	0.0
G. ruber pink	0.0	0.0	0.0	0.0	0.0	0.0	0.0	0.0	0.0	0.0	0.0	0.0
G. ruber white	3.6	7.8	27.9	14.7	5.9	5.4	5.1	23.1	5.1	12.7	5.5	13.0
G. tenellus	4.0	0.4	0.5	1.2	7.5	1.4	4.2	3.2	0.7	0.0	2.8	3.0
G.sacculifer	8.0	20.2	8.2	6.9	3.0	1.3	6.8	4.5	9.3	9.0	1.9	2.5
G. aequilateralis	2.9	1.9	2.4	5.8	1.1	3.0	0.8	3.6	4.9	2.3	0.6	0.0
G. calida	1.6	2.2	1.1	1.8	1.6	1.7	1.9	5.5	1.9	3.2	1.3	2.3
G. bulloides	5.3	5.7	1.3	10.1	11.7	10.1	5.1	5.0	8.8	8.3	14.3	11.4
G. falconensis	6.7	0.9	2.2	9.5	7.9	6.1	1.7	3.6	1.2	1.0	8.7	9.3
G. rubescens	3.6	0.7	0.3	0.4	1.9	2.8	0.8	2.5	1.6	0.3	2.2	0.7
G. quinqueloba	6.5	0.9	6.3	1.5	3.0	1.3	3.6	0.7	1.2	0.2	3.7	10.7
N. pachyderma sin.	0.4	0.1	0.5	0.1	2.6	0.7	0.0	0.9	0.2	0.1	0.0	0.5
N. pachyderma dex.	1.4	0.7	2.2	1.5	1.3	2.4	1.3	1.7	4.2	5.6	1.6	0.0
N. dutertrei	4.0	10.6	11.0	9.2	1.7	4.5	3.8	1.6	15.3	14.0	3.7	9.3
P. obliquiloculata	5.1	6.8	7.7	6.5	2.3	3.4	3.6	1.9	8.6	9.1	4.4	9.3
G. truncatulinoides sin.	0.0	0.0	0.0	0.0	0.0	0.0	0.4	0.0	0.0	0.0	0.0	0.0
G. truncatulinoides dex.	0.0	0.0	0.0	0.0	0.0	0.0	0.0	0.0	0.0	0.0	0.0	0.0
G. menardii	4.9	5.3	0.9	6.6	2.6	4.2	5.3	1.3	10.4	9.1	3.7	3.2
G. glutinata	37.5	28.1	14.7	16.2	36.7	43.5	42.4	35.0	15.0	9.9	32.9	17.5
Other	4.3	3.1	9.1	4.1	8.1	7.3	11.2	5.5	7.9	11.9	10.6	5.9

Species/**Depth in core** (cm)	510	520	530	540	550	560	570
O.universa	1.7	0.6	0.2	0.8	0.3	0.9	1.5
G. conglobatus	2.4	0.8	0.5	1.0	0.3	0.6	1.0
G. ruber pink	0.0	0.0	0.0	0.0	1.3	2.7	1.5
G. ruber white	15.9	5.4	12.4	25.4	15.6	23.9	21.4
G. tenellus	0.0	1.9	6.6	3.9	4.0	3.4	3.9
G.sacculifer	10.2	4.0	1.2	5.2	6.1	11.6	8.8
G. aequilateralis	1.4	0.8	1.2	2.9	2.1	5.4	5.2
G. calida	1.7	1.3	1.7	2.5	1.1	1.5	3.1
G. bulloides	7.4	16.0	13.3	4.3	7.1	3.2	2.3
G. falconensis	8.6	11.2	5.1	3.3	4.5	3.1	1.5
G. rubescens	3.5	2.5	5.1	2.7	1.3	2.0	1.5
G. quinqueloba	1.7	2.7	2.4	0.6	0.0	2.0	1.0
N. pachyderma sin.	0.0	0.0	1.0	0.0	0.0	0.3	0.0
N. pachyderma dex.	2.9	0.4	0.9	1.2	2.9	1.1	1.8
N. dutertrei	6.9	4.2	4.4	14.3	13.5	9.6	5.7
P. obliquiloculata	2.6	6.2	3.8	1.7	4.7	3.7	4.4
G. truncatulinoides sin.	0.0	0.0	0.0	0.0	0.0	0.0	0.0
G. truncatulinoides dex.	0.0	0.0	0.0	0.0	1.3	2.0	1.3
G. menardii	2.4	5.8	3.5	3.3	3.4	1.7	0.3
G. glutinata	20.7	32.0	27.2	22.1	26.9	14.8	25.5
Other	9.8	4.2	9.4	4.8	3.7	6.3	8.2

Table 10 Planktic foraminifera distribution in Core S-17666 from the Western Indian Ocean (%)

Species/**Depth in core** (cm)	0	10	20	30	40	50	60	70	80	90	100
G. conglobatus	0.6	4.3	2.3	2.5	2.2	2.3	0.7	2.7	0.0	0.0	0.5
G. ruber	56.2	64.8	45.8	52.3	49.5	39.0	57.3	44.4	69.7	34.3	75.0
G. tenellus	0.6	0.2	2.0	2.0	0.9	0.5	1.3	0.3	1.9	2.9	1.0
G. sacculifer	6.9	19.7	11.1	16.3	11.8	11.8	7.0	13.2	0.6	2.9	4.4
G. aequilateralis	7.7	1.5	4.2	3.5	0.6	2.3	2.0	1.5	0.0	0.0	0.2
G. calida	8.9	0.6	2.6	3.5	1.9	5.8	3.0	2.7	2.3	0.0	2.0
G. rubescens	0.0	0.4	1.6	1.2	1.6	0.3	1.0	1.5	1.9	5.7	2.7
N. pachyderma dex.	0.3	0.7	1.3	1.0	0.9	2.0	1.0	0.6	0.0	5.7	1.5
G. inflata	0.3	0.6	0.3	0.2	4.0	4.3	1.3	4.2	1.6	5.7	1.5
G. truncatulinoides dex.	0.0	0.2	2.3	1.7	2.5	2.8	2.3	4.5	1.9	0.0	1.0
G. menardii	0.0	2.0	5.6	5.2	6.2	11.3	4.3	6.0	4.2	17.1	2.0
G. glutinata	10.3	1.9	10.8	4.0	8.7	6.5	11.3	7.2	7.4	14.3	3.4
Other	8.3	3.2	10.1	6.7	9.0	11.1	7.3	11.1	8.4	11.4	4.9

Species/**Depth in core** (cm)	110	120	130	140	150	160	170	180	190	200	210
G. conglobatus	0.6	0.3	0.6	1.5	0.0	0.0	0.0	0.8	0.0	0.3	0.0
G. ruber	69.2	60.7	75.0	48.8	53.5	52.9	45.3	54.6	51.3	57.0	47.4
G. tenellus	0.4	2.0	0.6	4.5	3.2	2.9	4.7	3.1	2.6	0.3	0.0
G. sacculifer	5.0	4.6	1.8	6.6	3.9	6.9	4.7	10.9	10.5	6.7	2.6
G. aequilateralis	0.0	1.6	1.2	1.2	1.3	1.1	1.0	1.4	1.6	2.3	0.0
G. calida	2.6	1.3	3.4	4.8	1.0	2.4	1.7	2.9	2.9	2.7	2.6
G. rubescens	1.2	2.3	0.6	0.6	2.9	3.3	4.7	2.5	1.3	2.0	2.6
N. pachyderma dex.	2.0	1.3	1.2	0.9	0.3	1.1	2.0	0.8	2.4	2.3	1.3
G. inflata	3.4	2.0	1.2	1.8	1.3	0.9	3.3	1.2	0.8	1.7	1.3
G. truncatulinoides dex.	1.0	0.7	1.5	0.9	1.3	4.4	3.3	1.4	1.6	1.0	0.0
G. menardii	4.6	2.3	2.1	1.8	1.9	4.4	3.7	2.7	3.7	4.4	5.3
G. glutinata	3.4	11.1	5.8	12.9	21.6	12.4	15.0	9.7	13.6	9.7	27.6
Other	6.4	9.8	4.9	13.8	7.7	7.3	10.7	8.0	7.9	9.4	9.2

Species/**Depth in core** (cm)	220	230	240	250	260	270	280	290	300	310	320
G. conglobatus	3.0	0.4	0.3	0.5	0.3	2.0	0.3	0.9	0.8	1.6	1.2
G. ruber	53.3	49.6	45.1	62.7	49.1	57.4	44.1	51.5	59.6	44.2	39.1
G. tenellus	0.7	1.9	4.6	1.8	2.1	1.4	2.4	0.6	4.6	7.7	6.7
G. sacculifer	10.7	11.7	7.6	7.3	5.3	10.5	4.1	5.7	10.2	10.2	5.8
G. aequilateralis	3.7	2.1	0.9	1.0	1.8	2.1	1.2	3.9	2.5	1.6	3.6
G. calida	2.7	1.9	8.5	2.5	4.1	3.4	4.7	4.2	4.2	9.3	5.2
G. rubescens	0.3	4.9	3.0	1.0	2.1	1.2	3.6	0.3	0.8	3.3	3.6
N. pachyderma dex.	1.0	0.2	0.9	1.0	0.3	0.2	0.0	3.3	0.2	0.2	0.6
G. inflata	1.3	0.0	0.0	0.0	0.3	0.0	0.3	0.3	0.0	0.0	0.3
G. truncatulinoides dex.	0.3	0.6	0.9	1.5	3.2	0.4	0.6	0.0	0.0	0.2	1.2
G. menardii	4.3	4.7	2.1	2.3	2.9	1.6	9.2	1.8	0.5	1.4	2.4
G. glutinata	10.3	15.7	20.1	12.3	20.6	11.9	18.6	17.7	12.2	12.3	17.6
Other	8.3	6.4	5.8	6.0	7.9	7.8	10.9	9.9	4.6	7.9	12.7

Table 10 (continued)

Species/**Depth in core** (cm)	330	340	350	360	370	380	390	400	410	420	430
G. conglobatus	0.8	0.7	1.1	0.0	0.1	0.5	0.0	0.8	1.3	0.6	1.0
G. ruber	34.7	29.4	46.6	39.2	37.8	40.4	38.7	36.8	44.1	40.4	31.6
G. tenellus	8.2	6.6	4.6	4.8	6.8	4.1	1.1	5.9	3.7	0.6	6.1
G. sacculifer	4.5	8.0	7.9	5.1	8.9	14.2	14.0	12.6	10.6	9.9	12.9
G. aequilateralis	2.5	2.7	0.9	0.7	3.0	4.9	3.0	2.0	2.4	2.9	2.0
G. calida	4.0	6.6	1.8	4.0	6.1	3.3	5.2	4.5	4.0	5.8	10.2
G. rubescens	3.1	3.9	0.7	1.1	1.7	0.5	0.7	1.4	1.6	1.9	1.4
N. pachyderma dex.	1.4	1.7	0.7	0.7	1.1	1.4	1.5	2.8	0.8	2.2	0.3
G. inflata	0.3	0.5	1.1	0.0	1.0	0.8	1.1	0.6	1.8	2.2	0.0
G. truncatulinoides dex.	2.5	2.7	4.2	4.4	3.8	3.3	1.8	2.5	1.6	1.6	2.7
G. menardii	6.8	3.9	5.7	7.7	3.1	8.5	12.5	5.6	3.7	4.5	4.8
G. glutinata	16.1	18.0	11.4	16.5	15.0	7.7	8.5	14.3	14.5	14.7	16.0
Other	15.0	15.5	13.6	15.8	11.5	10.4	11.8	10.1	10.0	12.5	10.9

Species/**Depth in core** (cm)	440	450	460	470	480	490	500	510	520	530	540
G. conglobatus	1.8	0.0	1.0	0.0	1.8	0.3	1.9	0.7	1.3	2.4	2.3
G. ruber	47.6	38.5	34.5	37.9	46.5	54.9	49.4	54.5	28.9	38.9	35.7
G. tenellus	2.5	6.2	0.0	6.8	2.6	1.3	0.6	1.3	1.0	0.3	0.3
G. sacculifer	9.3	10.5	8.8	2.0	17.0	8.2	17.2	9.6	10.0	6.6	6.4
G. aequilateralis	1.3	11.4	7.2	2.0	1.8	4.2	2.2	1.7	3.3	1.7	1.6
G. calida	8.3	2.5	2.0	0.3	1.3	1.0	0.3	0.7	0.5	0.3	0.0
G. rubescens	0.3	0.6	1.0	5.1	1.5	0.3	0.6	0.7	0.3	0.3	0.0
N. pachyderma dex.	1.0	0.9	2.0	2.0	0.9	0.7	2.9	4.3	6.6	2.1	3.5
G. inflata	1.8	0.0	1.3	0.3	0.2	0.3	0.6	2.0	14.8	12.8	20.6
G. truncatulinoides dex.	2.0	0.0	0.3	3.8	4.2	2.9	2.9	0.3	1.8	4.2	1.9
G. menardii	3.0	5.2	9.1	6.5	1.8	2.6	4.1	6.3	10.7	10.4	10.6
G. glutinata	10.1	11.4	14.3	17.1	7.9	13.1	4.8	10.3	7.7	4.5	4.8
Other	11.1	12.9	18.6	16.0	12.6	10.1	12.4	7.6	13.0	15.3	12.2

Species/**Depth in core** (cm)	550	560	570	580	590	600	610
G. conglobatus	5.6	2.4	0.3	0.0	1.4	0.0	0.6
G. ruber	29.9	39.5	67.3	66.7	46.4	21.7	62.2
G. tenellus	0.0	0.3	0.3	0.0	0.3	0.0	1.8
G. sacculifer	1.9	10.2	2.4	16.7	13.6	8.7	9.6
G. aequilateralis	2.2	1.0	1.0	0.0	4.7	8.7	2.7
G. calida	0.3	1.4	0.0	0.0	1.7	17.4	2.1
G. rubescens	0.6	1.0	0.7	0.0	0.7	0.0	0.6
N. pachyderma dex.	0.3	1.0	1.7	0.0	1.7	0.0	0.0
G. inflata	33.0	10.2	9.1	0.0	9.5	0.0	0.3
G. truncatulinoides dex.	0.3	0.3	0.3	0.0	1.4	4.3	0.3
G. menardii	9.3	6.8	5.1	0.0	5.4	13.0	1.2
G. glutinata	4.9	14.3	6.4	0.0	2.7	0.0	9.3
Other	11.7	11.6	5.4	16.7	10.5	26.1	9.3

Table 11 Planktic foraminifera distribution in Core S-17954 from the South China Sea (%)

Species/**Depth in core** (cm)	5	15	25	35	45	55	65	75	85
G. ruber	18.4	25.0	32.8	21.8	18.4	22.8	27.7	24.2	18.8
G. sacculifer	18.6	23.0	21.3	22.0	24.7	19.4	22.6	16.9	18.8
G. aequilateralis	5.3	8.2	8.9	6.8	6.6	4.9	2.7	7.2	7.5
G. tenellus	0.0	1.3	0.6	0.5	1.3	1.7	1.4	3.0	1.8
G. rubescens	1.8	0.3	0.6	0.7	0.3	1.3	1.7	2.1	1.2
G. bulloides	2.8	2.0	1.8	1.8	1.3	3.6	1.0	2.1	2.7
N. pachyderma dex.	1.4	0.7	1.8	2.0	1.0	1.9	6.1	9.1	14.3
N. dutertrei	16.1	10.9	9.8	11.3	12.5	7.4	7.8	5.7	8.7
P. obliquiloculata	5.5	4.3	3.6	7.9	5.6	6.2	7.4	5.9	3.0
G. inflata	0.5	0.0	0.0	0.2	0.0	0.4	0.7	0.8	2.7
G. menardii	10.8	7.6	6.2	6.1	5.6	3.4	3.4	2.1	1.8
G. glutinata	7.4	5.3	6.5	9.5	14.5	10.9	7.1	11.6	5.4
G. truncatulinoides dex.	1.6	0.3	0.0	0.5	1.0	2.1	2.7	0.9	1.8
G. crassaformis	0.0	0.0	0.3	0.0	0.0	0.0	0.3	0.6	0.3
Other	9.9	11.2	5.9	8.8	7.2	14.0	7.4	8.0	11.3

Species/**Depth in core** (cm)	95	105	115	125	135	145	155	165	175
G. ruber	22.9	28.9	28.1	24.1	18.9	28.5	27.2	29.0	30.5
G. sacculifer	12.9	8.1	10.0	10.7	11.5	6.5	6.3	7.8	10.7
G. aequilateralis	7.9	7.7	7.2	7.6	8.1	7.6	6.8	6.8	6.5
G. tenellus	2.6	3.7	0.9	1.5	0.7	1.1	0.5	1.3	1.4
G. rubescens	1.3	2.7	1.9	4.9	0.5	2.1	2.2	1.5	3.1
G. bulloides	6.1	7.0	6.5	7.6	3.8	7.0	9.2	3.4	2.5
N. pachyderma dex.	14.2	15.4	13.7	7.9	14.8	17.1	14.4	11.2	12.1
N. dutertrei	10.7	5.7	11.2	7.3	15.6	6.1	4.1	5.7	9.0
P. obliquiloculata	3.3	2.0	1.9	4.6	6.7	4.2	4.9	5.5	5.1
G. inflata	2.4	1.0	0.2	1.5	0.0	0.2	1.1	1.3	1.4
G. menardii	2.2	2.0	3.3	3.4	1.4	2.5	2.4	2.1	0.6
G. glutinata	4.1	6.4	6.3	6.4	4.5	5.1	7.3	9.1	4.5
G. truncatulinoides dex.	0.2	0.0	0.2	0.6	0.5	0.2	0.0	0.0	0.0
G. crassaformis	0.4	1.0	0.7	1.8	4.8	2.3	0.3	1.3	3.7
Other	8.7	8.4	7.9	10.1	8.1	9.5	13.3	14.0	8.8

Depth in core (cm)	185	195	205	215	225	235	245	255	265
G. ruber	23.0	28.4	27.1	25.0	28.1	28.4	29.8	24.8	28.2
G. sacculifer	6.6	4.5	3.7	7.9	7.4	6.3	8.3	7.6	6.6
G. aequilateralis	6.1	5.8	5.9	5.4	7.8	6.1	5.7	5.3	4.1
G. tenellus	0.5	0.6	0.7	1.6	1.9	2.3	4.3	0.7	2.4
G. rubescens	2.8	3.2	3.7	1.6	4.3	4.0	1.7	1.7	2.4
G. bulloides	7.6	5.1	4.4	3.8	3.0	4.5	2.6	3.6	3.6
N. pachyderma dex.	13.7	15.3	16.3	10.1	13.2	12.7	12.9	15.2	15.0
N. dutertrei	5.1	9.3	5.9	6.3	3.0	5.6	5.7	6.6	7.3
P. obliquiloculata	10.4	5.1	7.6	12.0	5.0	3.8	7.6	7.3	5.8
G. inflata	2.0	2.9	5.7	5.4	2.2	1.6	1.7	3.3	1.5
G. menardii	5.6	3.8	3.4	3.2	0.9	4.0	3.1	2.0	3.4
G. glutinata	5.1	8.0	6.2	7.0	10.8	7.7	8.3	8.6	9.8
G. truncatulinoides dex.	0.0	0.0	0.0	0.0	0.0	0.0	0.0	0.3	0.0
G. crassaformis	2.5	2.9	2.0	3.5	1.9	2.3	2.9	6.3	3.6
Other	9.1	5.1	7.4	7.3	10.4	10.6	5.5	6.6	6.4

Table 11 (continued)

Species/**Depth in core** (cm)	275	285	295	305	315	325	335	345	355
G. ruber	28.8	22.8	24.7	24.5	30.1	28.0	27.7	28.3	30.1
G. sacculifer	5.8	9.1	6.9	8.5	7.2	7.8	10.5	13.5	9.4
G. aequilateralis	5.5	6.7	5.0	5.9	6.5	8.7	6.7	7.4	5.5
G. tenellus	3.3	0.7	1.1	2.1	1.2	1.7	2.0	0.6	2.0
G. rubescens	2.4	1.0	5.3	1.4	3.7	2.2	1.5	1.6	1.7
G. bulloides	2.1	3.0	2.2	4.3	4.2	3.4	2.3	4.2	3.1
N. pachyderma dex.	10.3	14.8	15.8	13.7	11.0	4.8	9.6	7.1	9.4
N. dutertrei	9.4	14.8	6.9	10.1	7.7	10.1	10.8	7.7	9.4
P. obliquiloculata	8.8	7.4	6.4	6.6	5.1	5.3	9.9	6.8	6.6
G. inflata	0.6	1.3	2.5	1.6	2.8	2.2	2.3	4.2	2.2
G. menardii	4.5	5.7	4.2	2.3	3.3	5.9	6.4	4.5	2.8
G. glutinata	8.2	5.4	9.7	8.2	9.1	9.8	4.7	6.4	7.6
G. truncatulinoides dex.	0.0	0.0	0.3	0.0	0.0	0.0	0.0	0.0	0.0
G. crassaformis	3.3	1.0	3.1	4.3	3.7	3.4	0.9	2.6	3.1
Other	7.0	6.4	5.8	6.4	4.4	6.7	4.7	5.1	7.2

Species/**Depth in core** (cm)	365	375	385	395	405	415	425	435	445
G. ruber	24.7	28.2	24.4	26.8	28.4	25.3	29.1	18.2	20.4
G. sacculifer	8.7	10.3	7.7	5.4	6.7	4.4	5.0	6.4	13.3
G. aequilateralis	8.0	10.6	5.7	7.7	4.0	4.8	5.0	4.1	3.8
G. tenellus	1.7	0.3	1.5	2.0	1.1	1.7	1.1	2.4	1.2
G. rubescens	0.7	0.5	1.2	1.7	3.8	3.7	2.8	3.7	0.6
G. bulloides	2.3	1.9	3.0	2.3	4.9	4.8	2.0	3.0	1.8
N. pachyderma dex.	9.3	6.5	5.4	6.0	12.0	9.4	12.3	16.6	11.5
N. dutertrei	12.0	7.6	10.1	10.1	7.8	8.7	7.8	11.5	11.8
P. obliquiloculata	5.0	7.0	7.9	7.4	6.7	10.0	8.1	6.1	5.3
G. inflata	3.7	4.3	4.7	3.0	5.3	6.5	5.3	1.4	5.3
G. menardii	6.7	4.1	9.9	7.7	3.5	2.2	2.0	5.1	5.9
G. glutinata	5.3	3.8	6.9	3.4	4.4	5.7	5.9	5.1	5.3
G. truncatulinoides dex.	0.0	0.0	0.2	0.3	0.2	0.2	0.0	0.0	0.0
G. crassaformis	7.3	6.2	6.4	7.7	4.0	4.8	6.4	10.5	7.1
Other	4.7	8.7	4.9	8.4	7.3	7.8	7.0	6.1	6.8

Species/**Depth in core** (cm)	455	465	475	485	495	505	515	525	535
G. ruber	25.8	19.9	28.9	26.2	22.7	16.4	24.2	19.2	17.8
G. sacculifer	10.0	6.9	8.5	6.6	11.7	14.9	12.2	15.7	9.3
G. aequilateralis	7.6	6.2	8.5	6.0	6.4	4.0	4.8	4.1	6.3
G. tenellus	1.1	1.3	1.2	2.0	0.7	1.3	0.8	0.9	1.9
G. rubescens	1.8	1.6	1.2	1.0	1.2	2.8	2.0	2.8	6.8
G. bulloides	2.4	5.2	1.5	3.0	0.5	0.5	3.1	2.8	3.0
N. pachyderma dex.	2.6	4.2	3.0	1.7	4.6	2.8	5.9	5.0	4.1
N. dutertrei	12.6	12.7	14.9	10.9	15.2	13.6	10.5	12.6	10.4
P. obliquiloculata	5.8	9.2	4.6	3.6	7.8	6.6	4.3	9.7	7.9
G. inflata	5.3	5.6	4.0	4.3	3.4	6.3	2.0	1.9	3.6
G. menardii	7.6	8.5	6.4	7.3	7.3	7.3	6.1	7.5	8.7
G. glutinata	6.3	9.8	7.0	13.2	9.0	8.1	10.2	5.0	6.8
G. truncatulinoides dex.	0.0	0.0	0.6	0.3	1.2	2.3	3.3	3.1	0.0
G. crassaformis	1.1	1.3	3.6	1.7	2.9	3.5	4.6	1.6	4.9
Other	10.0	7.5	6.1	12.3	5.1	9.6	5.9	7.9	8.5

Table 11 (continued)

Depth in core (cm)	545	555	565	575	585	595	605	615	625
G. ruber	24.0	21.9	16.0	17.6	15.2	20.3	15.3	22.4	22.3
G. sacculifer	8.4	9.9	6.5	6.7	7.9	10.8	16.7	11.4	15.5
G. aequilateralis	2.4	3.1	3.0	2.9	1.7	2.1	4.9	5.1	3.0
G. tenellus	4.2	1.9	1.2	1.3	2.6	0.8	2.0	3.7	1.6
G. rubescens	6.6	2.2	1.2	0.6	0.7	0.8	1.6	1.7	0.5
G. bulloides	3.6	3.4	3.9	2.2	3.0	1.0	0.8	2.3	0.5
G. menardii	9.0	8.6	8.4	7.4	10.2	6.9	7.7	6.5	9.3
G. glutinata	4.5	4.0	5.1	2.9	5.0	4.4	4.1	2.6	3.2
G. truncatulinoides dex.	0.0	0.6	1.2	0.6	0.0	0.0	0.0	0.6	0.0
G. crassaformis	3.9	4.0	5.6	3.8	1.7	2.6	0.6	0.6	0.5
Other	7.8	7.1	5.3	9.9	7.3	8.2	13.6	11.9	13.9

Species/**Depth in core** (cm)	635	645	655	665	675	685	695	705	715
G. ruber	18.5	16.8	18.4	15.3	14.8	19.5	22.1	16.9	18.3
G. sacculifer	12.1	14.4	12.5	12.0	15.4	14.1	6.4	6.8	6.3
G. aequilateralis	5.9	4.1	6.3	6.0	5.0	4.7	3.6	2.7	5.1
G. tenellus	1.0	3.3	0.9	0.5	0.7	1.3	0.0	1.6	0.3
G. rubescens	3.8	3.5	0.9	1.9	1.7	1.3	0.8	3.0	1.5
G. bulloides	1.0	0.5	1.6	3.0	2.3	3.7	3.4	3.9	4.5
N. pachyderma dex.	9.5	6.3	6.6	9.0	6.4	6.1	3.4	2.1	2.4
N. dutertrei	7.9	13.6	19.4	16.1	16.8	16.5	22.9	20.3	21.0
P. obliquiloculata	7.9	7.6	5.3	6.0	9.7	7.4	3.4	10.3	7.2
G. inflata	2.1	3.8	3.4	1.6	4.4	1.0	0.0	0.5	0.9
G. menardii	7.7	2.4	3.4	5.7	5.4	5.7	9.2	11.6	11.7
G. glutinata	9.5	6.0	5.3	4.9	4.4	4.7	10.3	5.0	6.9
G. truncatulinoides dex.	0.3	0.0	0.3	1.4	3.7	0.0	0.0	0.0	0.0
G. crassaformis	0.5	1.6	0.3	0.5	1.7	3.0	4.7	4.1	2.1
Other	12.3	16.0	15.3	15.8	7.7	10.8	9.8	11.4	12.0

Depth in core (cm)	725	735	745	755	765	775	785	795	805
G. ruber	20.6	18.5	20.0	19.2	25.1	20.7	21.7	19.7	18.7
G. sacculifer	7.6	11.0	15.3	14.9	18.5	20.2	17.8	14.8	20.1
G. aequilateralis	5.7	5.6	6.8	5.6	5.2	4.3	5.1	7.9	5.0
G. tenellus	1.1	0.8	1.6	1.0	1.7	1.2	1.0	5.6	1.0
G. rubescens	2.5	1.4	1.0	1.3	0.0	0.7	0.2	1.6	2.2
G. bulloides	4.0	1.7	1.8	2.3	1.1	2.7	4.1	0.7	2.6
N. pachyderma dex.	2.9	6.2	2.1	2.3	1.9	2.9	3.4	4.3	4.8
N. dutertrei	20.8	18.3	14.8	18.2	9.9	8.9	8.2	7.9	11.5
P. obliquiloculata	5.5	5.6	4.7	6.6	4.4	4.6	4.6	5.3	3.6
G. inflata	1.5	0.6	0.0	0.3	0.3	0.0	1.0	0.7	2.9
G. menardii	9.1	7.3	6.5	7.3	6.6	10.8	8.2	6.6	4.1
G. glutinata	4.6	8.4	9.4	5.3	13.5	12.0	13.5	10.2	6.7
G. truncatulinoides dex.	0.0	0.0	3.9	1.0	0.6	0.0	0.7	1.0	1.4
G. crassaformis	2.5	2.8	3.4	3.6	3.3	3.8	2.2	1.6	0.2
Other	11.6	11.8	8.8	10.9	7.7	7.2	8.4	12.2	15.1

Table 11 (continued)

Species/**Depth in core** (cm)	815	825	835	845	855	865	875	885	895
G. ruber	20.9	21.6	25.3	29.5	29.0	21.9	24.9	22.0	18.0
G. sacculifer	18.3	15.4	12.3	11.9	10.9	5.2	7.6	12.8	15.7
G. aequilateralis	7.1	5.9	9.9	8.1	9.4	6.3	8.3	5.7	5.0
G. tenellus	1.0	3.1	3.2	1.9	1.5	1.4	2.7	3.0	1.1
G. rubescens	1.3	2.2	1.1	0.5	1.5	2.2	1.0	3.6	2.3
G. bulloides	1.3	2.8	2.4	1.7	1.2	2.5	1.0	0.6	3.0
N. pachyderma dex.	3.4	2.2	3.2	1.9	2.1	11.8	8.0	8.9	5.2
N. dutertrei	15.4	14.5	15.5	21.6	20.8	13.7	13.0	11.9	15.2
P. obliquiloculata	5.2	5.6	5.6	0.7	0.9	7.1	5.0	1.8	2.0
G. inflata	1.0	2.2	2.1	1.7	2.4	2.2	4.0	2.1	2.0
G. menardii	3.4	0.6	0.5	1.0	0.9	1.6	3.3	4.5	6.4
G. glutinata	5.5	4.9	5.1	5.5	3.9	5.5	8.0	5.4	5.5
G. truncatulinoides dex.	2.1	0.3	0.0	0.0	0.0	0.0	0.0	0.0	0.0
G. crassaformis	1.3	5.9	3.7	1.7	2.7	4.9	2.3	0.9	4.5
Other	12.6	13.0	10.1	12.6	12.7	13.7	11.0	17.0	14.1

Species/**Depth in core** (cm)	905	915	925	935	945	955	965	975	985
G. ruber	17.1	20.9	26.8	27.1	14.9	23.8	24.5	23.2	27.1
G. sacculifer	15.9	14.4	16.1	6.4	14.9	12.0	12.1	10.5	15.6
G. aequilateralis	5.4	5.0	3.3	5.7	8.6	6.9	5.5	8.6	3.1
G. tenellus	2.1	2.2	2.0	3.3	1.7	2.0	1.7	1.3	1.4
G. rubescens	3.0	1.4	2.0	1.3	1.3	2.0	2.1	1.6	0.0
G. bulloides	1.5	4.1	1.3	2.7	6.0	2.0	2.5	1.6	0.7
N. pachyderma dex.	9.6	4.8	1.3	1.0	1.7	4.9	9.1	14.0	9.5
N. dutertrei	12.3	14.4	7.7	9.4	8.6	12.9	13.9	11.5	13.9
P. obliquiloculata	5.1	4.8	6.4	8.4	6.3	2.0	0.9	4.1	6.8
G. inflata	2.1	2.9	3.0	2.3	2.6	4.9	0.6	1.0	0.0
G. menardii	5.7	2.9	8.4	4.0	4.6	3.7	1.9	5.1	2.0
G. glutinata	5.4	6.7	7.0	8.4	5.3	6.3	7.4	2.2	2.0
G. truncatulinoides dex.	0.3	0.0	0.0	0.0	0.3	0.0	0.2	3.8	0.0
G. crassaformis	3.9	3.6	3.7	4.3	5.6	3.2	2.8	1.0	4.4
Other	10.5	11.8	11.0	15.7	17.5	13.5	14.8	10.5	13.6

Species/**Depth in core** (cm)	995	1005	1015	1025	1035	1045	1055	1065	1075
G. ruber	26.8	19.3	15.2	20.6	15.3	14.5	26.1	23.6	23.4
G. sacculifer	10.6	13.9	13.9	9.5	12.5	6.5	3.5	5.8	13.7
G. aequilateralis	6.5	6.1	7.1	7.4	7.2	3.8	5.7	6.4	3.2
G. tenellus	1.5	1.0	0.0	1.7	2.2	3.6	2.4	0.3	0.5
G. rubescens	0.0	1.7	2.0	0.7	0.6	0.3	1.3	0.3	0.7
G. bulloides	1.7	2.4	2.0	2.4	3.9	2.7	5.7	0.6	2.8
N. pachyderma dex.	9.1	11.1	12.8	9.8	13.3	28.4	4.3	1.9	0.2
N. dutertrei	15.3	16.9	10.5	17.6	13.9	12.1	11.6	12.8	13.0
P. obliquiloculata	5.0	5.4	10.8	9.5	7.8	2.1	3.5	6.1	7.4
G. inflata	0.4	0.0	2.0	0.0	0.3	0.0	0.3	4.5	2.6
G. menardii	3.5	5.1	8.4	4.7	5.3	3.3	8.1	10.9	5.8
G. glutinata	5.2	3.4	4.7	4.4	4.2	4.4	3.2	4.8	8.1
G. truncatulinoides dex.	0.0	0.7	0.7	0.0	0.0	3.8	8.9	9.9	7.9
G. crassaformis	3.7	1.0	0.3	2.4	1.7	1.8	2.2	4.2	2.3
Other	10.8	12.2	9.5	9.5	11.9	12.7	13.2	8.0	8.4

Abbreviations

accelerator mass spectrometry (AMS)
artificial neural networks technique (ANN)
Antarctic Bottom Water (AABW)
Antarctic Circumpolar Current (ACC)
Antarctic Intermediate Water (AAIW)
Arctic Oscillation (AO)
Atlantic meridional overturning circulation (AMOC)
B/A – Bølling/Allerød
cal. ka – calendar kilo-years
Carbonate Compensation Depth (CCD)
Dansgaard-Oeschger events/cycles (DO cycles)
eastern equatorial Pacific (EEP)
El Niño-Southern Oscillation (ENSO)
Foraminiferal productivity index (FP index, portion of *G. bulloides*+ *G. glutinata*+*N. dutretrei* in foraminiferal assemblage)
Glacial North Atlantic Intermediate Water (GNAIW)
Heinrich events (H-1, H-2, etc.)
ice rafted debris (IRD)
International Marine Past Global Change Study (IMAGES)
Indonesian Throughflow (ITF)
intertropical convergence zone (ITCZ)
ka – kilo-years
Last Glacial Maximum (LGM)
Ma – million years
Multiproxy Approach for the Reconstruction of the Glacial Ocean surface (MARGO)
Marine Isotope Stage (MIS)
Modern Analog Technique (MAT)
North Atlantic Deep Water (NADW)
North Atlantic Oscillation (NAO)
North Pacific Deep Water (NPDW)
North Pacific Intermediate Water (NPIW)
Older Dryas (OD)

PACE WG (Working Group on Reconstruction of Past Ocean Circulation)
Pacific Decadal Oscillation (PDO)
PDB standard (refers to the Cretaceous belemnite formation at Peedee in South Carolina, USA)
primary production (PP)
Revised Modern Analogue Technique (RAM)
Sea surface salinity (SSS)
Sea surface temperature (SST)
South China Sea (SCS)
Southern Oscillation Index (SOI)
standard oxygen isotope record (SPECMAP)
total organic carbon (TOC)
Tropical Ocean and Global Atmosphere Program (TOGA)
thermohaline circulation (THC)
Younger Dryas (YD)
western equatorial Pacific (WEP)
Western Pacific Warm Pool (WPWP)

References

Aagaard K, Carmack C (1989) The role of sea ice and other fresh water in the Arctic circulation. J. Geophys. Res. 94:14,485–14,498

Aagaard K, Coachman LK, Carmack C (1981) On the halocline of the Arctic Ocean. Deep-Sea Res. 28A(6):529–545

Aagaard K, Swift JH, Carmack EC (1985) Thermohaline circulation in the Arctic Mediterranean Seas. J. Geophys. Res. 90:4833–4846

Aagaard K, Fanrbach E, Meincke J, Swift JH (1991) Saline outflow from the Arctic Ocean: Its contribution to the deep waters of the Greenland, Norwegian and Iceland Seas. J. Geophys. Res. 96(C11):20,433–20,441

Adams CG, Gentry AW, Whybrow PJ (1983) Dating the terminal Tethyan event. Utrecht. Micropal. Bull. 30:273–293

Adkins JF, Boyle EA (1997) Changing atmospheric $\Delta^{14}C$ and the record of deep water paleoventilation ages. Paleoceanography 12:337–344

Adkins JF, Boyle EA, Keigwin L, Cortijo E (1997) Variability of the North Atlantic thermohaline circulation during the last interglacial period. Nature 390:154–156

Adkins JF, McIntyre K, Schrag DP (2002) The salinity, temperature, and $\delta^{18}O$ of the glacial deep ocean. Science 298:1769–1773

Adkins JF, Ingersoll AP, Pasquero C (2005) Rapid climate change and conditional instability of the glacial deep ocean from the thermobaric effect and geothermal heating. Quat. Sci. Rev. 24, doi:10.1016/j.quascirev.2004.11.2005:581–594

Aibulatov NA, Matyushenko VA, Shevchenko VP, Politova, NV, Potekhina EM (1999) New data on the transverse structure of lateral suspended matter fluxes in the Barents Sea periphery. Geoekologiya 6:526–540 (in Russian)

Akhrimenko NO, Ivanova EV (2005) Stratigraphy of postglacial sediments from the south-western branch of the Franz-Victoria Trough (Barents Sea) by foraminiferal assemblages. Abstracts of XVI International School on Marine Geology, V1:14–15. Shirshov Institute of Oceanology, RAS, Moscow (in Russian)

Alexander MA, Bladé I, Newman M, Lanzante JR, Lau N-C, Scott J (2002) The atmospheric bridge: the influence of ENSO teleconnections on air-sea interaction over the global oceans. J. Climate 15(16):2205–2231

Alexanderson H, Hjort Ch, Möller P, Antonov O, Pavlov M (2001) The North Taymyr Ice-marginal Zone, Arctic Siberia—a preliminary overview and dating. Glob. Planet Change 31:427–445

Allen MB, Armstrong HA (2008) Arabia–Eurasia collision and the forcing of mid-Cenozoic global cooling. Palaeogeogr., Palaeoclimatol., Palaeoecol. 265:52–58

Alley RB (1998) Icing the North Atlantic. Nature 392:335–337

Alley RB, Meese DA, Shuman CA, Gow AJ, Taylor KL, Grootes PM, White JWC, Ram M, Waddington ED, Mayewski PA, Zielinski GA (1993) Abrupt increase in Greenland snow accumulation at the end of the Younger Dryas event. Nature 362:527–529

Alley RB, Mayewski PA, Sowers T, Stuiver M, Taylor KC, Clark PU (1997) Holocene climatic instability: a prominent, widespread event 8200 yr ago. Geology 25:483–486

Alley RB, Anandakrishnan S, Jung P (2001) Stochastic Resonance in the North Atlantic. Paleoceanography 16(2):190–198

Almogi-Labin A, Hemleben C, Meischner D (1998) Carbonate preservation and climatic changes in the central Red Sea during the last 380 kyr as recorded by pteropods. Mar. Micropal. 33:87–107

Almogi-Labin A, Schmiedl G, Hemleben C, Siman-Tov R, Segl M, Meischner D (2000) The influence of the NE winter monsoon on productivity changes in the Gulf of Aden, NW Arabian Sea, during the last 530 kyr as recorded by foraminifera. Mar. Micropal. 40(3):295–319

Altabet MA, Higginson MJ, Murray DW (2002) The effect of millennial-scalechanges in Arabian Sea denitrification on atmospheric CO_2. Nature 415:159–162

Alve E, Murray JW (1995) Experiments to determine the origin and paleoenviromental significance of agglutinated foraminiferal assemblages. In: Kaminski MA, Geroch S, Gasinski MA (eds) Proceedings of the Fouth International Workshop on Agglutinated Foraminifera. Grzybowski Foundation

An Z, Kukla G, Porter SC, Xiao X (1991) Paleomonsoons of China over the last 130,000 years – paleomonsoon variation. Sci. China (B) 34:1016–1024

Anderson DM, Prell WL (1991) Coastal upwelling gradient during the Late Pleistocene. Proc. ODP. Sci. Res. 117:265–276

Anderson DM, Prell WL (1993) A 300 kyr record of upwelling off Oman during the Late Quaternary: evidence of the Asian southwest monsoon. Paleoceanography 8:193–208

Andreasen DG, Ravelo AC (1997) Tropical Pacific Ocean thermocline depth reconstructions for the last glacial maximum. Paleoceanography 12:395–413

Andersen C, Koc N, Jennings AE, Andrews JT (2004) Nonuniform response of the major surface currents in the Nordic Seas to insolation forcing: implications for the Holocene climate variability. Paleoceanography 19, PA200310.1029/2002PA000873

Andreassen K, Laberg JS, Vorren TO (2008) Seafloor geomorphology of the SW Barents Sea and its glaci-dynamic implications. Geomorphology 97(1–2):157–177

Andreev A, Klimanov VA (2000) Quantitative Holocene climatic reconstruction from Arctic Russia. J. Paleolimnology 24:81–91

Anisimov MV, Ivanov Ju A, Subbotina MM (2002) The global oceanic conveyor. Oceanology 42:1–5 (English translation)

Antoine D, Andre J-M, Morel A (1996) Oceanic primary production 2. Estimation at global scale from satellite (coastal zone color scanner) chlorophyll. Glob. Biogeochem. Cycles 10(1):57–69

Arnone RA, Gould RW, Kindle J, Martinolich P, Brink K (1998) Remote sensing of coastal upwelling and filaments off the coast of Oman. Oceanography 11(2):33

Arz HW, Patzold J, Wefer G (1998) Correlated millennial scale changes in surface hydrography and terrigenous sediment yield inferred from last-glacial marine deposits of northeastern Brazil. Quat. Res. 50:157–166

Arz HW, Lamy F, Ganopolski A, Nowaczyk N, Pätzold J (2007) Dominant Northern Hemisphere climate control over millennial-scale glacial sea-level variability. Quat. Sci. Rev. (26):312–321

Ashkenazy Y, Tziperman E (2006) Scenarios regarding the lead of equatorial sea surface temperature over global ice volume. Paleoceanography 21, PA2006 doi10.1029/2005PA001232

Austin WEN, Kroon D (2001) Deep sea ventilation of the northeastern Atlantic during the last 15,000 years. Glob. Planet Change 30:13–31

Banerjee SK (1995) Chasing the paleomonsoon over China: Its magnetic record. GSA Today 5:93–97

Barash MS (1964) Stratigraphy and foraminifera in North Atlantic sediments. MOIP Bulletin, Series of Geology 6:148–156 (in Russian)

Barash MS (1988) Quaternary Paleooceanology of the Atlantic Ocean. Nauka, Moscow (in Russian)

Barash MS, Blyum NS (1974) Distribution of planktonic foraminiferal species in the North Atlantic sediments depending on water temperature. In: Micropaleontology of seas and oceans, pp. 138–149. Nauka, Moscow (in Russian)

Barash MS, Yushina IG (1999a) Reconstruction of Quaternary paleooceanological parameters: new approaches. Oceanology 39:270–280 (in Russian)

Barash MS, Yushina IG (1999b) Reconstruction of the Quaternary North Atlantic paleohydrological variability by means of planktic foraminifera data (method of factor analysis and spline interpolation). Rep. Pol. Res. 306:5–34

Barash MS, Bulatov RP, Devdariani AS (1974) North Atlantic at the end of the glaciation. Oceanology 14:112–117 (in Russian)

Barash MS, Oskina NS, Ivanova EV (1980) Upwelling off Africa during the Late Pleistocene from planktonic foraminiferal data. Oceanology 20:62–68 (English Translation)

Barash MS et al. (1989) Neogene-Quaternary Paleooceanology by Micropaleontological Data. Nauka, Moscow (in Russian)

Barash MS, Yushina IG, Shpilhagen PX (2002) Reconstruction of quaternary paleogydrological variability by planctonic foraminifera (North Atlantic, Reykjanes ridge). Oceanology 42:744–756 (in Russian)

Barber DC, Dyke A, Hilaire-Marcel C, Jennings AE, Andrews JT, Kerwin MW, Bilodeau G, McNeely R, Southon J, Morehead MD, Gagnon JM (1999) Forcing of the cold event of 8,200 years ago by catastrophic drainage of Laurentide lakes. Nature 400:344–348

Barbieri R, Benjamini C, Monechi S, Reale V (2003) Stratigraphy and benthic foraminiferal events across the Middle–Late Eocene transition in the Western Negev, Israel. In: Prothero DR, Ivany LC, Nesbitt EA (eds) From Greenhouse to Icehouse: The Marine Eocene–Oligocene Transition. Columbia University Press, New York, pp. 453–471

Bard E (2002) Climate shock: abrupt changes over Millennial time scales. Physics Today 55:32–38

Bard E, Hamelin B, Fairbanks RG, Zindler A (1990) Calibration of the ^{14}C timescale over the past 30,000 years using mass spectrometric U-Th ages from Barbados corals. Nature 345:405–410

Bard E, Rostek F, Sonzogni C (1997) Interhemispheric synchrony of the last deglaciation inferred from alkenone palaeothermometry. Nature 385:707–710

Bard E, Ardold M, Hamelin B, Tismerat-Iabore N (1998) Comparison of U/Th and ^{14}C ages measured on corals: an update based on new samples from Pacific islands. Radiocarbon 40:1085–1092

Bard E, Rostek F, Turon J-L, Gendreau S (2000) Hydrological impact of Heinrich events in the subtropical Northeast Atlantic. Science 289:1321–1324

Barker S, Cacho I, Benway H, Tachikawa K (2005) Planktonic foraminiferal Mg/Ca as a proxy for past oceanic temperatures: a methodological overview and data compilation for the Last Glacial Maximum.Quat. Sci. Rev. 24:821–834

Barrows TT, Juggins S (2005) Sea-surface temperatures around the Australian margin and Indian Ocean during the Last Glacial Maximum. Quat. Sci. Rev. 24:1017–1047, doi:10.1016/j.quascirev.2004.07.020

Bartolacci DM, Luther ME (1999) Patterns of co-variability between physical and biological parameters in the Arabian Sea. Deep-Sea Res. 46(8–9):1933–1964

Bauch D, Bauch HA (2001) Last glacial benthic foraminiferal $\delta^{18}O$ anomalies in the polar North Atlantic: A modern analogue evaluation. J. Geophys. Res. 106(5):9135–9143

Bauch HA, Kandiano ES (2007) Evidence for early warming and cooling in North Atlantic surface waters during the last interglacial. Paleoceanography 22, PA1201, doi:10.1029/2005PA001252

Bauch D, Schlosser P, Fairbanks RG (1995) Freshwater balance and the sources of deep and bottom waters in the Arctic Ocean inferred from the distribution of $H_2{}^{18}O$. Progress in Oceanography 35:53–80

Bauch D, Carstens J, Wefer G (1997) Oxygen isotope composition of living Neogloboquadrina pachyderma (sin.) in the Arctic Ocean. Earth Planet Sci. Lett. 146:47–58

Bauch D, Erlenkeuser H, Stanovoy V, Simstich J, Spielhagen R (2003) Freshwater distribution and brine waters in the southern Kara Sea in summer 1999 as depicted by $\delta^{18}O$ results. In: Stein R, Fahl K, Fütterer DK, Galimov EM, Stepanets OV (eds) Siberian River Run-off in the Kara Sea: Characterisation, Quantification, Variability and Environmental Significance. Proceedings in Mar. Sci. 6:73–90

Bauch HA, Erlenkeuser H, Jung SJA, Thiede J (2000) Surface and deep water changes in the subpolar North Atlantic during Termination II and the last interglaciation. Paleoceanography 15:76–84

Bauch HA, Erlenkeuser H, Spielhagen RF, Struck U, Matthissen J, Thiede J, Heinemeier J (2001) A multiproxy reconstruction of the evolution of deep and surface waters in the subarctic Nordic seas over the last 30,000 yr. Quat. Sci. Rev. 20(4):659–678

Bauch H, Pavlidis YuA, Matishov GG et al. (2002) Pechora Sea coastal zone during Late Pleistocene and Holocene period (results in frame INTAS project N 1489–99). Fifth Workshop on Land Ocean Interactions in the Russian Arctic (LOIRA) devoted to memory of acad. I.S.Gramberg. Moscow, pp. 14–17

Bé AWH (1977) An ecological, zoogeographic and taxonomic review of recent planktonic foraminifera. In: Ramsay ATS (ed) Oceanic Micropaleontology 1:1–100

Bé AWH, Tolderlund DS (1971) Distribution and ecology of living planktonic foraminifera in surface waters of the Atlantic and Indian oceans. In: The micropaleontology of oceans. Cambridge University Press, London

Bearman G (ed) (1989) Ocean Circulation. Pergamon Press, Oxford

Beaufort L (1996) Dynamics of the monsoon in the Equatorial Indian Ocean over the last 260,000 years. Quat. Int. 31:13

Beaufort L, Lancelot Y, Camberlin P, Cayre O, Vincent E, Bassinot F, Labeyrie L (1997) Insolation cycles as a major control of Equatorial Indian Ocean primary production. Science 278: 1451–1454

Beaufort L, Bassinot F, Vincent E (1999) Primary production response to orbitally induced variations of the Southern oscillation in the Equatorial Indian Ocean. In: Abrantes F, Mix A (eds) Reconstructing Ocean History. Kluwer Academic/Plenum Publishers, New York

Beaufort L, de Garidel-Thoron T, Mix AC, Pisias NG (2001) ENSO-like forcing on oceanic primary production during the Late Pleistocene. Science 293:2440–2444

Beaufort L, Ivanova EV, Vidal L, Leduc G (2004) Dynamics of the Costa Rica Dome during the last climatic cycle: results from IMAGES core MD02–2529. 8th International Conference on Paleoceanography. Biarritz, France

Becquey S, Gersonde RA (2003) 0.55-Ma paleotemperature record from the Subantarctic zone: Implications for Antarctic Circumpolar Current development. Paleoceanography 18(1):1014, doi:10.1029/2000PA000576

Benway HM, Mix AC, Haley BA, Klinkhammer GP (2006) Eastern Pacific Warm Pool paleosalinity and climate variability: 0–30 kyr. Paleoceanography 21:3008, doi:10.1029/2005PA001208

Behl RJ, Kennet JP (1996) Evidence for brief interstadial events in the Santa Barbara Basin, NE Pacific during the past 60 Kyr. Nature 379:243–246

Belkin IM, Levitus S, Antonov J, Malmberg S-A (1998) Great Salinity Anomalies in the North Atlantic. Progress in Oceanology 41:1–68

Belyaeva NV and Burmistrova II (1984) Critical levels of carbonates accumulation in the Indian Ocean. Doklady of the USSR Academy of Sciences, Earth Science Section 277:652–655 (in Russian)

Bender M, Sowers T, Dickson M-L, Orchado, Grootes P, Mayewski P, Meese D (1994) Climate correlations between Greenland and Antarctica during the past 100,000 years. Nature 372:663–666

Bendle JAP, Rosell-Melé A (2007) High-resolution alkenone sea surface temperature variability on the North Icelandic Shelf: implications for Nordic Seas palaeoclimatic development during the Holocene. The Holocene 17(1):9–24, doi:10.1177/0959683607073269

Berger AL (1977) Support for the astronomical theory of climatic change. Nature 269:44–45

Berger AL (1978) Long-term variations of daily insolation and Quaternary climatic changes. J. Atmos. Sci. 35:2362–2367

Berger WH (1968) Planktonic foraminifera: selective solution and paleoclimatic interpretation. Deep-Sea Res. 15:31–43

Berger WH (1971) Sedimentation of planktonic foraminifera. Mar. Geol. 11:325–358

Berger WH, Jansen E (1995) Younger Dryas episode: Ice collapse and super-fjord heat pump, Versl. Gewone Vergad. Afd. Natuurkd. K. Ned. Akad. Wet. 44:61–105

Berger A, Loutre MF, Melice JL (2006) Equatorial insolation: From precession harmonics to eccentricity Frequencies. Clim. Past 2:131–136

Berger WH, Wefer G (1996) Expeditions in to the past: Paleoceanographic studies in the South Atlantic. In: Wefer G, Berger WH, Siedler G, Webb DJ (eds) The South Atlantic: Present and Past Circulation. Springer-Verlag

Berggren WA, Schnitker D (1983) Cenozoic marine environments in the North Atlantic and Norwegian-Greenland Sea. Structure and Development of the Greenland-Scotland Ridge. Plenum Publ. Corp.:495–548

Bianchi G, McCave N (1999) Holocene periodicity in the North Atlantic climate and deep-ocean flow south of Iceland. Nature 397:515–517

Bickert T, Mackensen A (2004) Late Glacial to Holocene changes in South Atlantic deep water circulation. In: Wefer G et al. (eds) The South Atlantic in the Late Quaternary: Reconstruction of Material Budget and Current Systems, Springer-Verlag, Berlin

Bjerknes J (1969) Atmospheric teleconnections from the equatorial Pacific. Mon. Weather Rev. 18:820–829

Blunier T, Brook EJ (2001) Timing of millennial-scale climate change in Antarctica and Greenland during the last glacial period. Science 291:109–112

Blunier T, Chappellaz J, Schwander J et al (1998) Acynchrony of Antarctic and Greenland climate change during the last glacial period. Nature 394(6695):739–743

Blunier T, Stocker TF, Chappellaz J, Raynaud D (1999) Phase Lag of Antarctic and Greenland Temperature in the Last Glacial and Link between CO_2 Variations and Heinrich Events. In: Abrantes F, Mix A (eds) Reconstructing Ocean History: A Window into the Future. Kluwer Academic/Plenum Publishers, New York

Blyum NS, Oskina NS, Ivanova EV (1987) Reconstruction of the climatic zonality in the Pliocene. In: Velichko AA, Chepalyga AL (eds) Earth Climates in Geological Past. Nauka, Moscow (in Russian), pp. 125–139

Blyum NS, Ivanova EV, Oskina NS (1988) Climatic zonality of the Atlantic, Indian and Pacific oceans in Middle and Late Miocene. Doklady of the USSR Academy of Sciences. Earth Science Section 301:1167–1170 (in Russian)

Boersma A, Mikkelsen N (1990) Miocene-age primary productivity episodes and oxygen minima in the Central equatorial Indian Ocean. Proc. ODP. Sci. Res. Wash. 115:589–609

Bol'shakov VA (2003) Modern climatic data for the Pleistocene: Implications for a new concept of the orbital theory of paleoclimate. Russian J. Earth Sci. 5(2):125–143 (English Translation)

Bol'shakov V. A. (2008) How long will the 'precession epoch' last in terms of Pleistocene glacial cycles? Russian J. Earth Sci. 10, ES3004, doi:10.2205/2008ES000299 (English Translation)

Bond G, Lotti R (1995) Iceberg dischanges into the North Atlantic on millenial time scales during the last glacition. Science 267:1005–1010

Bond G, Heinrich H, Broecker WS, Labeyrie L, MacManus J, Andrews J, Huon S, Jantschik R, Clasen S, Simet C, Tedesco K, Klas M, Bonani G, Ivy S (1992) Evidence for massive discharges of icebergs into the North Atlantic ocean during the last glacial period. Nature 360:245–251

Bond G, Broecker WS, Johnsen S, McManus J, Labeyrie L, Jouzel J, Bonani G (1993) Correlations between climate records from North Atlantic sediments and Greenland ice. Nature 365: 143–147

Bond G, Showers W, Cheseby M, Lotti R, Almasi P, deMenocal P, Priore P, Cullen H, Hajdas I, Bonani G (1997) A pervasive millenial-scale cycle in North Atlantic Holocene and glacial climates. Science 278:1257–1266

Boucsein B, Fahl K, Stein R (2000) Variability of river dischange and Atlantic-water inflow at the Laptev Sea continental margin during the past 15000 years: implications from maceral and biomarker records. Int. J. Earth Sci. 89:578–591

Boyle EA, Keigwin L (1987) North Atlantic thermohaline circulation during the past 20,000 years linked to high-latitude surface temperature. Nature 330:35–40

Bradley RS, Jones PD (1993) 'Little Ice Age' summer temperature variations: their nature and relevance to recent global warming trends. The Holocene 3:367–376

Brezgunov VS, Debolskiy VK, Nechaev VV, Ferronskiy VI, Yakimova TV (1982) Formation of oxygen isotope composition and salinity during sea water and river water mixing in the Barents and Kara seas. Vodnye Resursy 4:3–14 (in Russian)

Briffa KR (2000) Annual climate variability in the Holocene: Interpreting the message of ancient trees. Quat. Sci. Rev. 19:87–105

Briffa KR, Jones PD, Schweingruber FH, Shiyatov SG, Cook ER (1995) Unusual twentieth-century summer warmth in a 1,000-year temperature record from Siberia. Nature 376:156–159

Brock JC, McClain CR, Anderson DM, Prell WL, Hay WW (1992) Southwest monsoon circulation and environments of recent planktonic foraminifera in the northwestern Arabian Sea. Paleoceanography 7(6):799–813

Broecker WS, Peng TH (1982) Tracers in the Sea, Eldigo, New York

Broecker WS (1991) The great ocean conveyor. Oceanography 1:79–89

Broecker WS (1992) The Strenght of the Nordic Heat Pump. NATO ASI Series 12:173–180

Broecker WS (1996) Paleoclimatology. Geotimes, 41:40–41

Broecker WS (1997) Thermohaline Circulation, the Achilles Heel of our climate system: will man-made CO_2 upset the current balance. Science 278:1582–1588

Broecker WS (1998) Paleoocean circulation during the last deglaciaion: A bipolar seesaw. Paleoceanography 13:119–121

Broecker WS (1999) What If the Conveyor Were to Shut Down? Reflections on a Possible Outcome of the Great Global Experiment. GSA Today 9(1):2–7

Broecker WS (2000) Abrupt climate change: Causal constraints provided by the paleoclimate record. Earth-Sci. Rev. 51:137–154

Broecker WS (2001) Was the Medieval Warm Period Global? Science 291(6508):1498–1499

Broecker WS, van Donk J (1970) Insolation changes, ice volumes and the 0^{18} record in deep-sea cores. Geophysics and Spase Rhys. Rev. 1:169–198

Broecker WS, Denton GH (1989) The role of ocean atmosphere reorganizations in glacial cycles. Geoch. Cosmochim. Acta. 53:2465–2501

Broecker WS, Hemming S (2001) Climate Swings Come into Focus. Science 294:2308–2309

Broecker WS, Peteet DM, Rind D (1985) Does the ocean-atmosphere system have more than one stable mode of operation. Nature 315:21–26

Broecker WS, Andree M, Bonani G, Wolfli W, Mix A, Oeschger H (1988) Comparison between radiocarbon ages obtained on coexisting planktonic foraminefera. Paleoceanography 3: 647–657

Broecker WS, Bond G, Klas M (1990) A salt oscillator in the glacial Atlantic. The concept. Paleoceanography 5:469–477

Broecker WS, Sutherland S, Peng T-H (1999) A possible 20th-century slowdown of Southern Ocean deep water formation. Science 286:1132–1135

Brook EJ, White JWC, Schilla ASM, Bender ML, Barnett B, Severinghaus JP, Taylore KC, Alley RB, Steig EJ (2005) Timing of millennial-scale climate change at Siple Dome, West Antarctica, during the last glacial period. Quat. Sci. Rev. 24:1333–1343

Bruevich SV (ed) (1966) Chemistry of the Pacific Ocean. Nauka, Moscow (in Russian)

Bryden HL, Beal LM (2001) Role of the Agulhas Current in Indian Ocean circulation and associated heat and freshwater fluxes. Deep-Sea Res. 48:1821–1845

Burenkov VI, Vasilkov AP (1994) The influence of runoff from land on the distribution of hydrological characteristics in the Kara Sea. Oceanology 34:591–599 (English Translation)

Burenkov VI, Goldin YuA, Gureev BA, Sud'bin AI (1995) The basic notions of optical water properties of the Kara Sea. Oceanology 35:346–357 (English Translation)

Burkov VA (1980) General circulation in the World Ocean. Gidrometeoizdat, Leningrad (in Russian)

Burkov VA, Neyman VG (1977) General water circulation in the Indian ocean. In: Hydrology of the Indian ocean. Nauka, Moscow (in Russian)

Byshev VI, Galerkin LI, Grotov AS (2001a) Thermohaline structure of the Barents Sea waters in September 1988. In: Lisitzin AP, Vinogradov ME, Romankevich EA (eds) Experience of System Oceanologic Studies in the Arctic, Scientific World, Moscow (in Russian)

Byshev VI, Galerkin LI, Grotov AS (2001b) Mesoscale thermohaline structure of the Barents sea waters and its variability. In: Lisitzin AP, Vinogradov ME, Romankevich EA (eds) Experience of System Oceanologic Studies in the Arctic, Scientific World, Moscow (in Russian)

Cail Y, Anl Z, Cheng H, Edwards RL, Kelly MJ, Liu W, Wang X, Shen C-C (2006) High-resolution absolute-dated Indian Monsoon record between 53 and 36 ka from Xiaobailong Cave, southwestern China. Geology 34(8):621–624. DOI: 10.1130/G22567.1

Calov R, Ganopolski A, Petoukhov V, Claussen M, Greve R (2002) Large-scale instabilities of the Laurentide ice sheet simulated in a fully coupled climate-system model. Geophys. Rev. Lett. 29(24):2216, doi:10.1029/2002GL016078

Cane MA, Molnar P (2001) Closing of the Indonesian seaway as a precursor to east African aridification around 3–4 million years ago. Nature 411:157–162

Carstens J, Hebbeln D, Wefer G (1997) Distribution of planktic foraminifera at the ice margin in the Arctic (Fram Strait). Mar. Micropal. 29:257–269

Caulet JP, Vénec-Peyré M-T, Vergnaud-Grazzini C, Nigrini C (1992) Variation of South Somalian upwelling during the last 160 ka: radiolarian and foraminifera records in core MD 85674. In: Summerhayes CP, Prell WL, Emeis KC (eds) Upwelling Systems: Evolution Since the Early Miocene). Geol. Soc. Spec. Publ. 64:379–389

Cayre O, Bard E (1999) Planktonic foraminiferal and alkenone records of the last deglaciation from the Eastern Arabian Sea. Quat. Res. 52:337–342

Cayre O, Beaufort L, Vincent E (1999a) Paleoproductivity in the Equatorial Indian Ocean for the last 260,000 yr: atransfer function based on planktonic foraminifera. Quat. Sci. Rev. 18:839–857

Cayre O, Lancelot Y, Vincent E (1999b) Paleoceanographic reconstructions from planktonic foraminifera off the Iberian margin: Temperature, salinity and Heinrich events. Paleoceanography 14(3):284–396

Chapman MR, Shacklenton NJ, Duplessy J-C (2000) Sea surface temperature variability during the last glacial-interglacial cycle: assessing the magnitude and pattern of climate change in the North Atlantic. Palaeogeogr. Palaeoclim. Palaeoecol. 157:1–25

Chappell J, Omura A, Esat T, McCulloch M, Pandolfi J, Ota T, Pilans B (1996) Reconciliation of late Quaternary sea levels derived from coral terraces at Huon Peninsula with deep-sea oxygen isotope records. Earth Planet Sci. Lett. 141:227–236

Charles CD, Lynch-Stieglitz J, Ninnemann US, Fairbanks RG (1996) Climate connection between the hemisphere revealed by deep sea sediment core/ice core correlations. Earth Planet Sci. Lett. 142:19–27

Chen M-T, Farrell J (1991) Planktonic foraminifer faunal variations in the northeastern Indian Ocean: a high-resolution record of the past 800,000 years from site 758. Proc. ODP. Sci. Res. Wash. 121:121–140

Chelton DB, Freilich MH, Esbensen SK (2000) Satellite observations of the wind jets off the Pacific Coast of Central America: Part II. Relationships and dynamical considerations. Monthly Weather Rev. 128:2019–2043

Chistyakova NO, Ivanova EV, Risebrobakken B, Ovsepyan EA, Ovsepyan YaS. Reconstruction of Postglacial environments in the southwestern Barents Sea based on foraminiferal assemblages. Oceanology (in press)

Chizhov OP, Koryakin VS, Davidovich NV, Kamenevskiy ZM et al. (1968) Glaciation of the Novaya Zemlya. Nauka, Moscow (in Russian)

Churun V, Ivanov B (1998) Investigations of the hydrophysical structure in the mixing zone between fresh and saline waters. In Matthiessen J, Stepanets OV (eds) Scientific Cruise Report of the Kara Sea Expedition of RV "Akademik Boris Petrov" in 1997. Ber. Polarforsch. 266:11–18

Clemens SC, Prell WL (1991) Late Quaternary forcing of Indian Ocean summer monsoon winds: A comparison of Fourier Model and General Circulation Model results. J. Geophys. Res. 96(D12):22,683–22,700

Clemens SC, Prell WL (2007) The timing of orbital-scale Indian monsoon changes. Quat. Sci. Rev. 26:275–278

Clement AC, Seager R, Cane MA (1999) Orbital controls on the El Ninõ/Southern Oscillation and the tropical climate. Paleoceanography 14:441–456

Ciais P, Jouzel J, Lorius C, Barkov NI, Lipenkov V, Nikolaev V (1992) Evidence for an early Holocene climatic optimum in the Antarctic deep ice-core record. Clim. Dyn. 6:169–177

Clarke A, Church J, Gould J (2001) Ocrean Processes and Climate Phenomena. In: Siedler G, Church J, Gould J (eds) Ocean Circulation & Climate. Observing and Modelling the Global Ocean. Int. Geophysics Series 77:11–30. Academic Press, San Diego

Clark PU, Mitrovica JX, Milne GA, Tamisiea M (2002) Sea-level fingerprinting as a direct test for the source of global meltwater pulse IA. Science 295:2438–2441

Clemens SC, Prell WL (1996) Late Quaternary forcing of Indian Ocean summer monsoon winds: a comparison of Fourier Model and General Circulation Model results. J. Geophys. Res. 96(D12):22,683–22,700

Clemens SC, Prell WL (2003) A 350,000 year summer-monsoon multiproxy stack from the Owen Ridge, Northern Arabian Sea. Mar. Geol. 201:35–51

Clemens S, Prell W, Murray D, Shimmield G, Weedon G (1991) Forcing mechanisms of the Indian Ocean monsoon. Nature 353:720–725

Clemens SC, Murray DW, Prell WL (1996) Nonstationary phase of the Plio-Pleistocene Asian Monsoon. Science 274:943–948

Clement AC, Cane M (1999) A role for the tropical Pacific coupled Ocean-Atmosphere system on Milankovitch and millennial timescales. Part I. A modeling study of tropical Pacific variability. In: Clark PU, Webb RS, Keigwin LD (eds) Mechanisms of Global Climate Change at Millennial Time Scale. Geophys. Monograph. 112:363–371

Climap Project Members (1981) Seasonal Reconstruction of the Earth's surface at the last glacial maximum. Geol. Soc. Amer. Map and Chart series. Mc-36

Climap Project Members (1984) The Last Interglacial Ocean. Quat. Res. 21(2):123–224

Corliss BH (1979) Recent deep-sea benthonic foraminiferal distributions in the Southeast Indian Ocean; inferred bottom-water routes and ecological implications. Mar. Geol. 31(1–2):115–138

Cortijo E, Duplessy JC, Labeyrie L, Leclaire H, Duprat J, van Weering TCE (1994) Eemian cooling in the Norwegian Sea and North Atlantic Ocean preceding continental ice-sheet growth. Nature 372:446–449

Cortijo E, Labeyrie L, Vidal L, Vautravers M, Chapman M, Duplessy J-C, Elliot M, Arnold M, Turon J-L, Auffret G (1997) Changes in sea surface hydrology associated with Heinrich events 4 in the North Atlantic Ocean between 40° and 60°N. Earth Planet Sci. Lett. 146:29–45

Cortijo E, Balbon E, Elliot M, Labeyrie L, Turon J-L (1999a) Glacial and interglacial hydrological changes in th North Atlantic Ocean. In: Abrantes F, Mix A (eds) Reconstructing Ocean History: A Window into the Future. Kluwer Academic/Plenum Publishers, New York

Cortijo E, Lehman S, Keigwin LD, Chapman M, Paillard D, Labeyrie L (1999b) Changes in meridional temperature and salinity gradients in the North Atlantic Ocean (32° to 72°N) during the last interglacial period. Paleoceanography 14:23–33

Cortijo E, Labeyrie L, Elliot M, Balbon E, Tisnerat N (2000) Rapid climatic variability of the North Atlantic Ocean and global climate: a focus of the IMAGES program. Quat. Sci. Rev. 19:227–241

Craig H, Gordon LI (1965) Deuterium and Oxygen-18 Variations in the Ocean and the Marine Atmosphere. Proc. Spoleto Conf. on Stable Isotopes in Oceanographic Studies and Paleotemperatures 2:1–87

Crowley TJ (1992) North Atlantic deep water cools the southern hemisphere. Paleoceanography 7:489–497

Cullen JL (1981) Microfossil evidence for changing salinity patterns in the Bay of Bengal over the last 20,000 years. Palaeogeogr. Palaeoclim. Palaeoecol. 35(2–4):315–356

Cullen JL, Prell WL (1984) Planktonic foraminifera in the northern Indian Ocean: distribution and preservation in surface sediments. Mar. Micropal. 9(1):1–52

Curry WB, Oppo DW (1997) Synchronous, high-frequency oscillations in tropical sea surface temperatures and North Atlantic Deep Water production during the last glacial cycle. Paleoceanography 12:1–14

Curry WB, Oppo DW (2005) Glacial water mass geometry and the distribution of $\delta^{13}C$ of ΣCO_2in the western Atlantic Ocean. Paleoceanography 20, PA1017, doi:10.1029/2004PA001021

Cruz FW, Burns SJ, Karmann I, Sharp WD, Vuille M, Cardoso AO, Ferrari JA, Silva Dias PL, Viana O (2005) Insolation-driven changes in atmospheric circulation over the past 116,000 years in subtropical Brazil. Nature 434, doi:10.1038

Dansgaard WS, Johnsen SJ, Clausen HB, Langway CC Jr (1971) Climatic record revealed by the Camp Century ice core. In: Turekian KK (ed) The Late Cenozoic Glacial Ages. Yale University Press

Dansgaard WS, Johnsen SJ, Clausen HB, Dahl-Jensen D, Gundestrup NS, Hammer CU, Hvidberg CS, Steffensen JP, Sveinbjornsdottir AE, Jouzel J, Bond G (1993) Evidence for general instability of past climate from a 250-kyr ice-core record. Nature 364:218–220

Dansgaard W, Clausen HB, Gundestrup N, Hammer CU, Johnsen SJ, Kristinsdottir KM, Reeh N (1982) Science 218:1273–1277

Dansgaard W, White JWC, Johnsen S (1989) The abrupt termination of the Younger Dryas climate event. Nature 339:532–533

De Deckker P, Tapper NJ, van der Kaars S (2003) The status of the Indo-Pacific Warm Pool and adjacent land at the Last Glacial Maximum. Global and Planet Change 35(102):25–35

De Garidel-Thoron T, Beaufort L, Linsley BK, Dannenmann S (2001) Millenial-scale dynamics of the East Asian winter monsoon during the last 200,000 years. Paleoceanography 16(5):491–502

De Garidel-Thoron T, Rosenthal Y, Bassinot F, Beaufort L (2005) Stable sea surface temperatures in the western Pacific warm pool over the past 1.75 million years. Nature 433:294–298

De Menocal P, Bloemendal J, King J (1991) Rock-magnetic record of monsoonal dust deposition to the Arabian Sea: Evidence for a shift in the mode of deposition at 2.4 Ma. Proc. ODP. Sci. Results. (117):389–407.

De Menocal P, Ortiz J, Guilderson T, Sarnthein M (2000) Coherent high – and low latitude climate variability during the Holocene warm period. Science 288:2198–2202

Den Dulk M, Reichart GJ, Memon GM, Roelofs EMP, Zachariasse WJ, van der Zwaan GJ (1998) Benthic foraminiferal response to variations in surface water productivity and oxygenation in the northern Arabian Sea. Mar. Micropal. 35:43–66

Den Dulk M, Reichart GJ, van Heyst S, Zachariasse WJ, Van der Zwaan GJ (2000) Benthic foraminifera as proxies of organic matter flux and bottom water oxygenation? A case history from the northern Arabian Sea. Palaeogeogr., Palaeoclimatol., Palaeoecol. 161:337–359

De Vernal A, Pedersen TF (1997) Micropaleontology and palynology of core PAR87A-10. A 23,000 year record of paleoenvironal changes in the Gulf of Alaska, northeast North Pacific. Paleoceanography 12:821–830

De Vernal A, Hillaire-Marcel C (2000) Sea-ice cover, seasurface salinity and halo-thermocline structure of the northwest North Atlantic: Modern versus full glacial conditions. Quat. Sci. Rev. 19(1–5):65–85

De Vernal A, Eynaud F, Henry M, Hillaire-Marcel C, Londeix L, Mangin S, Matthiessen J, Marret F, Radi T, Rochon A, Solignac S, Turon JL (2005) Reconstruction of sea-surface conditions at middle to high latitudes of the Northern Hemisphere during the Last Glacial Maximum (LGM) based on dinoflagellate cyst assemblages. Quat. Sci. Rev. 24(7–9):897–924

Dickson RR, Meincke J, Malmberg S-A, Lee AJ (1988) The "Great Salinity Anomaly" in the northern North Atctic. 1968–1982. Prog.Oceanogr. 20:103–151

Ding X, Bassinot F, Guichard F, Li QY, Fang NQ, Labeyrie L, Xin RC, Adisaputra MK, Hardjawidjaksana K (2006) Distribution and ecology of planktonic foraminifera from the seas around the Indonesian Archipelago. Mar. Micropaleontol. 58:114–134

Dmitrenko IA, Holemann GA, Kirilov SA, Vegner S, Gribanov VA, Berezovskaya SL, Kassens X (2001) Thermal regime of the Laptev sea bottom water and controlling factors. Kriosphera Zemli 3:40–55 (in Russian)

Dobrolyubov SA (2004) Meridional and interoceanic circulation of water masses, heat and fresh water. In: Alexeevsky NI et al. (eds) Dynamics and interaction of the atmosphere and the hydrosphere. Geological Department, Moscow State University, Moscow (in Russian)

Dobrovol'skiy AD, Zalogin VS (1982) USSR Seas. MSU, Moscow (in Russian)

Dokken TM, Hald M (1996) Rapid climatic shifts during isotope stage 2–4 in Polar North Atlantic. Geology 24:599–602

Dokken TM, Jansen E (1999) Rapid changes in the mechanism of ocean convection during the last glacial period. Nature 401:458–461

Dokken T, Jansen E, Risebrobakken B (2001) High resolution deglacial records in the Nordic Seas: Phase relationship between different oceanic proxies during rapid transitions. ICP VII Program & Abstracts. Sapporo. Japan, pp. 97–98

Dowdeswell JA, Hagen JO, Bjornsson H, Glazovsky AF, Harrison WD, Holmlund P, Jania J, Koerner RM, Lefauconnier B, Ommanney CSL, Thomas RH (1997) The mass balance of circum-Arctic glaciers and recent climate change. Quat. Res. 48:1–14

Dürkop A, Holbourn A, Kuhnt W, Zuraida R, Andersen N, Grootes PM (2008) Centennial-scale climate variability in the Timor Sea during Marine Isotope Stage 3 Mar. Micropal. 66(3–4):208–221

Duplessy J-C (1982) Glacial to interglacial contrasts in the northern Indian Ocean. Nature 295(5849):494–498

Duplessy J-C (1996) Quand l'océan se fâche. Editions Odile Jacob, Paris

Duplessy J-C, Be AWH, Blanc PL (1981a) Oxygen and carbon isotopic composition and biogeographic distribution of planktonic foraminifera in the Indian Ocean. Palaeogeogr. Palaeoclimatol. Palaeoecol. 31:9–46

Duplessy J-C, Delibrias G, Turon JL, Pujol C, Duprat J (1981b) Deglacial warming of the Northeastern Atlantic Ocean: correlation with the paleoclimatic evolution of the European continent. Ralaeogeogr. Palaeoclimatol. Palaeoecol. 35(2/4):121–144

Duplessy J-C, Shackleton N, Fairbanks R, Labeyrie L, Oppo D, Kallel N (1988) Deepwater source variation during the last climatic cycle and their impact on the global deepwater circulation. Paleoceanography 3:343–360

Duplessy J-C, Arnold M, Bard E, Juillet-Leclerc A, Kallel N, Labeyrie L (1989) AMS ^{14}C study of transient events and of the ventilation rate of the Pacific intermediate water during the last deglaciation. Radiocarbon 31:493–502

Duplessy J-C, Bard E, Arnold M, Shackleton NJ, Duprat J, Labeyrie L (1991a) How fast did the ocean-atmosphere system run during the last deglaciation. Earth Planet Sci. Lett. 103:27–40

Duplessy J-C, Labeyrie L, Juillet-Leclerc A, Duprat J, Sarnthein M (1991b) Surface salinity reconstruction of the North Atlantic Ocean during the Last Glacial Maximum. Oceanol. Acta. 14:311–324

Duplessy J-C, Labeyrie L, Arnold M, Paterne M, Duprat J, van Weering TCE (1992) Changes in surface salinity of the North Atlantic Ocean during the last deglaciation. Nature 358:485–487

Duplessy J-C, Ivanova EV, Murdmaa IO, Paterne M, Labeyrie L (2001) Holocene paleoceanography of the Northern Barents Sea and variations of the northward heat transport by the Atlantic Ocean. Boreas 30(1):2–16

Duplessy J-C, Labeyrie L, Waelbroeck C (2002) Constraints on the ocean oxygen isotopic enrichment between the Last Glacial Maximum and the Holocene: Paleoceanographic implications. Quat. Sci. Rev. 21:315–330

Duplessy J-C, Cortijo E, Ivanova EV, Khusid T, Labeyrie L, Levitan M, Murdmaa IO, Paterne M (2005) Paleoceanography of the Barents Sea during the Holocene, doi:10.1029/2004PA001116

Duplessy J-C, Roche M, Kageyama M (2007) The Deep Ocean during the last interglacial period. Science 316:89–91

Eagles G, Livermore R, Morris P (2006) Small basins in the Scotia Sea: The Eocene Drake Passage gateway. Earth Planet Sci. Lett. 242:343–353

Elderfield H, Ganssen G (2000) Past temperature and $\delta^{18}O$ of surface ocean waters inferred from foraminiferal Mg/Ca ratios. Nature 405:442–445

Elliot M, Labeyrie L, Bond G, Cortijo E, Turon J-L, Tisnerat N, Duplessy J-C (1998) Millenial-scale iceberg discharge in the Irminger Basin during the last glacial period: Relationship with the Heinrich events and environmental setting. Paleoceanography 13(5):433–446

Elliot M, Labeyrie L, Duplessy J-C (2002) Changes in North Atlantic deep-water formation associated with the Dansgaard-Oeschger temperature oscillations (60–10 ka). Quat. Sci. Rev. 21:1153–1165

Elverhøi A, Dowdeswell JA, Funder S, Mangerud J, Stein R (1998) Glacial and oceanic history of the Polar North Atlantic margins: an overview. Quatern. Sci. Rev. 17(1–3):1–10

Emeis K-C, Anderson DM, Doose H, Kroon D, Schulz-Bull D (1995) Sea-surface temperatures and the history of monsoon upwelling in the northwest Arabian sea during the last 500,000 years. Quat. Res. 43:355–361

Emiliani C (1955) Pleistocene temperatures. J. Geol. 63:538–578

EPICA community members (2004) Eight glacial cycles from an Antarctic ice core. Nature 429:623–628

EPICA community members (2006) One-to-one coupling of glacial climate variability in Greenland and Antarctica. Nature 444:195–198, doi:10.1038/nature05301

Evans CDR, Brett CP, James JWC, Holmes R (1995) Shallow seismic reflection profiles from the waters of East and Southeast Asia: An interpretation manual and atlas. British Geological Survey, Tech. Rep. WC/94/60. Overseas Geology Series

Fairbanks RG (1989) A 17,000-year glacio-eustatic sea level record: Influence of glacial melting rates on the Younger Dryas event and deep-ocean circulation. Nature 342:637–642

Farrell JW, Janecek TR (1991) Late Neogene paleoceanography and paleoclimatology of the northeast Indian ocean (site 758). Proc. ODP. Sci. Res. 121, Washington, DC.

Farrell JW, Pedersen TF, Calvert SE, Nielsen B (1995) Glacial-interglacial changes in nutrient utilization in the equatorial Pacific Ocean. Nature 377:514–517

Fedorov AV, Dekens PS, McCarthy M, Ravelo AC, deMenocal PB, Barreiro M, Pacanowski RC, Philander SG (2006) The Pliocene Paradox (Mechanisms for a Permanent El Niño). Science 312:1485–1489

Feldberg MJ, Mix AC (2003) Planktonic foraminifera, sea surface temperatures, and mechanisms of oceanic change in the Peru and south equatorial currents, 0–150 ka BP. Paleoceanography 18(1):1016, doi:10.1029/2001PA000740

Fenchel T, Finlay BJ (1995) Ecology and Evolution in Anoxic Worlds. Oxford University Press, New York

Feyling-Hanssen RW (1964) Foraminifera in Late Quaternary deposits in the Oslofjord area. Norges Geologiske Undersokelse 225

Fichefet T, Hovine S, Duplessy J-C (1994) A model study of the Atlantic thermohaline circulation during the Last Glacial Maximum. Nature 372:252–255

Fiedler PC (2002) The annual cycle and biological effects of the Costa Rica Dome. Deep-Sea Res. 49:321–338

Fiedler PC, Talley LD (2006), Hydrography of the eastern tropical Pacific: A review. Prog. Oceanogr. 69(2–4):143–180

Filyushkin BN (1987) Temporal-spatial variability in the thermal structure of the oceanic active layer. Habilitation (Dr.Sc.) Thesis. Moscow State University, Moscow (in Russian)

Fischer G, Wefer G (eds) (1999) Use of Proxies in Paleoceanography: Examples from the South Atlantic, pp. 229–254. Springer-Verlag, Berlin Heidelberg

Fleitmann D, Burns SJ, Mangini A, Mudelsee M, Kramers J, Villa I, Neff U, Al-Subbary AA, Buettner A, Hippler D, Matter A (2007) Holocene ITCZ and Indian monsoon dynamics recorded in stalagmites from Oman and Yemen (Socotra). Quat. Sci. Rev. 26:170–188

Flückiger J, Knutti R, White JWC (2006) Oceanic processes as potential trigger and amplifying mechanisms for Heinrich events. Paleoceanography 21, PA2014, doi:10.1029/2005PA001204

Fluckiger J, Blunier T, Stauffer B, Chappellaz J, Spahni R, Kawamura K, Schwander J, Stocker TF, Dahl-Jensen D (2004) N_2O and CH_4 variations during the last glacial epoch: Insight into global processes. Global Biogeochem. Cycles 18, GB1020, doi:10.1029/2003GB002122

Fontugne MR, Duplessy J-C (1986) Variations of the monsoon regime during the upper Quaternary: evidence from carbon isotopic record of organic matter in north Indian Ocean sediment cores. Palaeogeogr. Palaeoclim. Palaeoecol. 56:69–88

Forman SL, Polyak L (1997) Radiocarbon content of pre-bomb marine mollusks and variations in the ^{14}C reservoir age for coastal areas of the Barents and Kara seas, Russia. Geophys. Res. Lett. 24:885–888

Forman SL, Lubinski DA, Miller GH, Matishov GG, Snyder J, Korsun S (1995) Post-glasial emergence and distribution of the late Weichselian ice sheet loads in the northern Barents and Kara Seas, Russia. Geology 23:113–116

Forman SL, Lubinski DA, Miller GH, Matishov GG, Korsun S, Snyder J, Herlihy F, Weihe R, Myslivets V (1996) Postglacial emergence of Western Franz Josef Land, Russia, and retreat of the Barents Sea ice sheet. Quatern. Sci. Rev. 15:77–90

Forman SL, Weihe R, Lubinski D, Tarasov G, Korsun S, Matishov G (1997) Holocene relative sea-level history of Franz Josef Land, Russia. Geol. Soc. Amer. Bull. 109:1116–1133

Forman SL, Lubinski DJ, Zeeberg JJ, Polyak L, Miller GH, Matishov G, Tarasov G (1999) Observations of Postglacial emergence and Late Quaternary glaciation at selected localities on northern Novaya Zemlya, Russia. Boreas 28:133–145

Fretzdorff S, Paterne M, Stoffers P, Ivanova EV (2000) Explosive activity of the Reuniun Island volcanoes through the past 260,000 years as recorded in deep-sea sediments. Bull. Volcanol. 62:266–277

Fronval T, Jansen E (1996) Rapid changes in ocean circulation and heat flux in the Nordic seas during the last interglacial period. Nature 383:806–810

Fronval TE, Jansen E, Bloemendal J, Jonsen S (1995) Oceanic evidence for coherent fluctuations in Fennoscandian and Laurentide ice sheets on millenium timescales. Nature 374:443–446

Galimov EM, Laverov NP, Siepanets OV, Kodina LA (1996) Preliminary results of ecological and geochemical studies in Russian Arctic seas. Geochemistry 34:1–18 (in Russian)

Ganopolski A, Rahmstorf S (2001) Rapid changes of glacial climate simulated in a coupled climate model. Nature 409:153–158

Ganopolski A, Rahmstorf S (2002) Abrupt glacial climate changes due to stochastic resonance. Phys. Rev. Lett. 88(3), N038501

Ganopolski A, Rahmstorf S, Petoukhov V, Claussen M (1998) Simulation of modern and glacial climates with a coupled global model of intermediate complexity. Nature 391:351–356

Gardner WD, Gundersen JS, Richardson MJ, Walsh ID (1999) The role of seasonal and diel changes in mixed-layer depth on carbon and chlorophyll distribution in the Arabian Sea. Deep-Sea Res. 46(II):1833–1858.

Gasse F (2000) Hydrogenical changes in the African tropics since the Last Glacial Maximum. Quat. Sci. Rev. 19:189–221

Gasse F, Van Campo E (1994) Abrupt post-glacial climate events in West Asia and North Africa monsoon domains. Earth Planet Sci. Lett. 126:435–456

Gataullin V, Mangerud J, Svendsen JI (2001) The extend of the Late Weichselian ice sheet in the southeastern Barents Sea. Glob. Planet Change 31:453–474

Gersonde R, Abelmann A, Brathauer U, Becquey S, Bianchi C, Cortese G, Grobe H, Kuhn G, Niebler H-S, Segl M, Sieger R, Zielinski U, and Fütterer DK (2003) Last glacial sea surface temperatures and sea-ice extent in the Southern Ocean (Atlantic-Indian sector): A multiproxy approach. Paleoceanography 18(3):1061, doi:10.1029/2002PA000809

Gordeev VV, Martin J-M, Sidorov IS, Sidorova MN (1996) A reassessment of Eurasian river input of water, sediment, major elements and nutrients to the Arctic Ocean. Amer. J. Sci. 296:655–695

Gordon AH (1986) Interocean exchange of thermocline water. J. Geophys. Res. 101:12,217–12,237

Gordon AL (1989) The Role of Thermohaline Circulation in Global Climate Change. Lamont-Doherty Annual Report, pp. 44–51

Gordon AL (2001) Interocean Exchange. In: Siedler G, Church J, Gould J (eds) Ocean Circulation & Climate. Observing and modelling the Global Ocean. Int. Geophysics Series. Academic Press, San Diego 77:303–314

Gordon AL (2005) Oceanography of the Indonesian Seas and their throughflow. Oceanography 18(4):14–27

Gordon AL, Fine RA (1996) Pathways of water between the Pacific and Indian Oceans in the Indonesian seas. Nature 379:146–149

Gordon AL, Zebiak SE, Bryan K (1992) Climate variability and the Atlantic Ocean. EOS Trans, AGU 79:161–165

GRIP project members (1993) Climatic instability during the last interglacial period revealed in the Greenland summit ice – core. Nature 364:203–207

Grootes PM, Stuiver M (1997) Oxygen 18/16 variability in Greenland snow and ice with 10^3- to 10^5-year time resolution. J. Geophys. Res. 102(C12):26455–26470

Grootes PM, Stuiver M, White JWC et al. (1993) Comparison of oxygen isotope records from the GISP2 and GRIP Greenland ice cores. Nature 366:552–554

Grootes PM, Steig EJ, Stuiver M, Waddington ED, Morse DL, Nadeau M-J (2001) The Taylor Dome Antarctic ^{18}O Record and Globally Synchronous Changes in Climate. Quat. Res. 56:289–298

Gross MG (1993) A View of Earth. Prentice Hall

Grousset FE, Cortijo E, Huon S, Herve L, Richter T, Burdloff D, Duprat J, Weber O (2001) Zooming in on Heinrich layers. Paleoceanography 16(3):240–259

Grousset FE, Labeyrie L, Sinko JA, Cremer M, Bond G, Duprat J, Cortijo E, Huon S (1993) Patterns of ice-rafted detritus in the glacial North-Atlantic (40–55 N). Paleoceanography 8:175–192

Guiot J, de Vernal A (2007) Transfer Functions: Methods for Quantitative Paleoceanography Based on Microfossils. Developments in Marine Geology, 1, ISSN 1572–5480, doi 10.1016/S1572–5480(07)01018–4

Gulev SK, Zolina O, Grigoriev S (2001) Extratropical Cyclone variability in the Northern Hemisphere winter from the NCEP/NCAR reanalysis data. Climate Dynamics 17:795–809

Gupta AK, Thomas E (1999) Latest Miocene-Pleistocene productivity and deep-sea ventillation in the northwestern Indian Ocean (Deep Sea Drilling Project Site 219). Paleoceanography 14(1):62–73

Gupta AK, Sundar Raj M, Mohan K, De S (2008) A major change in monsoon-driven productivity in the tropical Indian Ocean during ca 1.2–0.9 Myr: Foraminiferal faunal and stable isotope data. Palaeogeogr. Palaeoclimatol. Palaeoecol. 261:234–245

Hagelberg T, Mix AC (1991) Long-term monsoon regulators. Nature 353:703–704

Hald M, Steinsund PI (1996) Benthic foraminifera and carbonate dissolution in the surface sediments of the Barents and Kara Seas. In: Stein R, Ivanov G, Levitan M, Fahl K (eds) Surface sediment composition and sedimentary processes in the central Arctic Ocean and along the Eurasian Continental Margin. Ber. Polarforsch 212:285–307

Hald M, Aspeli R (1997) Rapid climatic shifts of the northern Norwegian Sea during the last deglaciation and the Holocene. Boreas 26:15–28

Hald M, Korsun S (1997) Distribution of modern benthic foraminifera from fjords of Svalbard, European Arctic. J. Foraminiferal Res. 27(2):101–122

Hald M, Korsun S (2008) The 8200 cal. yr BP event reflected in the Arctic fjord, Van Mijenfjorden, Svalbard. The Holocene 18(6):981–990

Hald M, Steinsund PI, Dokken T, Korsun S, Polyak L, Aspeli R (1994) Recent late Quaternary distribution of Elphidium excavatum f. clavatum in Arctic seas. Cushman Foundation Spec. Publ. 32:141–153

Hald M, Kolstad V, Polyak L, Forman SL, Herlihy FA, Ivanov G, Nescheretov A (1999) Late-glacial and Holocene paleoceanography and sedimentary environments in the St. Anna Trough, Eurasian Arctic Ocean margin. Palaeogeogr. Palaeoclim. Palaeoecol. 146:229–249

Hald M, Andersson C, Ebbesen H, Jansen E, Klitgaard-Kristensen D, Risebrobakken B, Salomonsen GR, Sarnthein M, Sejrup HP, Telford RJ (2007) Variations in temperature and extent of Atlantic Water in the northern North Atlantic during the Holocene. Quat. Sci. Rev.26:3423–3440

Hall IR, Bianchi GG, Evans JR (2004) Centennial to millennial scale Holocene climate-deep water linkage in the North Atlantic. Quat. Sci. Rev. 23:1529–1536

Hantoro WS (1997) Quaternary sea level variations in the Pacific-Indian Ocean Gateways: Response and impact. Quat. Intern 37:73–80

Haq BU, Hardenbol J, Vail P (1987) Chronology of fluctuating sea level since the Triassic. Science 235:1156–1167

Harms IH, Hübner U, Backhaus JO, Kulakov M, Stanovoy V, Stepanets OV, Kodina LA, Schlitzer R (2003) Salt intrusions in Siberian river estuaries: Observations and model experiments in Ob and Yenisei. In: Stein R, Fahl K, Fütterer DK, Galimov EM, Stepanets OV (eds) Siberian River Run-off in the Kara Sea: Characterisation, Quantification, Variability, and Environmental Significance. Proceedings in Mar. Sci. 6:27–45

Hastenrath S (1991) Climate Dynamics of the Tropics. Kluwer Academic Publishers, Dordrecht/Boston/London

Haug GH, Tiedemann R (1998) Effect of the formation of the Isthmus of Panama on Atlantic Ocean thermohaline circulation. Nature 393:673–676

Haug GH, Tiedemann R (2001) Role of the Central American landbridge on Late Neogene climate. 7th Intern. Conf. on Paleoceanography, Program&Abstracts, Sapporo, pp. 31–32

Haug GH, Maslin MA, Sarnthein M, Stax R, Tiedemann R (1995) Evolution of the northwest Pacific sedimentation patterns since 6 Ma (site 882). Proc. ODP. Leg 145:293–314

Haug GH, Sigman DM, Tiedemann R, Pedersen TF, Sarnthein M (1999) Onset of permanent stratification in the subarctic Pacific Ocean. Nature 401:779–782

Haug GH, Hughen KA, Sigman DM, Peterson LC, Rohl U (2001) Southward migration of the Intertropical Convergence Zone through the Holocene. Science 293:1304–1308, doi:10.1126/science.1059725

Hays JD, Imbrie J, Shackleton NJ (1976) Variations in the earth's orbit: pacemaker of the ice ages. Science 194:1121–1132

Hebbein D, Dokken TM, Andersen ES, Hald M, Elverhoi A (1994) Moisture supply for northern ice-sheet growth during the last glacial maximum. Nature 370:357–359

Hebbeln D, Henrich R, Baumann K-H (1998) Paleoceanography of the last interglacial/glacial cycle in the polar North Atlantic. Quat. Sci. Rev. 17:125–153

Heinrich H (1988) Origin and consequences of cyclic ice rafting in the northeast Atlantic Ocean during the past 130,000 years. Quat. Res. 29:142–152

Hemleben Ch, Spindler M, Anderson OR (1989) Modern planktonic foraminifera. Springer, New York, Berlin, Heidelberg, London, Paris, Tokyo

Hemleben Ch, Herrle JO, Koesser P (2001) Icehouse elements characteristics greenhouse world. 7th Intern. Conf. on Paleoceanography, Program&Abstracts, Sapporo

Hemming SR (2004) Heinrich events: massive late Pleistocene detritus layers of the North Atlantic and their global climate imprint. Rev. Geophys 42:RG1005

Hendy IL, Kennett JP (1999) Latest Quaternary North Pacific surface-water responses imply atmosphere-driven climate instability. Geology 27(4):291–294

Hendy IL, Kennett JP (2000) Dansgaard-Oeschger cycles and the California Current System: Planktonic foraminiferal response to rapid climate change in Santa Barbara Basin, Ocean Drilling Program hole 893A. Paleoceanography 15(1):30–42

Hendy IL, Kennett JP (2003) Tropical forcing of North Pacific Intermediate Water during late Quaternary climate change. Quat. Sci. Rev. 22:673– 689

Hendy IL, Kennett JP, Roark EB, Ingram BL (2002) Apparent synchroneity of submillennial scale climate events between Greenland and Santa Barbara Basin, California from 30–10 ka. Quat. Sci. Rev. 21:1167–1184

Hendy L, Pedersen TF, Kennett JP, Tada R (2004) Intermittent existence of a southern Californian upwelling cell during submillennial climate change of the last 60 kyr. Paleoceanography 19, PA3007, doi:10.1029/2003PA000965

Hermelin JO, Shimmield GB (1995) Impact of productivity events on the benthic foraminiferal fauna in the Arabian Sea over the last 150,000 years. Paleoceanography 10(1):85–116

Heusser L, Sirocko F (1997) Millenial Pulsing of vegetation change in southern California: A record of Indo-Pacific ENSO events from the past 24,000 years. Geology 25:243–246

Hewitt CD, Broccoli AJ, Mitchell JF, Stouffer RJ (2001) A coupled model study of the last glacial maximum: was part of the North Atlantic relatively warm? Geophys. Res. Lett. 28:1571–1574

Hillaire-Marcel C, de Vernal A (ed) (2007) Developments in Marine Geology. V1, Elsevier, Amsterdam

Holbourn A, Kuhnt W, Kawamura H, Zhimin J, Grootes P, Erlenkeuser H, Xu J (2005) Orbitally paced paleoproductivity variations in the Timor Sea and Indonesian Throughflow variability during the last 460 kyr. Paleoceanolography 20, PA3002, doi:10.1029/2004PA001094

Honda M, Nakamura H (2001) Interannual seesaw between the Aleutian and Icelandic lows. Part II: Its significance in the interannual variability over the wintertime Northern Hemisphere. J. Climate (14):4512–4529

Hong YT, Hong TB, Lin QH, Shibata Y, Hirota M, Zhua YX, Leng XT, Wang Y, Wang H, Yia L (2005) Inverse phase oscillations between the East Asian and Indian Ocean summer monsoons during the last 12,000 years and paleo-El Ninõ. Earth Planet Sci. Lett. 231:337– 346

Huang B, Jian Z, Cheng X, Wang P (2002) Foraminiferal responses to upwelling variations in the South China Sea over the last 220, 000 years. Mar. Micropal. 47:1–15

Hughen KA, Baillie MGL, Bard E, Bayliss A, Beck JW, Bertrand C, Blackwell PG, Buck CE, Burr G, Cutler KB, Damon PE, Edwards RL, Fairbanks RG, Friedrich M, Guilderson TP, Kromer B, McCormac FG, Manning S, Bronk Ramsey C, Reimer PJ, Reimer RW, Remmele S, Southon JR, Stuiver M, Talamo S, Taylor FW, van der Plicht J, Weyhenmeyer CE (2004) Marine04 marine radiocarbon age calibration, 0–26 Cal Kyr BP. Radiocarbon 46:1059–1086

Hunt AS, Corliss BH (1993) Distribution and microhabitats of living (stained) benthic foraminifera from the Canadian Arctic Archipelago. Mar. Micropal. 20:321–345

Hurrell JW, van Loon H (1997) Decadal variations in climate associated with the North Atlantic Oscillation. Clim. Change. 36:301–326

Ikeda M (1989) Decadal oscillations of the air-ice-ocean system in the Northern Hemisphere. Atmosph.-Ocean 28:106–139

Imbrie J, Kipp N (1971) A new micropaleontologic method for quantitative paleoclimatology. Application to a Late Pleistocene Caribbean core. In: Turekian KK (ed) The Late Cenozoic Glacial Ages. Yale University, New Haven

Imbrie J (1982) Astronomical theory of the Pleistocene Ice Ages: A brief historical review. Icarus 50:408, doi:10.1016/0019-1035(82)90132-4

Imbrie J, Hays J-D, Martinson D-G, McIntyre A, Mix AC, Morley JJ, Pisias NG, Prell WL, Shackleton NJ (1984) The orbital theory of Pleistocene climate: supports from a revised chronology of the marine ^{18}O record. In: Berger A et al. (eds) Milankovitch and Climate. Reidel, Dordrecht

Imbrie J, Boyle EA, Clemens SC, Duffy A, Howard WR, Kukla G, Kutzbach J, Martinson DG, McIntyre A, Mix AC, Molfino B, Morley JJ, Peterson LC, Pisias NG, Prell WL, Raymo ME, Shackleton NJ, Toggweiler JR (1992) On the structure and origin of major glaciation cycles: linear responses to Milankovitch forcing. Paleoceanography 7:701–738

Indermühle A, Monnin E, Stauffer B, Stocker TF, Wahlen M (2000) Atmospheric CO_2 concentration from 60 to 20 kyr BP from the Taylor Dome ice core, Antarctica. Geophys. Res. Lett. 27:735–738

IPCC (2007) Climate Change 2007 (Fourth Assessment Report of the Intergovernmental Panel on Climate Change) Cambridge University Press, Cambridge

Ivanenkov VN (1979) General distribution patterns of biogenic elements in the World Ocean. In: Bordovsky OK and Ivanenkov VN (eds) Chemistry of Oceans, Ser. Oceanology, Nauka, Moscow (in Russian)

Ivanov VV (2001) Atlantic waters in the western Arctic In: Lisitzin AP, Vinogradov ME, Romankevich EA (eds) Experience of System Oceanologic Studies in the Arctic, Scientific World, Moscow (in Russian)

Ivanov VV, Shapiro GI (2002) Dense water cascades around the Arctic Ocean. Fifth Workshop on Land Ocean Interactions in the Russian Arctic (LOIRA) devoted to memory of acad. I.S.Gramberg, Moscow

Ivanov VV, Shapiro GI (2005) Formation of a dense water cascade in the marginal ice zone in the Barents Sea. Deep-Sea Res. 52(1):1699–1717

Ivanova EV (1983) On paleotemperature analysis on planktonic foraminifers. Oceanology 23(3):456–463 (English Translation)

Ivanova EV (1985) Late Quaternary biostratigraphy and paleotemperatures of the Red Sea and the Gulf of Aden based on planktonic foraminifera and pteropods. Mar. Micropal. 9:335–364

Ivanova EV (1988) Late Quaternary paleoceanology of the Indian Ocean (based on planktonic foraminifers and pteropods). P.P. Shirshov Institute of Oceanology, USSR Ac. Sci., Moscow (in Russian)

Ivanova EV (1994) Reconstruction of Climatic Zonality in the Indian Ocean during the Neogene. In: Duplessy J-C, Spyridakis M-T (eds) NATO ASI Ser. 22. Springer-Verlag, pp. 267–275

Ivanova E (1998) Foraminiferal assemblages of the Norwegian and Barents seas: responses to post-glacial climatic changes. Reconstructing Ocean Hystory. 6 Intern. Conf. on Paleoceanography, Lisbon, pp. 131–132

Ivanova EV (1999) Foraminiferal assemblages of the Barents and Kara seas: response to Holocene environmental changes. In: Geodynamics and geoecology. Institute of Ecological Problems of the North, Urals Branch of RAS, Arkhangelsk, pp. 140–142

Ivanova EV (2000) Biogeographic zonality of the Indian Ocean in the Late Mesozoic Stratigraphy. Geological Correlation 5:461–469 (English Translation)

Ivanova EV (2001) Benthic foraminifera in sediments from the southern Kara Sea: Preliminary results. In: Stein R, Stepanets O (eds) The German-Russian Project on Siberian River Run-off (SIRRO): Scientific Cruise Report of the Kara-Sea Expedition "SIRRO 2000" of RV “Akademik Boris Petrov” and first results. Reports on Polar Res. 393:141–150

Ivanova EV (2002a) Holocene foraminiferal assemblages of Barents and Kara seas: a response to paleoceanographic changes. EMMM’2002, Program&Abstracts, Vienna, Austria, pp. 106–107

Ivanova EV (2002b) Migration of mixing zone in the South-Eastern Kara Sea during the Holocene: evidence from benthic foraminiferal data. Fifth Workshop on Land Ocean Interactions in the Russian Arctic (LOIRA), Moscow, pp. 46–47

Ivanova EV (2002c) Initiation and Evolution History of the Monsoon Circulation in the Indian Ocean during the Neogene-Quaternary. Oceanology 42(1):120–131 (English Translation)

Ivanova EV (2003) The role of global thermohaline circulation in paleoceanographic events of the last glacial cycle. DrSc (Habilitation) thesis. Shirshov Institute of Oceanology, RAS, Moscow (in Russian)

Ivanova EV (2004) Report of the chief scientist. In: Report of the R.V. “Professor Shtockman” cruise 63 Vol. 1:1–80, Shirshov Institute of Oceanology, RAS, Moscow (unpublished in Russian)

Ivanova EV (2006) The Global Thermohaline Paleocirculation. Scientific World, Moscow (in Russian)

Ivanova EV, Beaufort L, Vidal L. (in prep.) Sea-surface temperature, productivity and circulation changes off Costa Rica, Eastern Equatorial Pacific, during the last 140 kyr

Ivanova EV, Kiselev VI (1985) Late Quaternary variations in surface water temperature and salinity in the Red Sea and Gulf of Aden. Doklady USSR Academy of Sciences. Earth Sci. Section 282(5):1215–1218 (in Russian)

Ivanova EV, Beaufort L, Vidal L. (in prep.) Eastern Equatorial Pacific hydrography, productivity and planktic assemblages off Costa Rica during the last 140 kyr

Ivanova EV, Ivanova AA (1996a) Reconstructions of climatic zoning and surface circulation patterns in the Indian Ocean during the Neogene. Oceanology 36(4):622–633 (English Translation)

Ivanova EV, Ivanova AA (1996b) New data on the Neogene history of the Indian Ocean. Oceanology 36(5):1–8 (English Translation)

Ivanova EV, Murdmaa IO (2001) Postglacial paleoceanography of the Northern Barents Sea. In: Lisitzin AP, Vinogradov ME, Romankevich EA (eds) Experience of System Oceanologic Studies in the Arctic, Scientific World, Moscow (in Russian)

Ivanova EV, Shackleton NJ (1993) Middle and Late Quaternary Paleoceanology of the East Pacific Rise at 21 S. Oceanology 33(4):609–614 (English Translation)

Ivanova EV, Kiselev VI, Late Quaternary variations in surface water temperature and salinity in the Red Sea and Gulf of Aden. Doklady USSR Academy of Sciences, Earth Science Section 282(5): 1215–1218 (in Russian)

Ivanova EV, Oskina NS, Blyum NS (1986) Climatic zonality of the Atlantic, Indian and Pacific oceans during the Early Miocene. Doklady of the USSR Academy of Sciences, Earth Science Section 288(5):1185–1188 (English Translation)

Ivanova EV, Oskina NS, Blyum NS (1989) Migration of the World Ocean climatic zones during the Neogene. Oceanology 29(2):249–255 (English Translation)

Ivanova EV, Burmistrova II, Vaikmae RA, Dmitrenko OB, Martma TA (1990) New data on Middle – Late Pleistocene paleoceanology of the Western Equatorial Indian Ocean. Oceanology 30(4):598–605 (English Translation)

Ivanova EV, Paterne M, Duplessy J-C, Cortijo E, Labeyrie L, Stoffers P (1995) Sea-surface temperature and salinity oscillations in the subtropical gyre of the Indian Ocean during the last climatic cycle. 5th P.P. Zonenshain Mem. Conf. on Plate Tectonics, Progr. and Abstr, IORAS/GEOMAR:186

Ivanova EV, Heilig S, Wang L, Pflaumann U, Sarnthein M, Pelejero C (1997) Paleoceanography of the South China Sea off Vietnam during the last 250,000 years. Terra Nova 9:614

Ivanova EV, Murdmaa IO, Duplessy J-C, Paterne M (2002) Late Weichselian to Holocene Paleoenvironments of the Barents Sea. Glob. Planet Change 34(3–4):69–78

Ivanova EV, Polyak L, Murdmaa IO, Khromova N, Vetrov A (2003a) Little ice age and medieval warm period in the Russkaya Gavan' Fjord (NE Barents Sea): relations to NAO and THC oscillations. Program and Abstracts, IMAGES. Holocene Working Group Workshop. Hafslo, Norway, August 27–29, 2003

Ivanova E, Schiebel R, Singh AD, Schmiedl G, Niebler H-S, Hemleben Ch (2003b) Primary production in the Arabian Sea during the last 135,000 years. Palaeogeogr. Palaeoecol. Palaeoclim. 197:61–82

Ivanova EV, Akhrimenko N, Risebrobakken B (2007a) Centennial-to-millenial-scale variability of foraminiferal assemblages in the Barents Sea during the Holocene: controlling factors and implications. 37th International Arctic Worshop, Skatefell, Iceland, pp. 121–124

Ivanova EV, Murdmaa IO, Chepalyga AL, Cronin TM, Pasechnik IV, Levchenko OV, Howe SS, Manushkina AV, Platonova E (2007b) Holocene sea-level oscillations and environmental changes on the Eastern Black Sea shelf. Palaeogeogr. Palaeoclimatol. Palaeoecol. 246:228–259

Ivanova EV, Chistyakova NO, Risebrobakken B, Khusid TA (2008a) Barents Sea environments from Late Weichselian extreme to present. In: Long AJ, Evans DJA, OCofaigh C, Roberts DH (eds) Arctic Palaeoclimate and its Extremes (APEX), second international conference and workshop: Abstract volume. Department of Geography, Durham University, Durham, UK

Ivanova EV, Ovsepyan EA, Risebrobakken B, Vetrov AA (2008b) Downcore distribution of living calcareous foraminifera and stable isotopes in the Western Barents Sea. J. Foraminiferal Res. 38(4):337–356

Jain S, Collins LS (2007) Trends in Caribbean Paleoproductivity related to the Neogene closure of the Central American Seaway. Mar. Micropal. 63:57–74

Jansen E, Veum T (1990) Evidence for two-step deglaciation and its impact on North Atlantic deep-water circulation. Nature 343:612–616

Jansen E, Koç N (2000) Century to Decadal Scale Records of Norwegian Sea Surface Temperature Variations of the Past 2 Millenia. CLIVAR Exchanges 5(1):13–14

Jansen E, Fronval T, Rack F, Channell JET (2000) Pliocene-Pleistocene ice rafting history and cyclicity in the Nordic Seas during the last 3,5 Myr. Paleoceanography 15(6):709–721

Jansen E, Anderson C, Nesje A et al. (2001) Decadal scale resords of oceanic variability in the Eastern Nordic seas over the past 10,000 years. 7th Intern. Conf. on Paleoceanography, Program&Abstracts, Sapporo, pp. 36–37

Jansen E, Overpeck J, Briffa KR et al. (2007) Palaeoclimate. In: Solomon S, Qin D, Manning M et al. (ed) Climate Change 2007: The Physical Science Basis. Contribution of Working Group I to the Fourth Assessment Report of the Intergovernmental Panel on Climate Change, Cambridge University Press, Cambridge

Jansen E, Andersson C, Moros M, Nisancioglu KH, Nyland BF, Telford RJ (2008) The early to mid-Holocene thermal optimum in the northern North Atlantic and Nordic Seas: the role of seasonal orbital forcing and Holocene century to millennial scale climate events. In: Battarbee RW, Binney HA (ed) Global Warming and Natural Climate Variability: A Holocene Perspective. Blackwell, Oxford

Jian Z, Li B, Pflaumann U, Wang P (1996) Late Holocene cooling event in the western Pacific. Sci. China 39(D):552–560

Jian Z, Wang P, Cheng X, Tian J, Zhao Q, Huang B (2001) Evolution and variability of the East Asian monsoon: records of isotope and foraminifers from the South China Sea. 7th Intern. Conf. on Paleoceanography, Program&Abstracts, Sapporo, p. *33*

Johnsen SJ, Clausen HB, Dansgaard W, Fuhrer K, Gundestrup N, Hammer CU, Iversen P, Jouzel J, Stauffer B, Steffensen JP (1992) Irregular glacial interstadials recorded in a new Greenland ice core. Nature 359:311–313

Jones EP, Anderson LG, Wallace DWR (1991) Tracers of near-surface, halocline and deep waters in the Arctic Ocean: Implications for circulation. J. Mar. Systems 2: 241–255

Jones GA, Keigwin LD (1988) Evidence from Fram Strait (78°N) for early deglaciation. Nature 336:56–59

Jones PD, Osborn TJ, Briffa KR (2001) The evolution of climate over the last millenium. Science 292:662–667

Jouzel J, Vaikmae R, Petit JR et al. (1995) The two step shape and timing of the last deglaciation in Antarctica. Climate Dynamics 11:125–126

Jung SJA, Davies GR, Ganssen GM, Kroon D (2004) Synchronous Holocene sea surface temperature and rainfall variations in the Asian monsoon system. Quat. Sci. Rev. 23:2207–2218

Jung T, Hilmer M, Ruprecht E, Kleppek S, Gulev SK, Zolina O (2003) Characteristics of the Recent Eastward Shift of Interannual NAO Variability. J. Climate 16(20):3371–3382

Kandiano ES, Bauch HA (2003) Surface ocean temperatures in the north-east Atlantic during the last 500,000 years: evidence from foraminiferal census data. Terra Nova 15:265–271, doi:10.1046/j.1365-3121.2003.00488

Kanfoush SL, Hodell DA, Charles CD, Guilderson TP, Mortyn PG (2000) US Ninneman, millennial-scale instability of the Antarctic ice sheet during the last glacial. Science 288:1815–1818

Kaplin PA, Selivanov AO (1999) Modification of the Russian seas level and coasts development. GEOS, Moscow (in Russian)

Kaufman DS, Ager TA, Anderson NJ, Anderson PM, Andrews JT, Bartlein PJ, Brubaker LB, Coats LL, Cwynar LC, Duvall ML, Dyke AS, Edwards ME, Eisner WR, Gajewski K, Geirsdottir A, Hu FS, Jennings AE, Kaplan MR, Kerwin MW, Lozhkin AV, MacDonald GM, Miller GH, Mock CJ, Oswald WW, Otto-Bliesner BL, Porinchu DF, Ruhland K, Smol JP, Steig EJ, Wolfe BB (2004) Holocene thermal maximum in the western Arctic (0–180ºW). Quat. Sci. Rev. 23:529–560

Kawamura K, Parrenin F, Lisiecki L, Uemura R, Vimeux F, Severinghaus JP, Hutterli MA, Nakazawa T, Aoki S, Jouzel J, Raymo ME, Matsumoto K, Nakata H, Motoyama H, Fujita S, Goto-Azuma K, Fujii Y, Watanabe O (2007) Northern Hemisphere forcing of climatic cycles in Antarctica over the past 360,000 years. Nature 448(7156):912–917, doi:10.1038/nature06015

Keeling RF, Visbeck M. (2005) Northern ice discharges and Antarctic warming: Could ocean eddies provide the link? Quat. Sci. Rev. 24:1809–1820

Keigwin LD (2004) Radiocarbon and stable isotope constraints on Last Glacial Maximum and Younger Dryas ventilation in the western North Atlantic. Paleoceanography 19, PA4012, doi:10.1029/2004PA001029

Kennett JP (1982) Marine Geology. Prentice-Hall, Englewood Cliffs, NJ

Kennett JP, Keller G, Srinivasan MS (1985) Miocene planktonic foraminiferal biogeography and paleoceanographic development of the Indo-Pacific region. Geol. Soc. Amer. Mem. 163:197–236

Kerr RA (1994) Did the tropical Pacific drive the World's warming. Science 266(5185):544–545

Kerwin MW, Overpeck JT, Webb RS, DeVernal A, Rind DH, Healy RJ (1999) The role of oceanic forcing in mid-Holocene Northern Hemisphere climatic change. Paleoceanography 14(2):200–210

Khodri M, Ramstein G, Paillard D, Duplessy J-C, Kageyama M, Ganopolski A (2003) Modelling the climate evolution from the last interglacial to the start of the last glaciation: The role of Arctic Ocean freshwater budget. Geophys. Res. Lett. 30(12):1606, doi:10.1029/2003GL017108

Khromova NV, Ivanova EV (2002) Stratigraphic subdivision of the Holocene deposits from the eastern Barents Sea. In: Proceedings of the Second Yanshin Readings "Modern problems of Geology". Nauchniy Mir, Moscow, pp. 340–343 (in Russian)

Khusid TA (1996) Benthic foraminiferal assemblages in the Kara Sea. Oceanology 36(5):716–722 (English Translation)

Khusid TA, Korsun SA (1996) Modern benthic foraminiferal assemblages in the Kara Sea. In: Stein R, Ivanov G, Levitan M, Fahl K (eds) Surface sediment composition and sedimentary processes in the central Arctic Ocean and along the Eurasian Continental Margin. Ber. Polarforsch 212:308–314

Kiefer T, Kienast M (2005) Patterns of deglacial warming in the Pacific Ocean: a review with emphasis on the time interval of Heinrich event. Quat. Sci. Rev. 24:1063–1081

Kiefer T, Sarnthein M, Erlenkeuser H, Grootes PM, Roberts AP (2001) North Pacific response to millenial-scale changes in ocean circulation over the past 60 kyr. Paleoceanography 16:179–189

Kiefer T, Lorenz S, Schulz M, Lohmann G, Sarnthein M, Elderfield H (2002) Response of precipitation over Greenland and the adjacent ocean to North Pacific warm spells during Dansgaard–Oeschger stadials. Terra Nova 14:295–300

Kienast SS, McKay JL (2001) Sea Surface Temperature in the subarctic Northeast Pacific reflect millennial-scale Climate Oscillations during thr last 16 kyrs. Geophys. Res. Lett. 28(8):1563–1566

Kienast SS, Calvert SE, Pedersen TF (2002) Nitrogen isotope and productivity variations along the northeast Pacific margin over the last 120 kyr: Surface and subsurface paleoceanography. Paleoceanography 17(4):1055, doi:10.1029/2001PA000650

Kienast M, Kienast SS, Calvert SE, Eglinton TI, Mollenhauer G, Francois R, Mix AC (2006) Eastern Pacific cooling and Atlantic overturning circulation during the last deglaciation. Nature 443, doi:10.1038/nature05222

Kienast M, Steinke S, Stattegger K, Calvert SE (2001) Synchronous Tropical South China Sea SST Change and Greenland Warming During Deglaciation. Science 291:2132–2134

Kincaid E, Thunell RC, Le J, Lange CB, Weinheimer AL, Reis FMH (2000) Planktonic foraminiferal fluxes in the Santa Barbara basin: responce to seasonal and interannual hydrographic changes. Deep-Sea Res. 47(2):1157–1179

Kissel C, Laj C, Labeyrie L, Dokken T, Voelker A, Blamart D (1999) Magnetic Signatures of Rapid Climatic Variations in North Atlantic Sediments. In: Abrantes F, Mix A (eds) Reconstructing Ocean History: A Window into the Future. Kluwer Academic, Plenum Publishers, New York

Klimanov VA (1990) Quantitative characteristics of the northern Eurasian climate during the late glaciation. Proceedings of the USSR Academy of Sciences, Georg. Series 4:116–126 (in Russian)

Klimenko VV, Klimanov VA, Sirin AA, Sleptsov AM (2001) Climatic changes in the western European Russia during the Late Holocene. Doklady of the Russian Acad. of Sci. 376(5): 679–683 (in Russian)

Klitgaard-Kristensen D, Sejrup HP, Haflidason H (2001) The last 18 kyr fluctuations in Norwegian Sea surface conditions and implications for the magnitude of climatic change: Evidence from the North Sea. Paleoceanography 16:455–467

Klootwijk CT, Gee JS, Peirce JW, Smith GM (1992) Neogene evolution of the Himalayan-Tibetan region: constraints from ODP site 758, northern Ninetyeast Ridge; bearing on climatic change. Palaeogeogr. Palaeoclim. Palaeoecol. 95:95–110

Knaack JJ (1997) A new transfer function to reconstruct paleoproductivity from marine diatom assemblages. Ber.-Rep., Geol.-Palaeontol. Inst. Univ. Kiel 83:1–125 (in German)

Knies J, Stein R (1998) New aspects of organic carbon deposition and its paleoceanographic implications along the northern Barents Sea margin during the last 30,000 years. Paleoceanography 13(4):384–394

Knies J, Vogt C, Stein R (1999) Late Quaternary growth and decay of the Svalbard/Barents Sea ice sheet and paleoceanographic evolution in the adjacent Arctic Ocean. Geo-Marine Lett. 18:195–202

Knies J, Nowaczyk N, Müller C, Vogt C, Stein R. (2000) Amultiproxy approach to reconstruct the environmental changes along the Eurasian continental margin over the last 150,000 years. Mar. Geol. 163:317–344.

Knorr G, Lohmann G (2003) Southern Ocean origin for the resumption of Atlantic thermohaline circulation during deglaciation. Nature 424:532–536

Knorr G, Lohmann G (2004) The Southern Ocean as the Flywheel of the Oceanic Conveyor Belt Circulation. Pages News 12(1):11–13

Knutti R, Flückiger J, Stocker TF, Timmermann A (2004) Strong hemispheric coupling of glacial climate through freshwater discharge and ocean circulation. Nature 430:851–856

Koç N, Jansen E (1994) Response of high-latitude Northern Hemisphere to orbital climate forcing: Evidence from the Nordic Seas. Geology 22:523–526

Koç N, Jansen E, Haflidason H (1993) Paleoceanographic reconstructions of surface ocean conditions in the Greenland, Iceland and Norwegian seas through the last 14 ka based on diatoms. Quat. Sci. Rev. 12:115–140

Koerner RM, Fisher DA (1990) A record of Holocene summer climate from a Canadian high-Arctic ice core. Nature 343:630–631

Koerner RM (2001) Sudden climatic change. In: Watanabe O, Yamanouchi T (eds) Environmental Research in the Arctic 2000. Mem. Nat. Inst of Polar Res. Spec. Issue 54, Tokyo

Kohfeld KE, Fairbanks RG, Smith SL, Walsh ID (1996) Neogloboquadrina pachyderma (sinistral coiling) as paleoceanographic tracer in polar oceans: evidence from northeast water polynia plankton tows, sediment traps, and surface sediments. Paleoceanography 11:679–699

Koopman B (1981) Saharan dust deposition in the subtropical Atlantic during the last 25,000 years (in German). 'Meteor' Forschungsergeb C 5:23–54

Kopelevich OV, Burenkov VI, Ershova SV, Shcheberstov SV, Evdoshenko MA, Lukyanova EA. (2002) Biooptical characteristics of the Russian sea by the SeaWiFS Satellite Ocean Color. CD Shirshov Institute of Oceanology. Moscow

Korsun S (1998) Benthic foraminifera in the Ob and Yenisei Estuaries. In Matthiessen J, Stepanets OV (eds) Scientific Cruise Report of the Kara Sea Expedition of RV "Akademik Boris Petrov" in 1997. Ber. Polarforsch. 266:29–31

Korsun S (1999) Benthic foraminifera in the Ob estuary, West Siberia. In: Matthiessen J et al. (eds) The Kara Sea Expedition of RV "Akademik Boris Petrov" 1997: First Results of a Joint Russian German Pilot Study. Ber. Polarforsch. 300:59–70

Korsun S, Hald M (1998) Modern benthic foraminifera off Novaya Zemlya tidewater glaciers, Russian Arctic. Arctic Alpine Res. 30:61–77

Korsun S, Hald M (2000) Seasonal dynamics of benthic foraminifera in a glacially fed fjord of Svalbard, European Arctic. J. Foramin. Res. 30(4):251- 271

Korsun SA, Pogodina IA, Tarasov GA, Matishov GG (1994) Foraminifers of the Barents Sea. Murmansk Marine Biological Institute, Kola Scientific Centre of the Russian Acad. Sci., Apatity (in Russian)

Korsun SA, Pogodina IA, Forman SL, Lubinski DJ (1995) Recent foraminifera in glaciomarine sediments from three Arctic fjords of Novaya Zemlya and Svalbard. Polar Res. 14:15–31

Koshlyakov MN, Sagina TG, Goldin AYu (2001) Pacific Antarctic Cell of the Global Oceanic Conveyor. Izvestiya RAN, Atmospheric and Oceanic Physics 37(4):482–489 (English Translation)

Koutavas A, Lynch-Stieglitz J, Marchitto TM, Sachs JP (2002) El Ninõ-like pattern in ice age tropical sea surface temperature. Science 297:226–230

Krasheninnikov VA, Basov IA (1986) Ceinozoic Stratigraphy of the Southern Ocean. Nauka, Moscow (in Russian)

Krasheninnikov VA, Basov IA, Golovina LA et al (1999) Miocene of the North-East Atlantic and Eastern Mediterranean. Nauchniy Mir, Moscow (in Russian)

Krey J, Babenerd B (1976) Phytoplankton production. Atlas of the International Indian Ocean Expedition. Institut für Meereskunde an der Universität Kiel

Kroon D (1988) Distribution of extant planktic foraminiferal assemblages in Red Sea and northern Indian Ocean surface waters. In: Brummer GJA, Kroon D (eds) Planktonic Foraminifera as Tracers of Ocean Climate History. Free University Press, Amsterdam

Kroon D (1991) Distribution of extant planktic foraminiferal assemblages in Red Sea and northern Indian Ocean surface waters. Rev. Esp. Micropal. 23(1):37–74

Kroon D, Ganssen G (1989) Northern Indian Ocean upwelling cells and stable isotope composition of living planktonic foraminifers. Deep-Sea Res. 36:1219–1236

Kroon D, Steens T, Troelstra S (1991) Onset of monsoonal related upwelling in the western Arabian sea as revealed by planktonic foraminifers. Proc. ODP. Leg. 117:257–263

Kroopnick (1985) The distribution of ^{13}C of ΣCO_2 in the World Oceans. Deep-Sea Res. 32:57–84

Kucera M, Weinelt M, Kiefer T, Pflaumann U, Hayes A, Weinelt M, Chen M-T, Mix A, Barrows TT, Cortijo E, Duprat J, Juggins S, Waelbroeck C (2005) Reconstruction of sea-surface temperatures from assemblages of planktonic foraminifera: multi-technique approach based on geographically constrained calibration datasets and its application to glacial Atlantic and Pacific Oceans. Quat. Sci. Rev. 24:951–998

Kudrass HR, Erlenkeuser H, Vollbrecht R, Weiss W (1991) Global nature of the Younger Dryas cooling event inferred from oxygen isotope data from Sulu Sea cores. Nature 349:406–409

Kuhlbrodt T, Griesel A, Montoya M, Levermann A, Hofmann M, Rahmstorf S (2007) on the driving processes of the Atlantic meridional overturning circulation. Rev. Geophys. 45, RG2001. doi:10.1029/2004RG000166

Kukla G, Heller F, Ming X-L, Cxhun X-T, Liu T-S, An Z-S (1988) Pleistocene climates in China dates by magnetic susceptibility. Geology 16:811–814

Kutzbach J, Bonan G, Foley J, Harrison SP (1996) Vegetation and soil feedbacks on the response of the African monsoon to Orbital forcing in the early to middle Holocene. Nature 384:623–626

Kutzbach JE, Gallimore RG (1988) Sensitivity of a coupled atmosphere/mixed layer ocean model to changes in orbital forcing at 9,000 years B.P. J. Geophys. Res. 93(D1):803–821

Kutzbach JE, Gueter PJ, Ruddiman WF, Prell WL (1989) Sensivity of climate to late Cenozoic uplift in southern Asia and the American West: numerical experiments. J. Geophys. Res. 94(D15):18393–18407

Labeyrie L (2000) Glacial Climate Instability. Science 290:1905–1907

Labeyrie L, Elliot M (1999) Abrupt climatic changes-causes and consequences. In: Reconstructing Ocean History: Abrantes F, Mix A (eds) A Window into the Future. Kluwer Academic / Plenum Publishers, New York, pp. 73–81

Labeyrie L, Duplessy J-C, Blanc PL (1987) Variations in mode of formation and temperature of oceanic deep waters over the past 125,000 years. Nature 327:477–482

Labeyrie L, Duplessy J-C, Duprat J, Juillet-Leclerc A, Moyes J, Michel E, Kallel N, Shackleton NJ (1992) Changes in the vertical structure of the North Atlantic Ocean between glacial and modern times. Quat. Sci. Rev. 11:401–413

Labeyrie L, Leclaire H, Waelbroeck C, Cortijo E, Duplessy J-C, Vidal L, Elliot M, Le Coat B (1999) Mechanisms of Global Climate Change at Millennial Time Scales. Geophys. Monograph, American Geophys. Union 112:77–97

Lacombe H (1965) Cours d'Océanographie Physique. Gantier-Villard. Paris.

Laj C, Mazaud A, Duplessy J-C (1996) Geomagnetic intensity and 14C abundance in the atmosphere and ocean during the past 50 ka. Geophys. Res. Lett. 23:2045–2048

Lal DT, Jull AJT, Pollard D, Vacher L (2004) Evidence for large century time-scale changes in solar activity in the past 32 kyr, based on in-situ cosmogenic ^{14}C in ice at Summit, Greenland. Earth Planet Sci. Let. 234:335–349

Lambeck K (1995) Constrains on the late Weichselian ice sheet over the Barents sea from observations of raised shorelines. Quat. Sci. Rev. 14:1–16

Lambeck K, Yokoyama Y, Jonston P, Purcell A (2000) Global ice volumes at the Last Glacial Maximum and early Lateglacial. Earth Planet. Sci. Lett. 181:513–527

Landais A, Jouzel J, Masson-Delmotte V, Caillon N (2005) Large temperature variations over rapid climatic events in Greenland: A method based on air isotopic measurements. Comptes Rendus Geosci. 337(10–11):947–956

Landvik JY, Bondebik S, Elyerhoi A, Fjeldskaar W et al (1998) The Last Glacial Maximum of Svalbard and the Barents sea Area: Ice Sheet Extent and Configuration. Qutern. Sci. Rev. 17(1–3):43-76

Lappo SS (1984) On reasons of the northward heat advection across the Equator in the Atlantic ocean. In: Lappo SS (ed) Study of Ocean and Atmosphere Interaction Processes. Moscow Department of Gidrometeoizdat, Moscow, pp. 125–129 (in Russian)

Lappo SS (1995) The history of interoceanic heat and salt exchange. Priroda, 2:100–106 (in Russian)

Lappo SS (2004) Large-scale interaction in the ocean-atmosphere system and energy active zones of the World Ocean. In: Alexeevsky NI et al. (eds) Dynamics and interaction of the atmosphere and the hydrosphere. Geological Department, Moscow State University, Moscow (in Russian)

Lappo SS, Gulev SK, Rozhdestvenskiy AE (1990) Large-scale thermal interaction in the ocean-atmosphere system and energy-active zones of the World Ocean. Gidrometeoizdat, Leningrad (in Russian)

Laskar J (1990) The chaotic motion of the solar system: A numerical estimate of the chaotic zones. Icarus 88:266–291

Lawver AL, Gahagan LM, Coffin MF (1992) The Development of Paleoseaways around Antarctica. In: Kennet JP, Warnke DA (eds) The Antarctic Paleoenvironement: A Perspective on Global Change. Amer. Geophys. Union, Washington, DC

Le J, Thunell RC (1996) Modelling planktic foraminiferal assemblage changes and application to sea surface temperature estimation in the western equatorial Pacific Ocean. Mar. Micropaleontol. (28):211–229.

Lea DW, Pak DK, Spero HJ (2000) Climate impact of Late Quaternary Equatorial Pacific Sea surface temperature variations. Science 289:1719–1723

Lea DW, Pak DK, Belanger CL, Spero HJ, Hall MA, Shackleton NJ (2006) Paleoclimate history of Galapagos surface waters over the last 135,000 yr. Quat. Sci. Rev., doi:10.1016/j.quascirev.2005.11.010

Leduc G, Vidal L, Tachikawa K, Rostek F, Sonzogni C, Beaufort L, Bard E (2007) Moisture transport across Central America as a positive feedback on abrupt climatic changes. Nature 445, doi:10.1038/nature05578

Lehman SJ (1991) Initiation of Fennoscandian oce-sheet retreat during the last deglaciation. Nature 349:513–516

Leuschner DC, Sirocko F (2003) Orbital insolation forcing of the Indian monsoon—A motor for global climate changes? Palaeogeogr. Palaeoclimatol. Palaeoecol. 197:83–95

Levi C, Labeyrie L, Bassinot F, Guichard F, Cortijo E, Waelbroeck C, Caillon N, Duprat J, de Garidel-Thoron T, Elderfield H (2007) Low-latitude hydrological cycle and rapid climate changes during the last deglaciation. Geochem. Geophys. Geosyst. 8, Q05N12, doi:10.1029/2006GC001514

Levitan MA, Seidov DG (1989) On reconstruction of the paleocirculation of Indian ocean waters. Doklady of the USSR Academy of Sciences 304(4):937–977 (in Russian)

Levitan MA, Arnold M, Burtman MV, Ivanova EV, Marina MM (2000). Sedimentation History of the Eastern Kara Sea. Oceanology 40(4):614–620 (English Translation)

Levitan MA, Belyaev NA, Burtman MV, Duplessy JC, Husid TA (2003) T The history of the Holocene sedimentation in the Sothern Novaya Zemlya Trench. Lithology and Mineral Resources 6:660–672 (in Russian)

Levitus S (1982) Climatological atlas of the Worls oceans. NOAA Prof. Pap. 13, US Govt. printing office

Levitus S, Boyer TP (1994) World Ocean Atlas. Temperature. NOAA, U.S. Dep. Commerce Washington, DC 4:117

Little MG, Schneider RR, Kroon D, Price B, Summerhayes CP, Segl M (1997) Trade wind forcing of upwelling, seasonally, and Heinrich events as a response to sub-Milankovitch climate variability. Paleoceanography 12(4):568–576.

Lisiecki LE, Raymo ME (2005) Plio-Pleistocene stack of 57 globally distributed benthic $\delta^{18}O$ records. Paleoceanography, doi:10.1029/2004PA001071

Lisitzin AP (1974) Sedimentation in Oceans. Nauka, Moscow (in Russian)

Lisitzin AP (1995) The marginal filter of ocean. Oceanology 34(5):671–682 (English Translation)

Lisitzin AP (2001) Unsettled problems of Arctic oceanology. In: Lisitzin AP, Vinogradov ME, Romankevich EA (eds) Experience of System Oceanologic Studies in the Arctic, Scientific World, Moscow (in Russian)

Lisitzin AP, Shevchenko VP, Vinogradov ME, Severina OL, Vavilova VV, Mitzkevich IN (1995) Sedimentary material fluxes in the Kara Sea and in Ob and Yenisei estuaries. Oceanology 34(5):683–694 (English Translation)

Lisitzin AP, Shevchenko VP, Burenkov VI (2000) Hydrooptics and suspended matter of Arctic seas. Atmospheric and Oceanic Optics 13:61–70 (English Translation)

Liu T, Ding Z (1993) Stepwise coupling of monsoon circulation to global ice volume variation during the last Cenozoic. Glob. Planet Sci. 7:119–130

Liu Z, Alexander M (2007) Atmospheric bridge, oceanic tunnel, and global climatic teleconnections. Rev. Geophys. 45, RG2005, doi:10.1029/2005RG000172

Loeng H (1991) Features of the physical oceanographic conditions of the Barents Sea. Polar Res. 10:5–18

Lohmann G, Gerdes R (1998) Sea ice effects on the sensitivity of the Thermohaline Circulation in simplified atmosphere-ocean-sea ice models. J. Climate 11:2789–2803.

Lohmann G, Schulz M (2000) Reconciling Bølling warmth with peak deglacial meltwater discharge. Paleoceanography 15(5):537–540

Lototskaya A, Ganssen GM (1999) The structure of Termination II (penultimate deglaciation and Eemian) in the North Atlantic. Quat. Sci. Rev. 18:1641–1654

Loubere P, Bennett S (2008) Southern ocean biogeochemical impact on the tropical ocean: Stable isotope records from the Pacific for the past 25,000 years. Glob. Planet. Change, doi: 10.1016/j.gloplacha.2008.08.001

Loutre MF, Paillard D, Vimeix F, Cortijo E (2004) Does mean annual insolation have the potential to change the climate? Earth Planet. Sci. Lett. 221:1, doi:10.1016/S0012-821X(04)00108-6

Lubinski DJ, Korsun S, Polyak L, Forman SL, Lehman SJ, Herlihy FA, Miller GH (1996) The last deglaciation of the Franz Victoria Trough, northern Barents Sea. Boreas 25:89–100

Lubinski DJ, Forman SL, Miller GH (1999) Holocene glacier and climate fluctuation on Franz Josef Land, Arctic Russia, 80°N. Quat. Sci. Rev. 19:85–108

Lubinski DJ, Polyak L, Forman SL (2001) Freshwater and Atlantic water inflows to the deep northern Barents and Kara seas since ca 13 ^{14}C ka: foraminifera and stable isotopes. Quat. Sci. Rev. 20:1851–1879

Lund DC, Mix AC (1998) Millennial-scale deep water oscillations: Reflections of the North Atlantic in the deep Pacific from 10 to 60 ka. Paleoceanography 13(1):10–19

Lynch-Stieglitz J (2003) Tracers of past ocean circulation. In: Turekian KK, Holland HD (eds) Treatise on Geochemistry 6: The Oceans and Marine Geochemistry, Elsevier, pp. 433–451

Lynch-Stieglitz J, Curry WB, Slowey N, Schmidt GA (1999) The Overturning circulation of the Glacial Atlantic. A View from the Top. In: Abrantes F, Mix A (eds) Reconstructing Ocean History: A Window into the Future. Kluwer Academic/Plenum Publishers, New York, pp. 7–32

Lynch-Stieglitz J, Adkins JF, Curry WB, Dokken T, Hall IR, Herguera JC, Hirschi JJ-M, Ivanova EV, Kissel C, Marchal O, Marchitto TM, McCave IN, McManus JF, Mulitza S, Ninnemann U, Peeters F, Yu E-F, Zahn R (2007) Atlantic Meridional Overturning Circulation During the Last Glacial Maximum. Science 66(316), doi:10.1126/science.1137127

MacAyeal DR (1993) Binge/purge oscillations of the Laurentide ice sheet as a cause of the North Atlantic's Heinrich events. Paleoceanography 8:775–784

Macdonald A (1998) The global ocean circulation: a hydrographic estimate and regional analyses. Progress in Oceanography 41(3):281–382

MacDonald A, Wunsch C (1996) An estimate of global ocean circulation and heat fluxes. Nature 382:436–439

Mackensen A, Bickert T (1999) Stable carbon isotopes in benthic foraminifera: proxies for deep and bottom water circulation and new production. In: Fischer G, Wefer G (eds) Use of Proxies in Paleoceanography: Examples from the South Atlantic, pp. 229–254. Springer-Verlag, Berlin, Heidelberg

Madhupratap M, Kumar SP, Bhattathiri PMA, Kumar MD, Raghukumar S, Nair KKC, Ramaiah N (1996) Mechanism of the biological response to winter cooling in the northeastern Arabian Sea. Nature 384:549–552

Manabe S, Stouffer R (1995) Simulation of abrupt climate-change induced by fresh-water input to the North Atlantic Ocean. Nature 378(6553):165–167

Manabe S, Stouffer R (1997) Coupled ocean-atmosphere model response to freshwater input: comparison to Younger Dryas event. Paleoceanography 12(2):321–336

Mangerud J, Gulliksen S (1975) Apparent radiocarbon ages of recent marine shells from Norway, Spitsbergen, and Arctic Canada. Quat. Res. 5:273–296

Mangerud J, Dokken T, Hebbeln D, Heggen B, Ingolfsson O, Landvik JY, Mejdahl V, Svendsen JI, Vorren TO (1998) Fluctuations of the Svalbard-Barents Sea Ice Sheet during the last 150000 years. Quat. Sci. Rev. 17:11–22.

Mangerud J, Svendsen J-I, Astakhov VL (1999) The age and extent of the Barents and Kara Sea ice sheets in Northern Russia. Boreas 28:46–80

Mangerud J, Astakhov VL, Svendsen J-I (2002) The extent of the Barents Kara ice sheet during the Last Glacial Maximum. Quat. Sci. Rev. 21:111–119

Mangerud J, Jakobsson M, Alexanderson H, Astakhov V, Clarke GKC, Henriksen M, Hjort C, Krinner G, Lunkka J-P, Möller P, Murray A, Nikolskaya O, Saarnisto M, Svendsen JI (2004) Ice-dammed lakes and rerouting ofthe drainage of northern Eurasia during the Last Glaciation. Quat. Sci. Rev. 23:1313–1332

Mangerud J, Bondevik S, Gulliksen S, Hufthammer AK, Høisæter T (2006) Marine ^{14}C reservoir ages for 19th century whales and molluscs from the North Atlantic. Quat. Sci. Rev. 25:3228–3245

Manghnani V, Morrison JM, Hopkins TS, Böhm E (1998) Advection of upwelled waters in the form of plumes off Oman during the Southwest Monsoon. Deep-Sea Res. 45(2):2027–2052

Manighetti B, McCave IN (1995) Late glacial and Holocene palaeocurrents through South Rockall Gap, NE Atlantic Ocean. Paleoceanography 10:611–626

Marchuk GI (ed) (1983). The Program of atmosphere-ocean interaction studies for investigation of short-term climate changes (The "Cross-sections" program). Moscow (in Russian)

Marchitto TM, Lynch-Stieglitz J, Hemming SR (2005) Deep Pacific CaCO3 compensation and glacial–interglacial atmospheric CO_2. Earth Planet. Sci. Lett. 231:317–336

Marchitto TM, Broecker WS (2006) Deep water mass geometry in the glacial Atlantic Ocean: A review of constraints from the paleonutrient proxy Cd/Ca. Geochem. Geophys. Geosyst. 7, Q12003

Marchitto TM, Lehman SJ, Ortiz JD, Flückiger J, van Geen A (2007) Marine radiocarbon evidence for the mechanism of deglacial atmospheric CO_2 rise. Science, doi:10.1126/science.1138679

Marchuk GI, Sarkisyan AS (1988) Mathematic modelling of oceanic circulation. Nauka, Moscow (in Russian)

Martinez I, Keigwin L, Barrows TT, Yokoyama Y, Southon J (2003) La Ninã-like conditions in the eastern equatorial Pacific and a stronger Choco jet in the northern Andes during the last glaciation. Paleoceanography 18(2):1033, doi:10.1029/2002PA000877

Martinez-Mendez G, Zahn R, Hall IR, Pena LD, Cacho I (2007) 345,000 year long multi-proxy records off South Africa document variable contributions of Northern versus Southern component water to the deep South Atlantic. Earth Planet. Sci. Lett. (267):309–321

Martinson DG, Pisias NG, Hays JD, Imbrie J, Moore TC, Shackleton NJ (1987) Age dating and the orbital theory of the Ice Ages: Development of a high-resolution 0 to 300,000-year chronostratigraphy. Quat. Res. 27:1–29

Martrat B, Grimalt JO, Lopez-Martinez C, Cacho I, Sierro FJ, Flores JA, Zahn R, Canals M, Curtis JH, Hodell DA (2004) Abrupt temperature changes in the western Mediterranean over the past 250,000 years. Science 306:1762–1765

Martrat B, Grimalt JO, Shackleton NJ, de Abreu L, Hutterli MA, Stocker TF (2007) Four climate cycles of recurring deep and surface water destabilizations on the Iberian margin centennial climate variability over the last ice age. Science 317:502–507

Maslin MA, Ridgwell AJ (2005) Mid-Pleistocene revolution and the 'eccentricity myth'. Geological Society, London, Special Publications 247:19–34

Maslin M, Sarnthein M, Knaack JJ (1996) Subtropical eastern Atlantic climate during the Eemian. Naturwissenschaften 83:122–126

Maslin M, Shackleton NJ, Pflaumann U (1995) Suface water temperature, salinity, density changes in the northeast Atlantic during the last 45,000 years: Heinrich events, deep water formation, and climatic rebounds. Paleoceanography 10:527–544

Matishov GG (1984) Ocean floor during the glaciation. Nauka, Leningrad (in Russian)

Matishov GG (ed) (1995) Environments and ecosystems of the Novaya Zemlya (archipelago and shelf). Apatity (in Russian)

Matishov GG, Golubov V, Adrov N, Slobodin V, Levitus S, Smolyar I (1998) Climatic Atlas of the Barents sea: Temperature, salinity, oxygen. NOAA Atlas NESDIS 26, Silver Spring, Murmansk

Matishov GG, Druzhkov NV, Druzhkova EI, Larionov VV (1999) Phytoplancton of the northern Barents Sea (Franz-Victoria Trough) in the beginning of winter. Doklady of theUSSR Acad. Sci. 367(4):560–562 (in Russian)

Matul AG, Yushina IG, Yemelanov EM (2002) About Late Quaternary paleohydrological parameters of the Labrador Sea by radiolarians. Oceanology 42(2):247–251 (English Translation)

Mayewski, PA, Meeker, LD, Twickler, MS, Whitlow, S, Yang, Q, Lyons, WB, Prentice, M (1997) Major features and forcing of high-latitude northern hemisphere atmospheric circulation using a 110,000-year-long glaciochemical series. J. Geophys. Res. 102, 26, 345–26, 366

Mayewski PA, Rohling EE, Stager JC, Karlen W, Maasch KA, Meeker LD, Meyerson EA, Gasse F, van Kreveld S, Holmgren K, Lee-Thorp J, Rosqvist G, Rack F, Staubwasser M, Schneider RR, Steig EJ (2004) Holocene climate variability. Quat. Res. 62:243–255

McIntyre A, Molfino B (1996) Forcing of Atlantic Equatorial and subpolar millennial cycles by presession. Science 274:1867–1870

McManus JF, Oppo DW, Cullen JL (1999) A 0.5-Million-Year record of millennial-scale climate variability in the North Atlantic. Science 283:971–973

McManus JF, Francois R, Gherardi J-M, Jeigwin LD, Brown-Leger S (2004) Collapse and rapid resumption of Atlantic meridional circulation linked to deglacial climate changes. Nature 428(6985):834–837

McPhaden MJ, Busalacchi AJ, Cheney R, Donguy J-R et al. (1998) The Tropical Ocean-Global Atmosphere observing system: A decade of progress. J. Geophys. Res. 103(C7):14,169–14,240

Meeker LD, Mayewski PA (2002) A 1400-year high-resolution record of atmospheric circulation over the North Atlantic and Asia. The Holocene 12:257–266

Metzger EJ, Hulbert HE (1996) Coupled dynamics of the South China Sea, the Sulu Sea, and the Pacific Ocean. J. Geophys. Res 101:12331–12352

Miao Q, Thunell RC (1994) Glacial-Holoence carbonate dissolution and sea-surface temperatures in the South China and Sulu Seas. Paleoceanography 9:269–290

Midttun L (1985) Formation of dense bottom water in the Barents Sea. Deep Sea Res. 32:1233–1241

Mikolaewich U, Crowley TJ, Schiller A, Voss R (1997) Modelling teleconnections between the North Atlantic and North Pacific during the Younger Dryas. Nature 387:384–387

Miller KG, Wright JD, Brower AN (1989) Oligocene to Miocene stable isotope stratigraphy and planktonic foraminifer biostratigraphy of the Sierra Leone Rise (DSDP Site 366 and ODP Site 667). In: Ruddiman W et al. (eds) Proc. ODP. Sci. Res. 108, College Station: Ocean Drilling Program, pp. 279–294

Mix AC (2006) Running hot and cold in the eastern equatorial Pacific. Quat. Sci. Rev. 25:1147–1149

Mix AC, Bard E, Schneider R (2001) Environmental processes of the ice age: land, oceans, glaciers (EPILOG). Quat. Sci. Rev. 20:1–34

Monin AS, Shishkov YuA (1979) History of Climate. Gidrometeoizdat, Leningrad (in Russian)

Monin AS, Shishkov YuA (1998) About statistical parameters of the Little Ice Age. Doklady of the USSR Academy of Sciences 358(2):252–255 (in Russian)

Moore MD, Charles CD, Ruberstone JL, Fairbanks RG (2000) U/Th-dated sclerosponges from the Indonesian Seaway record subsurface adjustments to west Pacific winds. Paleoceanography 15(4):404–416

Moore TC, Walker JCG, Rea DK, Lewis CFM, Shane LCK, Smith AJ (2000) Younger Drays interval and outflow from the Laurentide ice sheet. Paleoceanography 15(1):4–18

Moreno A, Cacho I, Canals M, Grimalt JO, Sannchez-Goni MA, Shackleton NJ, Sierro FJ (2005) Links between marine and atmospheric processes oscillating on a millennial time-scale. A multi-proxy study of the last 50,000 yr from the Alboran Sea (Western Mediterranean Sea). Quat. Sci. Rev. 24:1623–1636

Morley JJ, Heusser LE (1997) Role of orbital forcing in East Asian monsoon climates during the last 350 kyr: evidence from terrestrial and marine climate proxies from core RC14–99. Paleoceanography 12(3):483–493

Moros M, Emeis K, Risebrobakken B, Snowball I, Kuijpers A, McManus J, Jansen E (2004) Sea surface temperatures and ice rafting in the Holocene North Atlantic: climate influences on northern Europe and Greenland. Quat. Sci. Rev. 23:2113–2126

Mosby H (1938) Svalbard waters. Geofysiske Publikasjoner 12:85

Müller A, Opdyke BN (2000) Glacial-interglacial changes in nutrient utilization and paleoproductivity in the Indonesian Throughflow sensitive Timor Trough, easternmost Indian Ocean. Paleoceanography 15(1):85–94

Müller C, Stein R (1999) Grain-size distribution and clay-mineral composition in surface sediments and suspended matter of the Ob and Yenisei Rivers. In: Matthiessen J, Stepanets OV, Stein R, Fütterer DK, Galimov EM (eds) The Kara Sea Expedition of RV "Akademik Boris Petrov" in 1997: First Results of a Joint Russian-German Pilot Study. Ber. Polarforsch. 300:179–187

Murdmaa IO, Ivanova EV (1985) Recent and last glacial deep-sea facies: Response to global climatic oscillations. Palaeogeogr. Palaeoclim. Palaeoecol. (50):285–290

Murdmaa IO, Ivanova EV (1987) Deep-Sea facies: Recent and of the Last Glacial Maximum. Lithology and Mineral Resources 1:3–16 (English Translation)

Murdmaa IO, Ivanova EV (1999) Postglacial sedimentation history in the shelf depressions of the Barents Sea. Lithology and Mineral Resources 6:576–595 (English Translation)

Murdmaa IO, Ivanova EV, Pimenov NV (1998a) Paleoceanographic changes in the Barents Sea: interrelations with global thermohaline circulation. Reconstructing Ocean History. 6 Intern. Conf. on Paleoceanography, Lisbon:71

Murdmaa IO, Ivanova EV, Pimenov NV (1998b). Periglacial marine sedimentation in the Barents Sea during the post-glacial times. In: Marine Periglacial and Glaciation of the Barents and Kara Seas Shelf in the Pleistocene. Kola Scientific Center, RAS, Apatity. 78–80 (in Russian)

Murdmaa IO, Polyak L, Ivanova EV, Korneeva GA (2000) A high-resolution sedimentary record from Russkaya Gavan', a Novaya Zemlya fjord: inferences for paleoclimatic conditions in the northeastern Barents Sea during the past millenium. CAPE/ICAPP abstr. "Sea Ice in the Climatic System. The Record of the North Atlantic Arctic", Iceland, pp. 46–47

Murdmaa IO, Ivanova EV, Belousov MA, Levitan MA, Bourtman MV, Duplessy J-C, Paterne M, Gordeev VYu, Korneeva GA (2001) Facies system of the Barents Sea since the Last Glaciation to Recent : A response to Paleoceanographic evolution. ICP VII Program & Abstracts, Sapporo, Japan, 158 pp

Murdmaa IO, Ivanova EV, Merklin LR, Polyak L, Vetrov AA, Korneeva GA, Lobkovskiy LI, Khromova NV (2003). Little Ice Age in the Russkaya Gavan' (the North Island of the Novaya Zemlya). In: Laverov NP (ed) Actual problems of Oceanology. Nauka, Moscow, (in Russian)

Murdmaa IO, Polyak L, Ivanova EV, Khromova N (2004) Paleoenvironments in Russkaya Gavan Fjord (NW Novaya Zemlya, Barents Sea) during the last millennium. Palaeogeogr. Palaeoclim. Palaeoecol. 209:141–154

Murdmaa IO, Ivanova EV, Risebrobakken B, Akhrimenko N, Yamskova E, Alekseeva T (2005) The Holocene Thermal Optimum in the Barents Sea. AGU Fall Meeting, San Francisco 2298

Murdmaa IO, Ivanova EV, Duplessy JC, Levitan MA, Khusid TA, Alekseeva TN, Bourtman MV, Belousov MA, Serova VV (2006) Facies system of the Central and Eastern Barents Sea since the Last Glaciation to Recent. Mar.Geol. 230:273–303

Murray JW (2006) Ecology and Applications of Benthic Foraminifera. Cambridge University Press, Cambridge

Murray DW, Prell WL (1991) Pliocene to Pleistocene variations in calcium carbonate, organic carbon, and opal on the Owen Ridge, Northern Arabian Sea. Proc. ODP. Sci. Res. 117, Washington, pp. 343–363

Murray DW, Prell WL (1992) Late Pliocene and Pleistocene climatic oscillations and monsoon upwelling recorded in sediments from the Owen Ridge, northwestern Arabian Sea. Upwelling Systems: Evolution since the Early Miocene. Geol. Soc. Spec. Publ. 64:301–321

Murtugudde R, Busalacchi AJ, Beauchamp J (1998) Seasonal-to-interannual effects of the Indonesian throughflow on the tropical Indo-Pacific Basin. J. Geophys. Res. 103(C10):21,425–21,441

Naidu PD (2006) Link between Western Arabian Sea Surface Temperature and Summer Monsoon Strength and High-Latitude Abrupt Climate Events. J. Geological Society of India 68:379–385

Naidu PD, Malmgren BA (1996a) A high-resolution record of late Quaternary upwelling along the Oman Margin, Arabian Sea based on planktonic foraminifera. Paleoceanography 11(1):129–140

Naidu PD, Malmgren BA (1996b) Relationship between Late Quaternary upwelling history and coiling properties of Neoglonoquadrina pachyderma and Globigerina bulloides in the Arabian Sea. J. Foraminiferal Res. 26(1):64–70

Neelin JD, Battisti DS, Hirst AC, Jin F-F et al. (1998) ENSO theory. J. Geophys. Res. 103(C7):14,261–14,290

Nees S (1997) High-resolution benthic foraminiferal records of the last glacial termination in the northern North Atlantic. In: Hass HC, Kaminski MA (eds) Contributions to the Micropaleontology and Paleoceanography of the Northern North Atlantic. Grzybowski Foundation Special Publication 5:167–197

Nesje A, Kvanne M, Rye N, Lovlie R (1991) Holocene glacial and climate history of the Jostedalsbreen region, western Norway: Evidence from lake sediments and terrestrial deposits. Quat. Sci. Rev. 10:87–114

Neiman VG, Burkov VA, Shcherbinin AD (1997) Water Dynamics of the Indian Ocean. Scientific World, Moscow (in Russian)

Niitsuma N, Oba T, Okada M (1991) Oxygen and carbon isotope stratigraphy at site 723, Oman margin. Proc. ODP. Sci. Res. 117, Washington, pp. 321–341

Nikiforov EG, Shpaykher AO (1980) Formation patterns of large-scale oscillations in the Arctic Ocean hydrology. Gidrometeoizdat, Leningrad (in Russian)

Nisancioglu KH, Raymo ME, Stone PH (2003) Reorganization of Miocene deep water circulation in response to the shoaling of the Central American Seaway. Paleoceanography 18(1):1006, doi:10.1029/2002PA000767

Nørgaard-Pedersen N, Spielhagen RF, Thiede J, Kassens H (1998) Central Arctic surface ocean environment during the past 80,000 years. Paleoceanography 13(2):193–204

Nørgaard-Pedersen N, Spielhagen RF, Erlenkeuser H, Grootes P, Heinemeier J, Knies J (2003) The Arctic Ocean during the Last Glacial Maximum: Atlantic and Polar domains of surface water mass distribution and ice cover. Paleoceanography 18(3). 1063. doi:10.1029/2002PA000781.

North Greenland Ice Core Project members (2004) High resolution climate record of the northern hemisphere reaching into the last interglacial period. Nature 431:147–151

Novitskiy VP (1961) Stationary currents of the Northern Barents Sea. Trudy GOIN 64:1–32 (in Russian)

Nürnberg D (2000) Taking the Temperature of Past Ocean Surfaces. Science 289(5485):1698–1699

Nűrnberg D, Müller A (2000) Paleo-sea surface temperature calculation in equatorial east Atlantic from Mg/Ca ratios in planktic foraminifera: A comparison to sea surface temperature estimates from U $^{k}{}_{37}$, oxygen isotopes, and foraminiferal transfer function. Paleoceanography 15(1):124–134

Oda M, Takemoto A (1992) Planktonic foraminifera and paleoceanography in the domain of Kuroshio current around Japan during the last 20,000 years (in Japanese). Quat. Res. 31:341–357

Ogilvie AEJ, Jonson T (2001) "Little Ice Age" Research: A Perspective from Iceland. Climate Change 48:9–52

Oppo DW, Keigwin LD, McManus JF, Cullen JL (2001) Persistent suborbital climate variability in marine isotope stage 5 and Termination II. Paleoceanography 16(3):280–292

Oppo DW, Lehman SJ (1993) Mid-depth circulation of the subpolar North Atlantic during the last Glacial Maximum. Science 259:1148–1152

Oppo DW, Lehman SJ (1995) Suborbital timescale variability of North Atlantic deep water during the past 200,000 years. Paleoceanography 10(5):900–910

Oppo DW, Linsley BK, Rosenthal Y, Dannenmann S, Beaufort L (2003) Orbital and suborbital climate variability in the Sulu Sea, western tropical Pacific. Geochemistry, Geophysics, Geosystems 4(1), doi:10.1029/2001GC000260

Oskina NS, Ivanova EV, Blyum NS (1982) Pliocene climatic zonality of the Atlantic, Indian and Pacific oceans. Doklady of the USSR Acad. of Sci., Earth Science Section 264(2):400–407 (English Translation)

Overpeck J, Anderson D, Trumbore S, Prell W (1996) The southwest Indian monsoon over the last 18 000 years. Climate Dynamics 12:213–225

Ovsepyan EA, Ivanova EV (2009) Benthic foraminiferal assemblages as indicators of paleoceanographic conditions in the Eastern Equatorial Pacific. Oceanology 49(1)121–129 (English Translation)

Pahnke K, Sachs JP (2006) Sea surface temperatures of southern midlatitudes 0–160 kyr B.P. Paleoceanography 21:2003, doi:10.1029/2005PA001191

Paillard D, Labeyrie L (1993) A "User Friendly" Macintosh Software for Rapid Correlations of Paleoclimatic Signals and Treatments. Centre des Faibles Radioactivités, Programme National d'Etudes de la Dynamique du Climat. Institut des Sciences de l'Univers, Paris

Paillard D, Labeyrie L (1994) Role of the thermohaline circulation in the abrupt warming after Heinrich events. Nature 372:162–164

Paillard D, Labeyrie L, Yiou P (1996) Macintosh program performs time-series analysis. EOS Trans, AGU 77(39):379

Patrick A, Thunell RC (1997) Tropical Pacific sea surface temperatures and upper water column thermal structure during the last glacial maximum. Paleoceanography 12(5):649–657

Paul A, Mulitza S (2008) The ocean circulation dirong the Last Glacial Maximum: A model-proxy data comparison, Abstracts of symposium honouring André Berger. Climate Change: From the geological past to the uncertain future. Louvain-la-Neuve, Belgium

Pavlidis YuA (1992) Shelf of the World Ocean in the Late Quaternary. Nauka, Moscow (in Russian)

Pavlidis YuA, Murdmaa IO, Ivanova EV, Artem'ev AV, Belousov MA (2001) Were ice shields of the Novaya Zemlya and Franz Josef Land connected 18 kyr ago? In: Lisitxin AP, Vinogradov

ME, Romankevich EA (eds) Experience of System Oceanologic Studies in the Arctic. Scientific World, Moscow (in Russian)

Pavlov VK, Pfirman SL (1995) Hydrographic structure and variability of the Kara Sea: implications for pollutant distribution. Deep-Sea Res. 42(2):1369–1390

Pavlov VK, Timokov LA, Baskakov GA, Kulakov MYu, Kurazhov VK, Pavlov V, Pivovarov SV, Stanovoy VV (1996) Hydrometeorological regime of the Kara, Laptev and East Siberian seas. Technical Memorandum APL-UW TM 1–96

Peeters F (2000) The distribution and stable isotope composition of living planktic foraminifera in relation to seasonal changes in the Arabian Sea. PhD Thesis. Vrije Universiteit, Amsterdam

Peeters FJC, Acheson R, Brummer G-J, de Ruijter WPM, Schneider RR, Ganssen GM, Ufkes E, Kroon D (2004) Vigorous exchange between the Indian and Atlantic oceans at the end of the past five glacial periods. Nature 430:661–665

Pelejero C, Grimalt JO (1997) The correlation between the U^k_{37} index and sea surface temperature in the warm boundary: the South China Sea. Geochim. Cosmochim. Acta. 61:4789–4797

Pelejero C, Grimalt JO, Sarnthein M, Wang L, Flores J-A (1999) Molecular biomarker record of sea surface temperature and climatic change in the South China Sea during the last 130,000 years. Mar. Geol. 156:109–121

Pena LD, Cacho I, Ferretti P, Hall MA (2008) El Nin~o–Southern Oscillation–like variability during glacial terminations and interlatitudinal teleconnections. Paleoceanography 23:3101, doi:10.1029/2008PA001620

Peterson LC, Haug GH, Hughen KA, Rohl U (2000) Rapid changes in the hydrologic cycle of the tropical Atlantic during the Last Glacial. Science 290:1947–1951

Pfirman SL, Bauch D, Gammelsrod T (1994) The Northern Barents Sea: Water mass distribution and Modification. In: The Polar oceans and their role in shaping the global environment. AGU Geophys. Monogr. 85:77–94

Pflaumann U, Duprat J, Pujol C, Labeyrie LD (1996) SIMMAX: a Modern Analog Technique to deduce Atlantic sea surface temperatures from planktonic foraminifera in deep-sea sediments. Paleoceanography 11:15–35

Pflaumann U, Jian Z (1999) Modern distribution pattents of planktonic foraminifera in the South China Sea and West Pacific: A new transfer technique to estimate regional sea-suface temperature. Mar. Geol. 156:41–83

Pflaumann U, Sarnthein M, Chapman M, d'Abreu L, Funnell B, Huels M, Kiefer T, Maslin M, Schulz H, Swallow J, van Kreveld S, Vautravers M, Vogelsang E, Weinelt M (2003) Glacial North Atlantic: Sea-surface conditions reconstructed by GLAMAP 2000. Paleoceanography 18(3):1065, doi:10.1029/2002PA000774

Pfuhl HA, Shacklenton NJ, Hall IR, Schellenberg SA, Kelly DC (2001) Paleoceanographic change in the australo-antarctic gateway focussing on 23 Ma – evidence for opening of the Drake Passage. 7th Intern. Conf. on Paleoceanography, Program&Abstracts, Sapporo, 171 pp

Pickard GL, Emery WJ (1990) Descriptive Physical Oceanography. Pergamon Press.

Piotrowski AM, Goldstein SL, Hemming SR, Fairbanks RG (2004) Intensification and variability of ocean thermohaline circulation through the last deglaciation. Earth Planet Sci. Lett. 225:205–220

Piotrowski AM, Goldstein SL, Hemming SR, Fairbanks RG (2005) Temporal relationships of carbon cycling and ocean circulation at glacial boundaries. Science 307(5717):1933–1938

Pivovarov S, Schlitzer R, Novikhin A (2003) River run-off influence on the water mass formation in the Kara Sea. In: Stein R, Fahl K, Fütterer DK, Galimov EM, Stepanets OV (eds) Siberian River Run-off in the Kara Sea: Characterisation, Quantification, Variability, and Environmental Significance. Proceedings in Marine Sciences 66:9–25

Pogodina IA, Tarasov GA (2002) Sedimentation processes and evolution of foraminifera in the Barents Sea dirung the last deglaciation. Oceanology 42(5):156–160 (in Russian)

Polyak L, Mikhailov V (1996) Post-glacial environments of the southeastern Barents Sea: Foraminiferal evidence. In: Andrews JT, Austin WEN, Bergsten H, Jennings AE (eds) Late Quaternary paleoceanography of the North Atlantic margins. Geological Society Special Publication, Edinburgh, pp. 323–337

Polyak L, Solheim A (1994) Late- and postglacial environments in the northern Barents Sea, west of Franz Josef Land. Polar Res. 13:197–207

Polyak L, Lehman SJ, Gataullin V, Jull AJT (1995) Two-step deglaciation of the southeastern Barents Sea. Geology 23:567–571

Polyak L, Forman SL, Herlihy FA, Ivanov G, Krinitsky P (1997) Late Weichselian deglacial history of the Svyataya (Saint) Anna Trough, northern Kara Sea, Arctic Russia. Mar. Geol. 143:169–188

Polyak L, Gataullin V, Okuneva O, Stelle V (2000a) New constraints on the limits of the Barents-Kara ice sheet during the Last Glacial Maximum based on borehole stratigraphy from the Pechora Sea. Geology 28:611–614

Polyak L, Levitan M, Gataullin V, Khusid T, Mikhailov V, Mukhina V (2000b) The impact of glaciation, river discharge, and sea-level change on Late Quaternary environments in the southwestern Kara Sea. Int. J. Earth Sci. 89:550–562

Polyak L, Korsun S, Febo L, Stanovoy V, Khusid T, Hald M, Paulsen BE, Lubinski DA (2002a) Benthic foraminiferal assemblages from the Southern Kara Sea, a river-influenced arctic marine environment. J. Foraminiferal Res. 32:252–273

Polyak L, Levitan M, Khusid T, Merklin L, Mukhina V (2002b) Variations in the influence of riverine discharge on the Kara Sea during the last deglaciation and the Holocene. Glob. Planet Change 32:291–309

Polyak L, Gataullin V, Gainanov V, Gladysh V, Goremykin Y (2002c) Kara Sea expedition yields insight into extent of LGM ice sheet. EOS 83:525- 529

Polyak L, Murdmaa IO, Ivanova EV (2004) A high-resolution, 800-year glaciomarine record from Russkaya Gavan, a Novaya Zemlya fjord, eastern Barents Sea. The Holocene 14(4):638–644

Polyakova E, Stein R (2002) Holocene variations in the Eurasian Arctic shelf surface water salinity and sea ice regime: Evidences from diatom assemblages of the Siberian Arctic shelf, Kara Sea. EMMM'2002, Program&Abstracts, Vienna, Austria, pp. 164–165

Polyakova E, Stein R (2004) Holocene paleoenvironmental implications of diatom and organic carbon records from the southeastern Kara Sea (Siberian Margin). Quat. Res. 62:256–266

Polyakova EI, Bauch HA, Klyuvitkina TS (2005) Early to Middle Holocene changes in Laptev Sea water masses deduced from diatom and aquatic palynomorph assemblages. Glob. Planet Change 48:208–222

Poore PZ, Osterman L, Curry WB, Phillips RL (1999) Late Pleistocene and Holocene meltwater events in the western Arctic Ocean. Geology 27:759–762

Porter SC, An Z (1995) Correlation between climate events in the North Atlantic and China during the last glaciation. Nature 375:305–308

Potemra JT, Lukas R, Mitchum GT (1997) Large-scale estimation of transport from the Pacific to the Indian Ocean. J. Geophys. Res. 102(C13):27,795–27,812

Prahl FG, Dymond J, Sparrow MA (2000) Annual biomarker record for export production in the central Arabian Sea. Deep-Sea Res. 47:1581–1604

Prell WL (1985) The stability of low-latitude sea-surface temperature: An evaluation of the CLIMAP reconstruction with emphasis on the positive SST anomalies. Tech. Tep. TR025, Department of Energy, Washington D.C, USA

Prell WL, Kutzbach JE (1987) Monsoon variability over the past 150,000 years. J. Geophys. Res. 92:8411–8425

Prell WL, Kutzbach JE (1992) Sensitivity of the Indian monsoon to forcing parameters and implications for its evolution. Nature 360:647–652

Prell WL, Niitsuma N, Emeis K-C et al (1989) Site reports. Proc. ODP. Init. Rep. 117 Washington, pp. 157–654

Prell WL, Murray DW, Clemens SC, Anderson DM (1992) Evolution and variability of the Indian Ocean summer monsoon: Evidence from the western Arabian Sea drilling program. Synthesis of results from Scientific Drilling in the Indian Ocean. Geophys. Monogr. 70, AGU:447–469

Prell W, Martin A, Cullen J, Trend M (1999) The Brown University Foraminiferal Data Base. IGBP PAGES/World Data Center-A for Paleoclimatology. Data Contribution Series. 1999-2027. NOAA/NGDC Paleoclimatology Program. Boulder CO, USA

Prentice IC, Sarnthein M (1993) Self-regulatory processes in the biosphere in the face of climate change. In: Eddy JA, DOeschger H (eds) Global Changes in the Perspective of the Past. Dahlem Konferenzen, Wiley, New York, pp. 29–38

Proshutinsky AY, Johnson M (1997) Two circulation regimes of the wind-driven Arctic Ocean. J.Geophys. Res. (102):12,493–12,514

Pye K, Zhou L (1989) Late Pleistocene and Holocene aeolian dust deposition in north China and the northwest Pacific Ocean. Palaeogeogr., Palaeoclim., Palaeoecol. 73:11–23

Quahdi R (1997) Variations de la productivité au nord-ouest de l'océan Indien lors des derniers 70,000 ans dans l'upwelling de Socotra et de Somalie: enregistrements géochimique. Bull. Soc. Geol. Fr. 168(1) :93–107

Qasim SZ (1982) Oceanography of the northern Arabian Sea. Deep-Sea Res. 29(9A):1041–1068

Rahman A, Roth PH (1990) Late Neogene paleoceanography of the Gulf of Aden Region based on calcareous nannofossils. Paleoceanography 5(1):91–107

Rahman A, Vernal D (1994) Surface oceanographic changes in the eastern Labrador Sea: Nannofossil record of the last 31,000 years. Mar. Geology. 121:247–263

Rahmstorf S (1995) Bifurcations of the Atlantic thermohaline circulation in response to changes in the hydrological cycle. Nature 378:145–149

Rahmstorf S (2002) Ocean circulation and climate during the past 120,000 years. Nature 419:207–214

Rahmstorf S (2003) The current climate. Nature 421:699

Rahmstorf S (2006) Thermohaline Ocean Circulation. In: Elias SA (ed) Encyclopedia of Quaternary Sciences. Elsevier, Amsterdam

Rahmstorf S, Crucifix M, Ganopolski A, Goosse H, Kamenkovich I, Knutti R, Lohmann G, Marsh R, Mysak LA, Wang Z, Weaver AJ (2005) Thermohaline circulation hysteresis: A model intercomparison. Geophys. Res. Lett. 32, L23605, doi:10.1029/2005GL023655

Rao RR, Molinari RL, Festa JF (1989) Evolution of the climatological near-surface thermal structure of the tropical Indian Ocean. 1. Description of mean monthly mixed layer depth, and sea surface temperature, surface current, and surface meteorological fields. J. Geophys. Res. 94:10801–10815.

Rashid H, Flower BP, Poore RZ, Quinn TM (2007) A $\sim$ 25 ka Indian Ocean monsoon variability record from the Andaman Sea. Quat. Sci. Rev. 26(19–21):2586–2597, doi:10.1016/j.quascirev.2007.07.002

Rasmussen TL, Thomsen E (2004) The role of the North Atlantic Drift in the millennial timescale glacial climate fluctuations. Palaeogeogr. Palaeoclim. Palaeoecol. 210:101–116

Rasmussen TL, Thomsen E, Labeyrie L, van Weering TCE (1996) Circulation changes in the Faeroe-Shetland Channel correlation with cold events during the last glacial period (58–10 ka). Geology 24:937–940

Rasmussen TL, van Weering TCE, Labeyrie L (1997) Climatic instability, ice sheets and ocean dynamics at high northern latitudes during the last glacial period (58–10 ka BP). Quat. Sci. Rev. 16:71–80

Rasmussen TL, Thomsen E, Ślubowska MA, Jessen S, Solheim A, Koç N (2007) Paleoceanographic evolution of the SW Svalbard margin (76°N) since 20,000 ^{14}C yr BP. Quat. Res.67:100–114

Ravelo AC, Dekens PS, McCarthy M (2006) Evidence for El Niño-like conditions during the Pliocene. GSA Today 16:4–11

Raymo ME, Ganley K, Carter S, Oppo DW, McManus J (1998) Millennial-scale climate instability during the early Pleistocene epoch. Nature 392(6677):699–702

Rea DK, Basov IA, Krissek LA and the Leg 145 Scientific Party (1995) Scientific Results of Drilling the North Pacific transect. Proc. ODP. Leg 145:577–596

Reichart GJ, den Dulk M, Visser HJ, van der Weijden CH, Zachariasse WJ (1997) A 225 kyr record of dust supply, paleoproductivity and the oxygen minimum zone from the Murray Ridge (northern Arabian Sea). Palaeogeogr. Palaeoclim. Palaeoecol. 134(1–4):149–169

Reichart GJ, Lourns LJ, Zachariasse WJ (1998) Temporal variability in the northern Arabian Sea Oxygen Minimum Zone (OMZ) during the last 225,000 years. Paleoceanography 13(6):607–621

Richardson PL (2008) On the history of meridional overturning circulation schematic diagrams. Prog. Oceanogr. 76:466–486

Risebrobakken B, Jansen E, Andersson C, Mjelde E, Hevrøy K (2003) A high-resolution study of Holocene paleoclimatic and paleoceanographic changes in the Nordic Seas. Paleoceallography 18(1):1017, doi:10.1029/2002PA000764

Risebrobakken B, Moros M, Ivanova E, Chistyakova N, Rosenberg R. Holocene climate changes in the SW Barents Sea – effects of sea ice variability. The Holocene (submitted)

Rixen T, Haake B, Intekkot V, Guptha MVS, Nair RR, Schlüssel P (1996) Coupling between SW monsoon-related surface and deep-ocean processes as discerned from continuous particle flux measurements and correlated satellite data. J. Geophys. Res. 101:28,569–28,582

Robinson LF, Adkins JF, Keigwin LD, Southon J, Fernandez DP, Wang S-L, Scheirer DS (2005) Radiocarbon variability in the western North Atlantic during the last deglaciation. Science 310(5753):1469–1473

Rodwell MJ, Rowell DP, Folland CK (1999) Oceanic forcing of the wintertime North Atlantic Oscillation and European climate. Nature 398(6725):320–323

Rohling EJ, Pälike H (2005) Centennial-scale climate cooling with a sudden cold event around 8,200 years ago. Nature 434:975–979

Rohling EJ, Marsh R, Wells NC, Siddall M, Edwards NR (2005) Similar meltwater contributions to glacial sea level changes from Antarctic and northern ice sheets. Nature 430:1016–1021

Romanov YuA (1994) Atmospheric circulation patterns in the tropical zone of oceans. Gidrometeoizdat, St-Petersbourg (in Russian)

Rooth C (1982) Hydrology and ocean circulation. Prog. Oceanogr. 7:131–149

Rosenthal Y, Oppo DW, Linsley BK (2003) The amplitude and phasing of climate change during the last deglaciation in the Sulu Sea, western equatorial Pacific. Geophys. Res. Lett. 30(8):1428, doi:10.1029/2002GL016612

Rossignol-Strick M (1983) African monsoons, an immediate climate response to orbital insolation. Nature 304:46–49

Rostek F, Ruhland G, Bassinot FC, Müller PJ, Labeyrie LD, Lancelot Y, Bard E (1993) Reconstructing sea surface temperature and salinity using $\delta^{18}O$ and alkenone records. Nature 364:319–321

Rostek F, Bard E, Beaufort L, Sonzogni C, Ganssen G (1997) Sea surface temperature and productivity records for the past 240 kyr in the Arabian Sea. Deep-Sea Res. 44(6–7):1461–1480

Ruddiman WF (ed) (1997) Tectonic Uplift and Climatic Change. Plenum, New York

Ruddiman WF (1998) Early uplift in Tibet. Nature 394(6695):723–724

Ruddiman WF (2006a) Ice-driven CO_2 feedback on ice volume. Climate of the Past 2:1–13

Ruddiman WF (2006b) Orbital changes and climate. Quat. Sci. Rev. 25:3092–3112, doi:10.1016/j.quascirev.2006.09.001

Ruddiman WF (2006c) What is the timing of orbital-scale monsoon changes? Quat. Sci. Rev. 25:657–658

Ruddiman WF, McIntyre A (1981) The North Atlantic during the last deglaciation. Palaeogeogr., Palaeoclim., Palaeoecol. 35:145–214

Ruddiman WF, Kutzbach JE (1989) Forcing the late Cenozoic Northern hemisphere climate by plateau uplift in Southern Asia and the American West. J. Geophys. Res. 94(D15):18,409–18,427

Rudels B, Jones EP, Anderson LG, Kattner G (1994) On the intermediate depth waters of the Arctic Ocean. In: Johannessen OM, Meunch RD, Overland JE (eds) The Polar Oceans and their Role in Shaping the Global Environments. AGU, Washington, DC:33–46

Rühlemann C, Mulitza S, Müller PJ, Wefer G, Zahn R (1999) Warming of the tropical Atlantic Ocean and slowdown of thermohaline circulation during the last deglaciation. Nature 402(6761):511–514

Sachs JP (2005) Off the grid. Quat. Sci. Rev. 24:1779–1780

Saidova KhM (1976) Benthic foraminifera of the World Ocean. Nauka, Moscow (in Russian)

Saji NH, Goswami BN, Vinayachandran PN, Yamagata T (1999) A dipole mode in the tropical Indian Ocean. Nature 401:360–363

Salvigsen O, Forman S, Miller GH (1992) Thermophilus molluscs on Svalbard during Holocene and their paleoclimatic implications. Polar Res. 11:1–10

Samoylovich YuG, Kagan LYa, Ivanova LV (1993) Quaternary sediments of the Barents Sea. Apatity, 73 pp (in Russian)

Sancetta C (1978) Neogene Pacific microfossils and paleoceanography. Mar. Micropal. 3:347–376

Sapozhnikov VV and Chernyaev AM (1967). Oxygen and phosphate distribution in the Pacific Ocean and some problems related to detection of their extremal concentrations. Proceedings of the Shirshov Institute of Oceanology, USSR Acad. Sci. 83 pp (in Russian)

Sarnthein M, Winn K, Duplessy J-C, Fontugne MR (1988) Global variations in surface ocean productivity in low and mid latitudes: influence on CO_2 reservoirs of the deep ocean and atmosphere during the last 21,000 years. Paleoceanography 3:361–399

Sarnthein M, Jansen E, Arnold M, Duplesyy J-C, Erlenkeuser H, Flatoy A, Veum T, Vogelsang E, Weinelt MS (1992) $\delta^{18}O$ time-slice reconstruction of meltwater anomalies at termination I in the North Atlantic between 50 and 80 °N. In: Bard E, Broecker WS (eds) The last Deglaciation: Absolute and Radiocarbon chronologies. NATO ASI Series 12:183–200. Springer-Verlag, Berlin

Sarnthein M, Pflaumann U, Wang PX, Wong HK (1994a) Preliminary Report on SONNE-95 Cruise " Monitor Monsoon" to the South China Sea. Ber.-Rep. Geol.-Paleontol. Inst. Univ. Kiel 48:225

Sarnthein M, Tetzlaff G, Koopmann B, Wolter K, Pflaumann U (1981) Glacial and interglacial wind regimes overthe eastern subtropical Atlantic and NW Africa. Nature 293:193–196

Sarnthein M, Winn K, Jung SJA, Duplessy J-C, Labeyrie L, Erlenkeuser H, Ganssen G (1994b) Changes in east Atlantic deepwater circulation over the last 30,000 years: Eight time slice reconstructions. Paleoceanography 9:209–267

Sarnthein M, Jansen E, Weinelt M, Arnold M, Duplessy JC, Erlenkeuser H, Flatøy A, Johannessen G, Johannessen T, Jung S, Koç N, Labeyrie L, Maslin M, Pflaumann U, Schulz H (1995) Variations in Atlantic surface ocean paleoceanography, 50–80°N: a time-slice record of the last 30,000 years. Paleoceanography 10:1063–1094

Sarnthein M, Kennett JP, Chappel J, Crowley T, Curry W, Duplessy JC, Grootes P, Hendy I, Laj C, Negendank J, Schulz M, Shackleton NJ, Voelker A, Zolitschka B et al (2000a) Exploring Late Pleistocene Climate Variations. EOS, Transactions. AGU 81(51):625, 629–630

Sarnthein M, Stattegger K, Dreger D, Erlenkeuser H et al (2000b) Fundamental Modes of Abrupt Changes in North Atlantic Circulation and Climate over the last 60 ky – Concepts, Reconstruction and Numerical Modeling. In: Schäfer P, Ritzrau W, Schlüter M, Thiede J (eds) The Northern North Atlantic: A Changing Environment, pp. 265–410. Springer, Berlin

Sarnthein M, Grootes PM, Kiefer T, Schulz M (2001) Spatial and temporal variability of Dansgaard-oeschger cycles. 7th Intern. Conf. on Paleoceanography, Program & Abstracts, Sapporo, pp. 179–180

Sarnthein M, Pflaumann U, Weinelt M (2003a) Past extent of sea ice in the northern North Atlantic inferred from foraminiferal paleotemperature estimates. Paleoceanography 18(2):1047, doi:10.1029/2002PA00077I

Sarnthein M, van Kreveld S, Erlenkeuser H, Grootes PM, Kucera M, Pflaumann U, Schulz M (2003b) Centennial-to-millennial-scale periodicities of Holocene climate and sediment injections off the western Barents shelf, 75°N. Boreas 32:447–461

Sato K, Oda M, Chiyonobu S, Kimoto K, Domitsu H, Ingle JC (2008) Establishment of the western Pacific warm pool during the Pliocene: Evidence from planktic foraminifera, oxygen isotopes, and Mg/Ca ratios. Palaeogeogr., Palaeoclimatol., Palaeoecol. 265:140–147

Scher HD, Martin EE (2008) Oligocene deep water export from the North Atlantic and the development of the Antarctic Circumpolar Current examined with neodymium isotopes. Paleoceanography 23:1205, doi:10.1029/2006PA001400

Schiebel R, Waniek J, Bork M, Hemleben Ch (2001) Planktic foraminiferal production stimulated by chlorophyll redistribution and entrainment of nutrients. Deep-Sea Res. 48(1):721–740

Schlitzer R (2000) Ocean Data View, http://www.awi-bremerhaven.de/GEO/ODV

Schlosser P, Bauch D, Fairbanks R, Bönisch G (1994) Arctic river-runoff: mean residence time on the shelves and in the halocline. Deep-Sea Res. 41(7):1053–1068

Schmidt MW, Spero HJ, Lea DW (2004) Links between salinity variation in the Caribbean and North Atlantic thermohaline circulation. Nature 428:160–163

Schmiedel G, Leuschner DC (2005) Oxygenation changes in the deep western Arabian Sea during the last 190,000 years: Productivity versus deepwater circulation. Paleoceanography 20, PA2008, doi:10.1029/2004PA001044

Schmittner A, Clement AC (2002) Sensitivity of the thermohaline circulation to tropical and high latitude freshwater forcing during the last glacial-interglacial cycle. Paleoceanography 17(2):1017, doi:10.1029/2000PA000591

Schmittner A, Yoshimori M, Weaver AJ (2002) Instability of glacial climate in a model of the ocean–atmosphere–cryosphere system. Science 295:1489–1493

Schmittner A, Galbraith ED, Hostetler SW, Pederson TF, Zhang R (2007) Large fluctuations of dissolved oxygen in the Indian and Pacific oceans during Dansgaard-Oeschger oscillations caused by variations of North Atlantic Deep Water subduction. Paleoceanography 22:3207, doi:10.1029/2006PA001384

Schmitz WJ (1996) On the world ocean circulation: Volume I - Some global features. North Atlantic circulation. Woods Hole Oceanographic Institution, Technical Report WHOI-96–03:141

Schneider B, Schmittner A (2006) Simulating the impact of the Panamanian seaway closure on ocean circulation, marine productivity and nutrient cycling. Earth Planet Sci. Lett. 246:367–380

Schott FA, McCreary JP (2001) The monsoon circulation of the Indian Ocean. Progress in Oceanography 51:1–123

Schulte S, Müller PJ (2001) Variations of sea surface temperature and primary productivity during Heinrich and Dansgaard-Oeschger events in the northeastern Arabian Sea. Geo-Marine Lett. 21:168–175

Schulz H, von Rad U, Erlenkeuser H (1998) Correlation between Arabian Sea and Greenland climate oscillations of the past 110,000 years. Nature 393:54–57

Schulz M, Berger WH, Sarnthein M, Grootes PM (1999) Amplitude variation of 1470-years climate oscillation during the last 100,000 years linked to fluctuation of continental ice mass. Geophys. Res. Lett. 26(22):3385–3388

Schulz M, Seidov D, Sarnthein M, Stattegger K (2001) Modeling ocean-atmosphere carbon budgets during the Last Glacial Maximum-Heinrich 1 meltwater event-Bølling transition. Int. J. Earth Sciences 90:412–425

Seidov D, Haupt B (1997) Global ocean thermohaline conveyor at present and in the late Quat. Geophys. Res. Lett. 24:2817–2820

Seidov D, Haupt B (1999) Last glacial and meltwater interbasin water exchanges and sedimentation in the world ocean. Paleoceanography 14(6):760–769

Seidov D, Sarnthein M, Stattegger K, Prien R, Weinelt M (1996) North Atlantic ocean circulation during the last glacial maximum and subsequent meltwater event : a numerical model. J. Geophys. Res. 101:16305–16332

Seidov D, Stouffer RJ, Haupt BJ (2005) Is there a simple bi-polar ocean seesaw? Glob. Planet Change 49:19–27

Serebryanny L, Andreev A, Malyasova E, Tarasov P, Romanenko F (1998) Late glasial and early-Holocene environments of Novaya Zemlya and the Kara Sea region of the Russian Arctic. The Holocene 8:323–330

Serreze MC, Box JE, Barry RG, Walsh JE (1993) Characteristics of arctic synoptic activity, 1952–1989. Meteorol. Atmos. Phys. 51:147–164

Shackleton NJ (1969) The last interglacial in the marine and terrestrial record. Proc. Roy. Soc. London. B. 174(1034):135–154

Shackleton NJ (1974) Attainment of isotopic equilibrium between ocean water and benthonic foraminifera genus Uvigerina: isotopic changes in the ocean during the last glacial. In: Les méthodes quantitatives d'étude des variations du climat au cours du Pleistocène, pp. 203–209

Shackleton NJ (2000) The 100,000-year ice-age cycle identified and found to lag temperature, carbon dioxide, and orbital eccentricity. Science 289(5486):1897–1902

Shackleton NJ (2001) Climate change across the hemispheres. Science 291:58–59

Shackleton NJ, Opdyke ND (1973) Oxygen isotope and paleomagnetic stratigraphy of Equatorial Pacific core V28–238. Oxygen isotope temperatures and ice volumes on a 10^5 year scale. Quat. Res. 3:39–55

Shackleton NJ, Kennett JP (1975) Paleotemperature history of the Cenozoic and the initiation of Antarctic glaciation: oxygen and carbon analyses in DSDP Sites 277 and 281. Init. Rep. DSDP 29:743–755, Wash

Shackleton NJ, Hall MA, Vincent E (2000) Phase relationships between millennial-scale events 64,000–24,000 years ago. Paleoceanography 15:565–569

Shackleton NJ, Chapman M, Sanchez-Goñi MF, Paillard D, Lancelot Y (2002) The classic Marine Isotope Substage 5e. Quat. Res. 58:14–16

Shimmield GB, Mowbray SR (1991) The inorganic geochemical record of the northwest Arabian Sea: a history of productivity variation over the last 400 k.y. from sites.722 and 724. Proc. ODP. Sci. Res. 117:409–429, Wash

Sicre MA, Ezat U, Mazaud A, Turon JL, (2006) A 27 kyr terrestrial biomarker record in the southern Indian Ocean. Geochem. Geophys. Geosyst. 7(7) Q07014, doi:10.1029/2005GC001234

Shi Y, Kong Z, Wang S, Tang L, Yao T, Zhao X, Zhang P, Shi S (1993) Mid-Holocene climates and environments in China. Glob. Planet. Change 7:219–233

Shriver JF, Hurlburt HE (1997) The contribution of the global thermohaline circulation to the Pacific to Indian Ocean throughflow via Indonesia. J. Geophys. Res. 102(C3):5491–5511

Simstich J (2001) Is water from the rosette sampler a suitable approximation for bottom water? – Comparison of salinity in rosette and multicorer. In: Stein R, Stepanets O (eds) The German-Russian Project on Siberian River Run-off (SIRRO): Scientific Cruise Report of the Kara-Sea Expedition "SIRRO 2000" of RV "Akademik Boris Petrov" and first results. Reports on Polar Res, 393

Simstich J, Ivanova EV, Stanovoy V, Bauch D, Erlenkeuser H, Spielhagen R (2002) Hydrographic Variability at the Bottom of the Southern Kara Sea During the Last 8000 Years. Sixth QUEEN Workshop Programme, Spiez, Switzerland

Simstich J, Stanovoy V, Bauch D, Erlenkeuser H, Spielhagen RF (2004) Holocene variability of bottom water hydrography on the Kara Sea shelf (Siberia) depicted in multiple single-valve analyses of stable isotopes in ostracods. Mar. Geol. 206:147–164

Simstich J, Erlenkeuser H, Harms I, Spielhagen RF, Stanovoy V (2005) Modern and Holocene hydrographic characteristics of the shallow Kara Sea shelf (Siberia) as reflected by stable isotopes of bivalves and benthic foraminifera. Boreas 34:1–12

Sirocko F (1996) Past and Present Subtropical summer monsoons. Science 274:937–938

Sirocko F, Sarnthein M, Erlenkeuser H, Lang H, Arnold M, Duplessy J-C (1993) Century-scale events in monsoonal climate over the past 24,000 years. Nature 364:322–324

Sirocko F, Garbe-Schonberg D, McIntyre A, Molfino B (1996) Teleconnections between the subtropical monsoons and high-latitude climates during the last glaciation. Science 272(5261):526–529

Sirocko F, Leuschner D, Staubwasser M, Maley J, Heusser L (1999) High-frequency Oscillations of the Last 70,000 Years in the tropical/Subtropical and Polar Climates. Mechanisms of Global Climate Change at Millenial Time Scales. Geophys. Monogr. 112:113–126, AGU

Skinner LC, Shackleton NJ (2005) An Atlantic lead over Pacific deep-water change across Termination I: implications for the application of the marine isotope stage stratigraphy. Quat. Sci. Rev. 24:571–580

Ślubowska MA, Koç N, Rasmussen TL, Klitgaard-Kristensen D (2005) Changes in the flow of Atlantic water into the Arctic Ocean since the last deglaciation: Evidence from the northern Svalbard continental margin, 80°N. Paleoceanolography 20:4014, doi:10.1029/2005PA001141

Ślubowska-Woldengen M, Rasmussen TL, Koç N, Klitgaard-Kristensen D, Nilsen F, Solheim A (2007) Advection of Atlantic Water to the western and northern Svalbard shelf since 17,500 cal yr BP. Quat. Sci. Rev. 26:463–478

Sonzogni C, Bard E, Rostek F (1998) Tropical sea-surface temperatures during the last glacial period: a view based on alkenones in Indian Ocean sediments. Quat. Sci. Rev. 17:1185–1201

Sorokhtin OG and Ushakov SA (2003) Evolution of the Earth. MSU, Moscow (in Russian)

Sowers K, Bender M (1995) Climate records covering the last deglaciation. Science 269:210–214

Spero H, Lea D (2002) The cause of carbon isotope minimum events on glacial terminations. Science 296:522–525

Spero HJ, Mielke KM, Kalve EM, Lea DW, Pak DK (2003) Multispecies approach to reconstructing eastern equatorial Pacific thermocline hydrography during the past 360 kyr. Paleoceanography 18(1):1022, doi:10.1029/2002PA000814

Spielhagen RF, Erlenkeuser H (1994) Stable oxygen and carbon isotopes in planktic foraminifers from Arctic ocean surface sediments: reflection of the low salinity surface water layer. Mar. Geology 119:227–250

Spielhagen RF, Erlenkeuser H, Siegert C (2005) History of freshwater runoff across the Laptev Sea (Arctic) during the last deglaciation. Glob. Planet Change 48(1–3):187–207

Spooner MI, Barrows TT, De Deckker P, Paterne M (2005) Palaeoceanography of the Banda Sea, and Late Pleistocene initiation of the Northwest Monsoon. Global Planet. Change. 49(1–2):28–46.

Stattegger K, Kuhnt W, Wong HK et al. (1997) Cruise Report SONNE 115 SUNDAFLUT. Sequenzstratigraphie, spätpleistozän-holozäne Meeresspiegelschwankunger und hochauflösende Rekonstruktion der postpleistozänen Transgression am Sunda Schelf. Ber.-Rep. Geol.-Palaeontol. Inst. Univ. Kiel 86:1–211

Steens TNF, Ganssen G, Kroon D (1992) Oxygen and carbon isotopes in planktonic foraminifera as indicators of upwelling intensity and upwelling-induced high productivity in sediments from the northwestern Arabian Sea. In: Summerhays CP, Prell WL, Emeis KC (eds) Upwelling Systems: Evolution Since the Early Miocene. Geol. Soc. London Spec. Publ. 64107–64119

Stein R (2001) Lithostratigraphy of gravity cores and correlation with sediment echograph profiles ("Akademik Boris Petrov" Kara Sea expeditions 1999 and 2000). In: Stein R, Stepanets O (eds) The German-Russian Project on Siberian River Run-off (SIRRO): Scientific Cruise Report of the Kara-Sea Expedition "SIRRO 2000" of RV "Akademik Boris Petrov" and first results. Reports on Polar Res. 393:120–140

Stein R, Nam S-I, Schubert C, Vogt C, Futterer D, Heinemaeier J (1994) The last deglasiation events in the eastern central Arctic Ocean. Science 264:692–696

Stein R, Levitan M (2001) Grain-size and sediment composition of sediment cores based on lithological core description and smear-slide estimates. In: Stein R, Stepanets O (eds) The German-Russian Project on Siberian River Run-off (SIRRO): Scientific Cruise Report of the Kara-Sea Expedition "SIRRO 2000" of RV "Akademik Boris Petrov" and first results. Reports on Polar Res. 393:110–119

Stein R, Stepanets O (eds) (2001) The German-Russian Project on Siberian River Run-off (SIRRO): Scientific Cruise Report of the Kara-Sea Expedition "SIRRO 2000" of RV "Akademik Boris Petrov" and first results. Reports on Polar Res., 393

Stein R, Niessen F, Dittmers K, Levitan M, Schoster F, Simstich J, Steinke T, Stepanets OV (2002) Siberian river run-off and Late Quaternary glaciation in the southern Kara Sea, Arctic Ocean: Preliminary results. Polar. Res. 2002 21(2):1–8

Stein R, Dittmers K, Fahl K, Kraus M, Matthiessen J, Niessen F, Pirrung M, Polyakova Y, Schoster F, Steinke T, Futterer DK (2004) Arctic (palaeo) river discharge and environmental change: evidence from the Holocene Kara Sea sedimentary record. Quat. Sci. Rev. 23:1485–1511

Steinke S, Yu P-S, Kucera M, Chen M-T (2008) No-analog planktonic foraminiferal faunas in the glacial southern South China Sea: Implications for the magnitude of glacial cooling in the western Pacific warm pool. Mar. Micropal. 66:71–90

Steinsund PI, Hald M (1994) Recent calcium carbonate dissolution in the Barents Sea, Paleaoceanographic applications. Mar. Geol. 117:303–316

Stepanov VN (1974) The World Ocean. Znanie, Moscow (in Russian)

Stephantsev L, Shmelkov B (2001) Some peculiarities of the hydrological structure on a meridional section Kara Sea - Yenisei Estuary – Yenisei River. In: Stein R, Stepanets O (eds) The German-Russian Project on Siberian River Run-off (SIRRO): Scientific Cruise Report of the Kara-Sea Expedition "SIRRO 2000" of RV "Akademik Boris Petrov" and first results. Reports on Polar Res. 393:6–14

Stocker TF (1998) The seasaw effect. Science 282:61–62

Stocker TF (2002) North–south connections. Science 297:1814–1815

Stocker TF, Wright DG, Broecker WS (1992) The influence of high-latitude surface forcing on the global thermohaline circulation. Paleoceanography 7:529–541

Stommel HM (1958) The abyssal circulation. Deep-Sea Res. 5:80–82

Stommel HM (1961) Thermohaline convection with two stable regimes of flow. Tellus 13:224–230

Stott L, Poulsen C, Lund S, Thunell R (2002) Super ENSO and global climate oscillations at millennial time scales. Science 279:222–226

Stott L, Timmermann A, Thunell R (2007) Southern hemisphere and deep-sea warming led deglacial atmospheric CO_2 rise and tropical warming. Science 318(5849):435–438, doi: 10.1126/science.1143791

Stouffer RJ, Yin J, Gregory JM, Dixon KW, Spelman MJ, Hurlin W, Weaver AJ, Eby M, Flato GM, Hasumi H, Hu A, Jungclaus JH, Kamenkovich IV, Levermann A, Montoya M, Murakami S, Nawrath S, Oka A, Peltier WR, Robitaille DY, Sokolov AP, Vettoretti G, Weber N (2006) Investigating the Causes of the Response of the Thermohaline Circulation to Past and Future Climate Changes. J. Climate 19:1365–1387

Street FA, Grove AT (1979) Global maps of lake-level fluctuations since 30,000 yr. B.P. Quatern. Res. 12(1):83–118

Stuiver M, Braziunas TF (1993) Modelling atmosphere ^{14}C influences and ^{14}C ages of marine samples to 10,000 BC. Radiocarbon 35:137–189

Stuiver M, Reimer PJ (1993) Extended ^{14}C data base and revised CALIB 3.0 ^{14}C age calibration program. Radiocarbon 35:215–230

Stuiver M, Grootes PM (2000) GISP2 oxygen isotope ratios. Quat. Res. 53:277–284

Stuiver M, Grootes PM, Braziunas TF (1995) The GISP2 $\delta^{18}O$ climate record of the past 16,500 years and the role of the Sun, ocean, and volcanoes. Quat. Res. 44:341–354

Sukhoruk VJ, Tokarev VG, Vlasova LN, Kodina LA (2001) Nutrient distribution along the Yenisei River - Kara Sea transect in September 2000. Some peculiarities of the hydrological structure on a meridional section Kara Sea – Yenisei Estuary – Yenisei River. In: Stein R, Stepanets O (eds) The German-Russian Project on Siberian River Run-off (SIRRO): Scientific Cruise Report of the Kara-Sea Expedition "SIRRO 2000" of RV "Akademik Boris Petrov" and first results. Reports on Polar Res.393:21–28

Sun X, Li X (1999) Pollen records of the last 37 ka in deep-sea core 17940 from the northern slope of the South China Sea. Mar. Geol. 156:227–244

Svendsen JI, Mangerud J (1992) Paleoclimatic inferences from glacial fluctuations on Svalbard during the last 20,000 years. Climatic Dynamics 6:213–220

Svendsen JI, Mangerud J (1997) Holocene glacial and climatic variation on Spitsbergen, Svalbard. The Holocene 7:45–57

Svendsen JI, Astakhov VI, Bolshiyanov DYu, Dowdeswell JA, Gataullin V, Hjort C, Hubberten HW, Larsen E, Mangerud J, Melles M, Moller P, Saarnisto M, Siegert MJ (1999) Maximum

extent of the Eurasian ice sheets in the Barents and Kara Sea region during the Weichselian. Boreas 28:234–242
Svendsen JI, Alexanderson H, Astakhov VI, Demidov I, Dowdeswell JA, Funder S, Gataullin V, Henriksen M, Hjort C, Houmark-Nielsen M, Hubberten HW, Ingolfsson O, Jakobsson M, Kjaer KH, Larsen E, Lokrantz H, Lunkka J, Lysa A, Mangerud J, Matiouchkov A, Murray A, Möller P, Niessen F, Nikolskaya O, Polyak L, Saarnisto M, Siegert C, Siegert MJ, Spielhagen RF, Stein R (2004) Late Quaternary ice sheet history of northern Eurasia. Quat. Sci. Rev. 23:1229–1271
Taldenkova E, Bauch HA, Stepanova A, Strizh A, Demyankov S, Ovsepyan Y (2008) Postglacial to Holocene history of the Laptev Sea continental margin: Palaeoenvironmental implications of benthic assemblages. Quat. Int. 183:40–60
Talley LD (1999) Some aspects of ocean heat transport by the shallow, intermediate and deep overturning circulations. In: Clark PU, Webb RS, Keigwin LD (eds) Mechanisms of Global Climate Change at Millenial Time Scales. Geophys. Monograph 112:1–22
Talley LD, Reid JL, Robbins PE (2003) Data-based meridional overturning stream functions for the global ocean. J. Clim. 16:3213–3226
Tantsyura AI (1959). On the Barents Sea currents. In: Hydrological Studies in the Barents and Greenland Seas. Proceedings of PINRO 11:35–54. Pishchepromizdat, Murmansk (in Russian)
Tarasov GA, Pogodina IA, Khasankaev VB, Gritsenko II, Matishov GG (1999) About the Late Quaternary sedimentation in the area of the Central Barents Sea Rise. Doklady of the USSR Acad. Sci. 367(6):792–795 (in Russian)
Tarasov GA, Pogodina IA, Khasankaev VB, Kukina NA, Mityaev MV (2000) Sedimentation processes on the glacial shelves. Apatity, 473 pp (in Russian)
Tchernia P (1978) Océanographie Régional. Description physique des océans et des mers. École Nationale Supérieure de Techniques avancées
Tchernia P (1980) Descriptive Regional Oceanography. Marine Series 3:253. Pergamon Press, Oxford
Thiede J, Mangerud J (1999) New Map Revises Extent of Last Ice Sheet Over Barents and Kara Seas. EOS Trans. 80(42):493–494
Thiede JA, Winkler T, Wolf-Welling O, Eldholm A, Myhre M, Baumann KH, Henrich R, Stein R (1998) Late Cenozoic history of the Polar North Atlantic: Results from ocean drilling. Quat. Sci. Rev. 17:185–208
Thompson DWJ, Wallace JM (1998) The Arctic Oscillation signature in the wintertime geopotential height and temperature fields. Geophys. Res. Lett. 25:1297–1300
Thompson PR (1981) Planktonic foraminifera in the Western North Pacific during the past 150,000 years: comparison of modern and fossil assemblages. Paleoceanogr. Paleoclimatol. Paleoecol. 35:241–279
Thompson PR, Bé AWH, Duplessy J-C, Shackleton NJ (1979) Disappearance of pink-pigmented *Globigerinoides ruber* at 120,000 yr. B.P. in the Indian and Pacific oceans. Nature 280:554–558
Thunell R, Anderson D, Gellar D, Miao Q (1994) Sea-surface temperature estimates for the tropical western Pacific during the last glaciation and their implication for the Pacific Warm Pool. Quat. Res. 41:255–264
Thunell RC, Miao Q (1996) Sea surface temperature of the western equatorial Pacific ocean during the Younger Dryas. Quat. Res. 46:72–77
Tian J, Wang P, Chen R, Cheng X (2005) Quaternary upper ocean thermal gradient variations in the South China Sea: Implications for east Asian monsoon climate. Paleoceanography 20, doi:10.1029/2004PA001115
Timofeeva VT (1960) Water masses of the Arctic Basin. Gidrometeoizdat, Leningrad (in Russian)
Toggweiler J, Dixon D, Broecker W (1991) The Peru Upwelling and the ventilation of the South Pacific thermocline. J. Geophys. Res. 96:20467–20497
Toggweiler JR, Samuels B (1995) Effect of drake passage on the global thermohaline circulation. Deep-Sea Res. 42:477–500
Toggweiler JR (2005) Climate change from below. Quat. Sci. Rev. 24:511–512
Tolderlund DS, Be AWH (1971) Seasonal distribution of planktonic foraminifera in the western North Atlantic. Micropaleontology 17:297–329

Trenberth KE, Branstator GW, Karoly D, Kumar A et al (1998) Progress during TOGA in understanding and modeling global teleconnections associated with tropical sea surface temperatures. J. Geophys. Res. 103(C7):14,291–14,325

Treshnikov AF (1985) Arctic Atlas. AARI, Moscow (in Russian)

Treshnikov AF and Baranov GI (1972) Water Column Structure and Circulation in the Arctic Basin (in Russian). Gidrometeoizdat, Leningrad (in Russian)

Ujiié HA (2003) 370-ka paleoceanographic record from the Hess Rise, central North Pacific Ocean, and an indistinct 'Kuroshio Extension'. Mar. Micropal. 49:21–47

Van Campo E, Duplessy J-C, Rossignol-Strick M (1982) Climatic conditions deduced from a 150 kyr oxygen – pollen record from the Arabian sea. Nature 296:56–59

Van Geel B, Raspopov OM, Renssen H, Van der Plitcht J, Dergachew VA, Meijer HAJ (1999) The role of solar forcing upon climate change. Quat. Sci. Rev. 18:331–338

Van Kreveld S, Sarnthein M, Erlenkeuser H, Grootes P, Jung S, Nadeau MJ, Pflaumann U, Voelker A (2000) Potential links between surging ice sheets, circulation changes, and the Dansgaard-Oeschger cycles in the Irminger Sea, 60–18 kyr. Paleoceanography 15(4):425–442

Vedernikov VI and Gagarin VI (1998) Primary production and chlorophyll in the Barents sea in September-October 1997. Oceanology 38:642–649 (English Translation)

Vedernikov VI, Demidov AB, Sud'bin AI (1994) Primary production and chlorophyll in the Kara Sea in September 1993. Oceanology 34:693–704 (in Russian)

Velichko AA (ed) (1999) Climate and landscape changes during the last 65 million years (Cenozoic: from Paleocene to Holocene). GEOS, Moscow (in Russian)

Velichko AA, Dolukhanov PM, Rutter NW, Catto NR (eds) (1997) Quaternary of Northern Eurasia: Late Pleistocene and Holocene landscapes, stratigraphy and environments. Pergamon, Quat. International, INQUA/Elsevier Science Ltd.

Vénec-Peyré M-T, Caulet J-P (2000) Paleoproductivity changes in the upwelling system of Socotra (Somali Basin, NW Indian Ocean) during the last 72,000 years: evidence from biological sinatures. Mar. Micropal. 40:321–344

Vénec-Peyré M-T, Caulet J-P, Vergnaud-Grazzini C (1995) Paleoceanographic changes in the Somali Basin (5°N upwelling and equatorial areas) during the last 160 kyr, based on correspondence analysis of foraminiferal and radiolarian assemblages. Paleoceanography 10(3):473–491

Vénec-Peyré M-T, Caulet J-P, Vergnaud-Grazzini C (1997) Glacial/interglacial changes in the equatorial part of the Somali Basin (NW Indian Ocean) during the last 355 kyr. Paleoceanography 12(5):640–648

Vergnaud-Grazzini C, Caulet J-P, Vénec-Peyré M-T (1995) Index de fertilité et mousson dans le bassin de Somalie. Evolution au Quaternaire supérieur. Bull. Soc. Geol. France 166(3):259–270

Vetrov AA, Romankevich EA (2003) The Barents Sea: Distribution, sources, variability and burial of organic carbon. In: Stein R and Macdonald RW (eds) The Arctic Ocean Organic Carbon Cycle: Present and Past. Springer

Veski S, Seppä H, Ojala AEK (2004) Cold event at 8200 yr B.P. recorded in annually laminated lake sediments in eastern Europe. Geology 32(8):681–684, doi:10.1130/G20683.1

Veum T, Jansen E, Arnold M, Beyer I, Duplessy J-C (1992) Water mass exchange between the North Atlantic and the Norwegian Sea during the past 28,000 years. Nature 356:783–785

Via RK, Thomas DJ (2006) Evolution of Atlantic thermohaline circulation: Early Oligocene onset of deep-water production in the North Atlantic. Geology 34:441–444

Vidal L, Arz H (2004) Oceanic climate variability at millennial time scales: models of climate connections. In: Battarbee RW et al. (eds) Past Climate Variability Through Europe and Africa. Springer

Vidal L., Labeyrie L, Cortijo E, Arnold M, Duplessy JC, Michel E, Becque S, van Weering TCE (1997) Evidence for changes in the North Atlantic Deep Water linked to meltwater surges during the Heinrich events. Earth Planet Sci. Lett. 146:13–27

Vidal L, Labeyrie L, van Weering TCE (1998) Benthic $\delta^{18}O$ records in the North Atlantic over the last glacial period (60–10 kyr) : Evidence for brine formation. Paleoceanography 13(3):245–251

Vidal L, Schneider RR, Marchal O, Bickert T, Stocker TF, Wefer G (1999) Link between the North and South Atlantic during the Heinrich events of the last glacial period. Clim. Dyn. 15:909–919

Vincent E, Berger WH (1981) Planktic foraminifera and their use in paleoceanography. In: Emiliani C (ed) The Oceanic Lithosphere: The Sea, 7, Chapter 25. New York, pp. 1025–1119

Vincent E, Toumarkine M (1990) Neogene planctonic foraminifers from the western tropical Indian Ocean, leg 115. Proc.ODP. Sci. Res. 115:795–838, Washington

Vinje T (1997) On the variation during the past 400 years of the Barents Sea ice edge position and Northern Hemisphere temperatures. Polar Processes and Glob. Climate. Conf. Proc. Pt. 2:269–271

Vinje T (2001) Anomalies and trends of sea ice extent and atmospheric circulation in the Nordic seas during the period 1864–1998. J. Climate 14(3):255–267

Vinje T, Kvambekk ES (1991) Barents Sea drift ice characteristics. Polar Res. 10:59–68

Vinogradov ME, Shushkina EA, Lebedeva LP, Gagarin VI (1994) Mesoplankton of the easthern Kara Sea and the Ob and Yenisey river estuaries. Oceanology 34:653–661 (English Translation)

Vinogradov ME, Vedernikov VI, Romankevich EA, Vetrov AA (2000) Components of the carbon cycle in Russian Arctic seas. Primary production and flux of C_{org} from the photic layer. Oceanology 40:204–215 (English Translation)

Voelker AHL, Lebreiro SM, Schönfeld J, Cacho I, Erlenkeuser H, Abrantes F (2006) Mediterranean outflow strengthening during northern hemisphere coolings: A salt source for the glacial Atlantic? Earth Planet. Sci. Lett. 245(1–2):39–55

Vogelsang E (1990) Palaeo-Ozeanographie des Europaischen Nordmeeres an Hand stabiler Kohlenstoff- und Sauerstoffisotope. PhD Thesis. University of Kiel

Vogt C, Knies J, Spielhagen RF, Stein R (2001) Detailed mineralogical evidence for two nearly identical glacial/deglacial cycles and Atlantic water advection to the Arctic Ocean during the last 90,000 years. Global Planet. Change 31:23–44

Vőlker A, Sarnthein M, Grootes PM, Erlenkeuser H, Laj C, Mazaud A, Nadeau MJ, Schleicher M (1998) Correlation of marine ^{14}C ages from the Nordic Seas with the GISP2 isotope record: implication for radiocarbon calibration beyond 25 ka BP. Radiocarbon 40:517–534

Voelker AHL, WS members (2002) Global distribution of centennial-scale records for Marine Isotope Stage (MIS) 3: a database. Quat. Sci. Rev. 21:1185–1212

Von Grafenstein U, Erlenkeuser H, Müller J, Jouzel J, Johnsen S (1998) The cold event 8,200 years ago documented in oxygen isotope records of precipitation in Europe and Greenland. Climate Dynamics 14:73–81

Von Rad U, Schulz H, Riech V, den Dulk M, Berner U, Sirocko F (1999) Multiple monsoon-controlled breakdown of oxygen-minimum conditions during the past 30,000 years documented in laminated sediments off Pakistan. Palaeogeogr., Palaeoclim.,Palaeoceol. 152(1–2):129–161

Voronina E, Polyak L, De Vernal A, Peyron O (2001) Holocene variations of sea-surface conditions in the southeastern Barents Sea, reconstructed from dinoflagellate cyst assemblages. J. Quat. Sci. 16(7):717–726

Vorren TO, Laberg JS (1996) Late glacial air temperature, oceanographic and ice sheet interactions in the southern Barents Sea region. In: Andrews JT, Austin WEN, Bergsten H, Jennings AE (eds) Late Quaternary paleoceanography of the North Atlantic margins. Geo. Soc. Spec. Publ. 111:275–287, Boulder

Waelbroeck C, Labeyrie L, Duplessy J-C, Guiot J, Labracherie M, Leclaire H, Duprat J (1998) Improving past sea surface temperature estimates based on planctonic fossil faunas. Paleoceanography 13(3):272–283

Waelbroeck C, Labeyrie L, Michel E, Duplessy J-C, McManus JF, Lambeck K, Balbon E, Labracherie M (2002) Sea-level and deep water temperature changes derived from benthic foraminifera isotopic records. Quat. Sci. Rev. 21:295–305

Waelbroeck C, Mulitza S, Spero H, Dokken T, Kiefer T, Cortijo E (2005) A global compilation of late Holocene planktonic foraminiferal $\delta^{18}O$: relationshipbetween surface water temperature and $\delta^{18}O$. Quat. Sci. Rev. 24.:853–868

Waelbroeck C, Levi C, Duplessy J-C, Labeyrie L, Michel E, Cortijo E, Bassinot F, Guichard F (2006) Distant origin of circulation changes in the Indian Ocean during the last deglaciation. Earth Planet Sci. Lett. 243:244–251

Walker GT (1924) Correlation in seasonal variation of weather, IX. Mem. Indian Meteor. Dep. 24(9):275–332

Wallace JM, Zhang Y, Renwick JA (1995) Dynamic contribution to hemispheric mean temperature trends. Science 270:780–783

Wallace JM, Rasmusson EM, Mitchell TP, Kousky VE et al. (1998) On the structure and evolution of ENSO-related climate variability in the tropical Pacific: Lessons from TOGA. J. Geophys. Res. 103(C7):14,241–14,259

Wang L (1992) The late Quaternary oxygen isotope record and its implications for the ventilation state og the South China Sea. (In Chinese, with English abstr.). In: Ye Z, Wang P (eds) Contributions to Late Quaternary Paleoceanography of the South China Sea, Qingdao Ocean. University Press, Qingdao

Wang L, Wang P (1990) Late Quaternary paleoceanography of the South China Sea: glacial/interglacial contrasts in an enclosed basin. Paleoceanography 5:77–90

Wang L, Sarnthein M (1999) Long-Short- Term Variation of Monsoon Climate and Its Tele-Connection to Global Change: High Resolution Grain Size Records of Siliciclastic Sediments from the South China Sea. In: Abrantes F, Mix A (eds) Reconstructing Ocean History: A Window into the Future. Kluwer Academic/Plenum Publishers, New York

Wang Z, Mysak LA (2006) Glacial abrupt climate changes and Dansgaard-Oeschger oscillations in a coupled climate model. Paleoceanography 21, PA2001, doi:10.1029/2005PA001238

Wang L, Wang P, Bian Y, Jian Z (1995a) Late Quaternary paleoceanography of the South China Sea: surface circulation and carbonate cycles. Mar. Geol. 127:145–166

Wang L, Sarnthein M, Duplessy J-C, Erlenkeuser H, Jung S, Pflaumann U (1995b) Paleo sea surface salinities in the low-latitude Atlantic: The $\delta^{18}O$ record of *Globigerinoides ruber* (white). Paleoceanography 10(4):749–761

Wang L, Sarnthein M, Erlenkeuser H, Grimalt J, Grootes P, Heilig S, Ivanova EV, Kienast M, Pelejero C, Pflaumann U (1999a) East Asian monsoon climate during the Late Pleistocene: High resolution sediment records from the South China Sea. Mar. Geology 156(1–4):243–282

Wang L, Sarnthein M, Grootes P, Erlenkeuser H (1999b) Millennial reoccurrence of century-scale abrupt events of East Asian monsoon: A possible heat conveyor for the global deglaciation. Paleoceanography 14(6):725–731

Wang P (1992) West Pacific marginal seas in the last glaciation: a paleoceanographic comparison. (In Chinese, with English abstr.). In: Ye Z, Wang P (eds) Contributions to Late Quaternary Paleoceanography of the South China Sea, Qingdao Ocean. University Press, Qingdao

Wang P, Clemens S, Beaufort L, Braconnot P, Ganssen G, Jian Z, Kershaw P, Sarnthein M (2005) Evolution and variability of the Asian monsoon system: state of the art and outstanding issues. Quat. Sci. Rev. 24:595–629

Wang YJ, Cheng H, Edwards RL, An ZS, Wu JY, Shen C-C, Dorale JA (2001) A high-resolution absolute-dated late Pleistocene Monsoon record from Hulu Cave, China. Science 294:2345–2348, doi:10.1126/science.1064618

Wang Z, Mysak LA (2006) Glacial abrupt climate changes and Dansgaard-Oeschger oscillations in a coupled climate model. Paleoceanography 21, PA2001, doi:10.1029/2005PA001238

Washburn L, Armi L (1988) Observation of frontal instabilities on an upwelling filament. J. Phys. Oceanogr. 18:1075–1092

Weaver AJ (1995) Driving the ocean conveyor. Nature 378:135–136

Weber SL, Drijfhout SS, Abe-Ouchi A, Crucifix M, Eby M, Ganopolski A, Murakami M, Otto-Bliesner B, Peltier WR (2007) The modern and glacial overturning circulation in the Atlantic ocean in PMIP coupled model simulations. Climate of the Past 3:51–64

Webster PJ, Magaña VO, Palmer TN, Shukla J et al. (1998) Monsoons: Processes, predictibility, and the prospects for prediction. J. Geophys. Res. 103(C7):14,451–14,510

Webster PJ, Moore AM, Loschnigg JP, Leben RR (1999) Coupled ocean-atmosphere dynamics in the Indian ocean during 1997–1998. Nature 401:356–360

Wei K-Y, Lee M-Y, Duan W, Chen C, Wang CH (1998) Paleoceanographic change in the northeastern South China Sea during the last 15,000 years. J. Quat. Sci. 13:55–64

Weinelt MS, Sarnthein M, Vogelsang E, Erlenkeuser H (1991) Early decay of the Barents Shelf Ice Sheet -spread of stable isotope signals across the eatern Norwegian Sea. Norsk Geologisk Tidsskrift 71:137–140

Weinelt M, Vogelsang E, Kucera M, Pflaumann U, Samthein M, Völker A, Erlenkeuser H, Malmgren BA (2003) Variability of North Atlantic heat transfer during MIS 2. Paleoceallography 18, doi:10.1029/ 2002RA000772

Wiesner MG, Zhen L, Wong HK, Wang Y, Chen W (1996) Fluxes of particulate matter in the South China Sea. In: Itterkkot V, Schaefer P, Honjo S, Depetris PJ (eds) Particle Flux in the Ocean. Wiley, New York

Wilke I, Bickert T, Peeters FJC (2006) The influence of seawater carbonate ion concentration $[CO_3{}^{2-}]$ on the stable carbon isotope composition of the planktic foraminifera species Globorotalia inflate Mar. Micropal. 58:243–258

Williamowski C, Zahn R (2000) Upper ocean circulation in the glacial North Atlantic from benthic foraminiferal isotope and trace element fingerprinting. Paleoceanography 15:515–527.

Winguth AME, Archer D, Duplessy J-C, Maier-Reimer E, Mikolajewicz U (1999) Sensitivity of paleonutrient tracer distributions and deep-sea circulation to glacial boundary conditions. Paleoceanography 14(3)304–323

Winn K, Sarnthein M, Erlenkeuser H (1991) δ^{18}O stratigraphy and chronology of Kiel sediment cores from the East Atlantic. Ber.-Rep. Geol.-Palaeontol. Inst. Univ. Kiel 45:1–99

WOA (1998) World Ocean Atlas 1998, Version 2, http://www.nodc.noaa.gov/oc5/woa98.html. Tech. rep., National Oceanographic Data Center, Silver Spring, Maryland

WOA (2001) World Ocean Atlas 2001 http://www.nodc.noaa.gov/OC5/WOA01F/prwoa01f.html

WOA (2005) World Ocean Atlas 2005 http://www.nodc.noaa.gov/OC5/WOA05/pr_woa05.html

Wollenburg JE, Mackensen A (1998) Living benthic foraminifers from the central Arctic Ocean: faunal composition, standing stock and diversity. Mar. Micropal. 34:153–185

Wollenburg JE, Kuhnt W, Mackensen A (2001) Changes in Arctic Ocean paleoproductivity and hydrography during the last 145 kyr: The benthic foraminiferal record. Paleoceanography 16(1):65–77

Wollenburg JE, Knies J, Mackensen A (2004) High-resolution paleoproductivity fluctuations during the past 24 kyr as indicated by benthic foraminifera in the marginal Arctic Ocean. Palaeogeogr. Palaeoclim. Palaeoecol. 204:209–238

Wyrtki K (1961) Physical oceanography of the southern Asian waters. Scientific Results of Marine Investigations of the South China Sea and the Gulf of Thailand 1959–1961. NAGA Rep. 2. Scripps Institution of Oceanography, La Jolla

Wyrtki K, Bennett EB, Rochfeld DJ (1971) Oceanographic atlas of the international Indian ocean expedition. National Science Foundation, Washington, DC

Wyrtki K (1973) Physical oceanography of the Indian Ocean. In: Zeitschel B (ed) Biology of the Indian Ocean. Springer-Verlag, New York

Wyrtki K (1975) El-Niño the dynamic response of the equatorial Pacific Ocean to atmospheric forcing. J. Phys. Oceanogr. 5:572–584

Wyrtki K (1982) The Southern Oscillation ocean – atmosphere interaction and El-Nino. Marine Thechnol. Soc. J. 16(1):3–10

Yiou P, Baert E, Loutre MF (1996) Spectral analysis of climate data. Surveys of Geophys. 17:619–663

Yokoyama Y, Lambeck K, De Deckker P, Jonston P, Fifield L (2000) Timing of the last glacial maximum from observed sea-level minima. Nature 406:713–716

Yu G, Harrison SP (1996) An evaluation of the simulated water balance of Eurasia and northern Africa at 6000 yr BP using lake status data. Clim. Dyn. 12:723–735

Zachariasse WJ (1992) Neogene planktonic foraminifers from sites 761 and 762 off Northwest Australia. Proc. ODP. Sci. Res. 122:665–675, Washington

Zachos JC, Pagani M, sloan ET, Billups K (2001) Trends, rhythms, and aberrations in global climate 65 Ma to present. Science 292:686–694

Zahn R (1994) Fast flickers in the tropics. Nature 372:621–622

Zahn R (1997) North Atlantic thirculation during the last glasial period: evidence for coupling between meltwater events and convective instability. Geomar Report, Kiel 63:133

Zahn R (1999) Polar Tropical and Interhemispheric linkages. An Introduction. In: Abrantes F, Mix A (eds) Reconstructing Ocean History: A Window into the Future. Kluwer Academic. Plenum Publishers, New York

Zahn R (2003) Monsoon linkages. Nature 421:324–325

Zahn R, Pedersen TF (1991) Late Pleistocene evolution of surface and mid-depth hydrography at the Oman margin: planktonic and benthic isotope records at site 724. Sci. Res. ODP. Leg 117:291–308

Zahn R, Stüber A (2002), Suborbital intermediate water variability inferred from paired benthic foraminiferal Cd/Ca and $\delta^{13}C$ in the tropical West Atlantic and linking with North Atlantic Climates. Earth Planet Sci. Lett. 200:191–205

Zahn R, Schonfeld J, Kudrass H-R, Park M-H, Erlenkeuser H, Grootes P (1997) Thermohaline instability in the North Atlantic during meltwater events: Stable isotope and ice-rafted detritus records from core SO75–26KL, Portuguese margin. Paleoceanography 12:696–710

Zeeberg JJ (2001) Climate and glacial history of the Novaya Zemlya Archipelago, Russian Arctic. PhD thesis, University Illinois in Chicago. Rozenberg, Amsterdam

Zeeberg J, Forman SL (2000) Changes in glacier extent on north Novaya Zemlya in the twentieth century. The Holocene 11(2):161–175

Zeeberg J, Lubinski DJ, Forman SL (2001) Holocene Relative Sea-Level History of Novaya Zamlya, Russia, and Implications for Late Weichselian Ice-Sheet Loading. Quat. Res. 56(2)218–230

Zenkevich LA (1963) Biology of the USSR seas. Izd. Acad. Sci. USSR. Moscow (in Russian)

Zhang R, Delworth TL (2005) Simulated tropical response to a substantial weakening of the Atlantic thermohaline circulation. J. Clim. 18:1853–1860

Zhao M, Beveridge NAS, Shacklenton NJ, Sarntein M, Eglinton G (1995) Molecular stratigraphy of cores off northwest Africa: Sea-surface temperature history over the last 80 ka. Paleoceanography 10:661–675

Zhou B, Zhao Q (1999) Allochthonous ostracods in the South China Sea and their significance in indicating downslope sediment contamination. Mar. Geol. 156:187–195

Zinke J, Dullo W-Chr, Heiss GA, Eisenhauer A (2004) ENSO and Indian Ocean subtropical dipole variability is recorded in a coral record off southwest Madagascar for the period 1659 to 1995. Earth Planet Sci. Lett. 228:177–194

Zubakin GK (1990) Ice and ice edge position. In: The seas of the USSR. Gydrometeorologiya i Gidrokhimiya. V. 1, The Barents Sea. GMI, Leningrad (in Russian)

Colour Plates

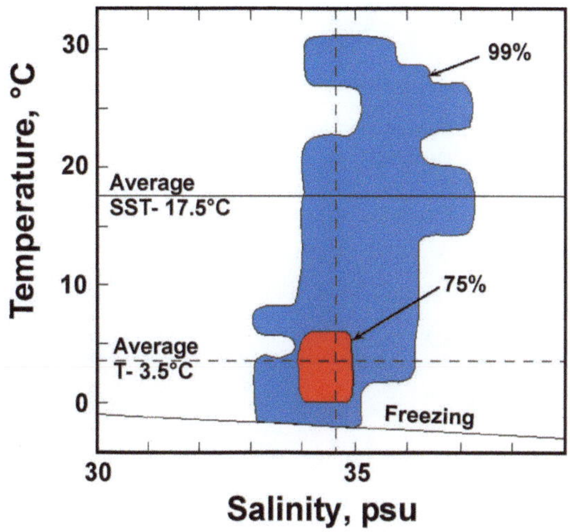

Plate 1.1 Volumertic TS diagram (modified after Gross 1993). Blue figure delineates 99% and red figure delineates 75% of the global ocean water. T = temperature, SST = sea surface temperature

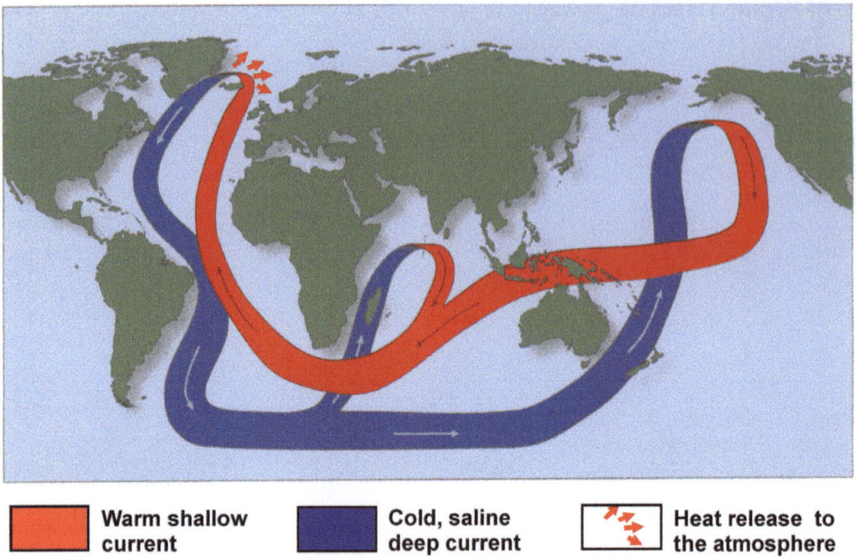

Plate 1.2 The global ocean conveyor belt (Broecker 1991)

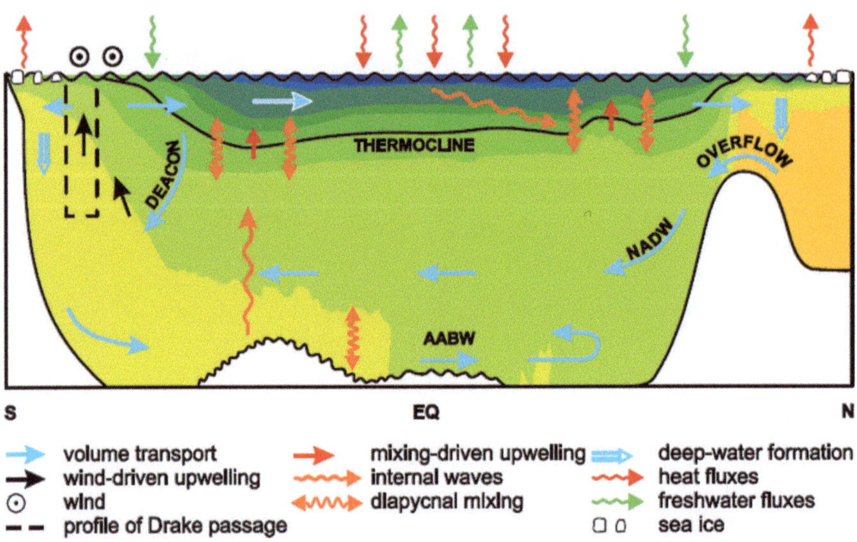

Plate 1.3 A schematic meridional section of the Atlantic Ocean representing a zonally averaged picture (from Kuhlbrodt et al. 2007, with AGU permission). The AMOC is denoted by *straight blue arrows*. The background color shading depicts zonally averaged density stratification from observational data with lightest waters in *blue* and densest in *orange*. The *blue double arrows* subsume the different deep-water formation sites in the North Atlantic (Nordic Seas and Labrador Sea) and in the Southern Ocean (Ross Sea and Weddell Sea). NADW = North Atlantic Deep Water, AABW = Antarctic Bottom Water. AMOC = Atlantic Meridional Overturning Circulation

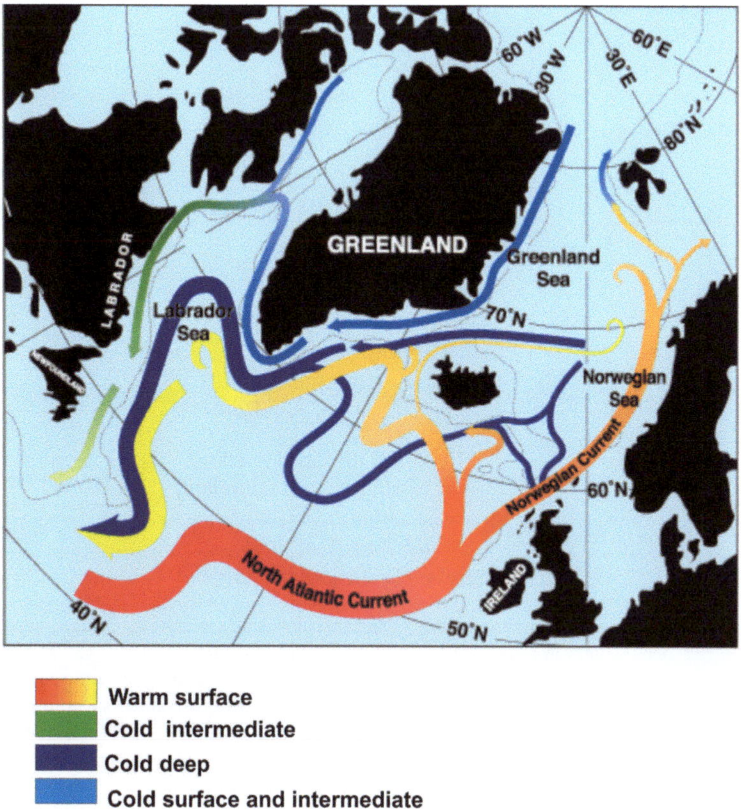

Plate 1.4 Schematic representation of the thermohaline circulation in the North Atlantic (modified after Schmitz 1996)

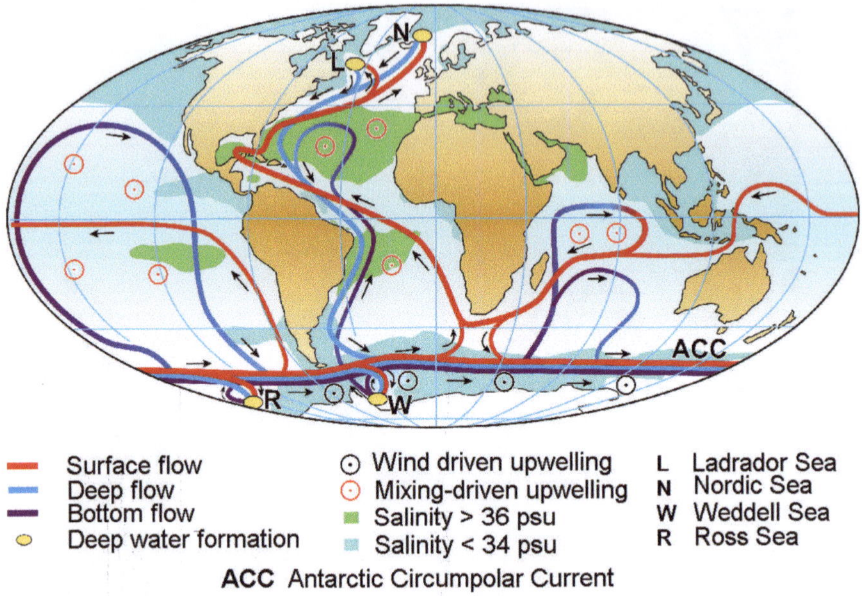

Plate 1.5 Schematic representation of the global thermohaline circulation. Surface currents are shown in *red*, deep waters in *light blue* and bottom waters in *dark blue*, *black flashed* denote the direction of water flows. The main deep water formation sites in the North Atlantic and Antarctic are shown in *yellow*. At the southern end of the Atlantic, the AMOC connects with the Antarctic Circumpolar Current (ACC) (after Rahmstorf 2002, Kuhlbrodt et al. 2007 with Nature permission)

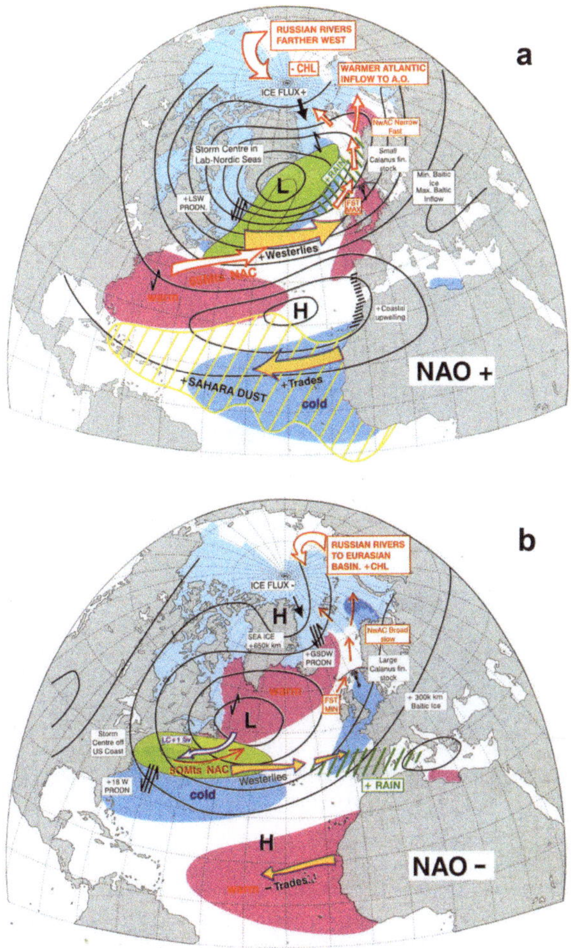

Plate 1.6 Positive (**a**) and negative (**b**) NAO index (modified after Dickson 1999, reproduced from PAGES Newsletter, Vol. 8, No. 1, 2000, with R. Dickson and PAGES permission). During the NAO+, westerlies become stronger and related cyclones shift northward; the Sakhara dust outblow to the North Atlantic by trade winds intensifies. During the NAO, trajectories of westerlies and cyclones shift to the south that results in weakening of the Atlantic water transport to the Arctic and climate humidization in the Western Mediterranean region

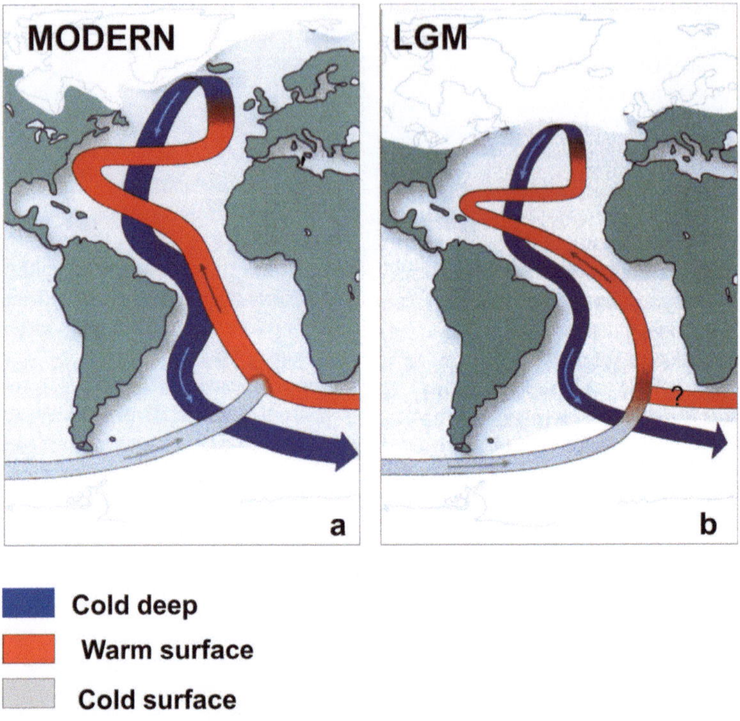

Plate 3.1 Simplified cartoon of the Atlantic Meridional Overturning Circulation (**a**) at present and (**b**) at the LGM (modified after Knorr and Lohmann (2004) suggesting the existence of Agulhas linkage at least at the end of the LGM)

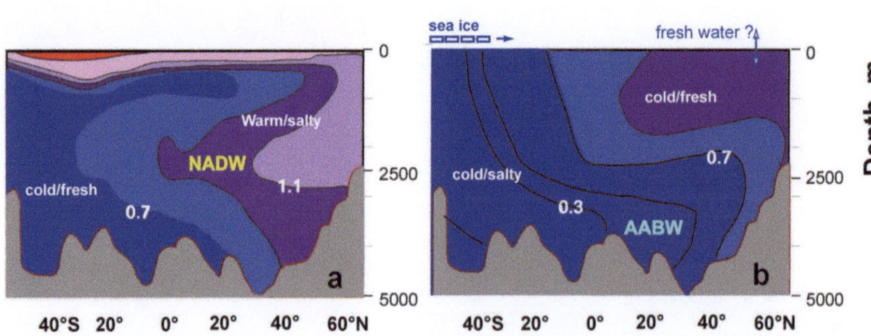

Plate 3.2 Circulation of modern (**A**) and glacial (**B**) Atlantic, as reconstructed by $\delta^{13}C$ tracer data (‰ vs. PDB) along the meridional transect (after Duplessy et al. 1988; Labeyrie et al. 1992; Adkins et al. 2005). Note the shallower flow of NADW and the upward and northward expansion of AABW (*darker blue*) in glacial times. The increase in AABW salinity by 0.4 psu due to intensified sea ice formation and export in the Southern Ocean maintained the meridional overturning, whereas the meltwater discharge into the North Atlantic weakened the deep water convection

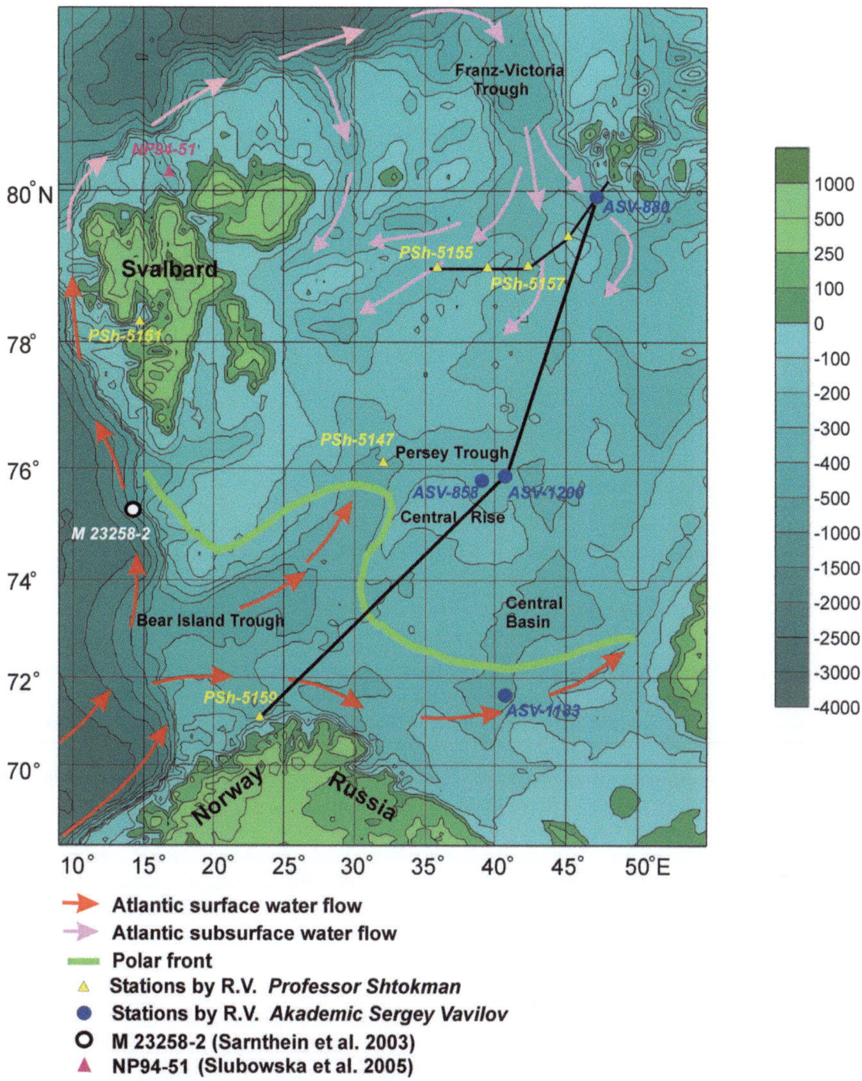

Plate 4.1 Location of two transects and sediment cores and two transects considered in Chapter 4 and in section 7.3 (see Color Figs. 4.3 and 7.2) plotted on the bathymetric map of the Barents Sea. Present mean position of the Polar front (*solid green line*) and direction of the surface (*red arrows*) and subsurface (*pink arrows*) Atlantic water inflow into the Barents Sea are shown

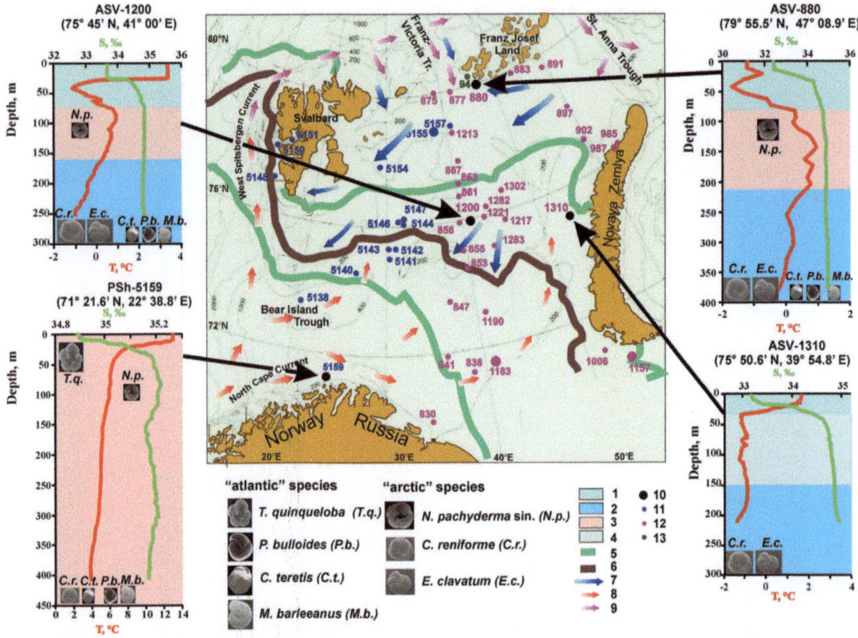

Plate 4.2 Distribution patterns of planktic and benthic foraminifera in Holocene sediments of the Barents Sea depending on water masses (data by the author). Four TS diagrams characterize typical structures of the water column in different Barents Sea regions (location of stations is shown by *arrows*) measured in August-September during expeditions of R.V. *Professor Shtokman* (Cruise 63, 2004), and *Akademik Sergey Vavilov* (cruises 11 and 14, 1997 and 1998). Size of the foraminiferal test images in the diagrams reflects qualitative abundance estimates. Legend: 1 - surface Arctic or Polar water, 2 - bottom Barents seawater, 3 - Atlantic water, 4 - subsurface Barents seawater, 5 - seasonal sea ice margins (after Vinje and Kvambekk 1991), 6 - Polar Front, 7 - Arctic water input to the Barents Sea, 8 - surface Atlantic water input to the Barents Sea, 9 - subsurface Atlantic water input to the Barents Sea, 10 - stations with TS profiles shown, 11 - stations by R.V. *Professor Shtokman*, 12 - stations by R.V. *Akademik Sergey Vavilov*, 13 - stations by R.V. *Akademik Alexandre Karpinski*. Surcuface circulation after (Tantsyura 1959)

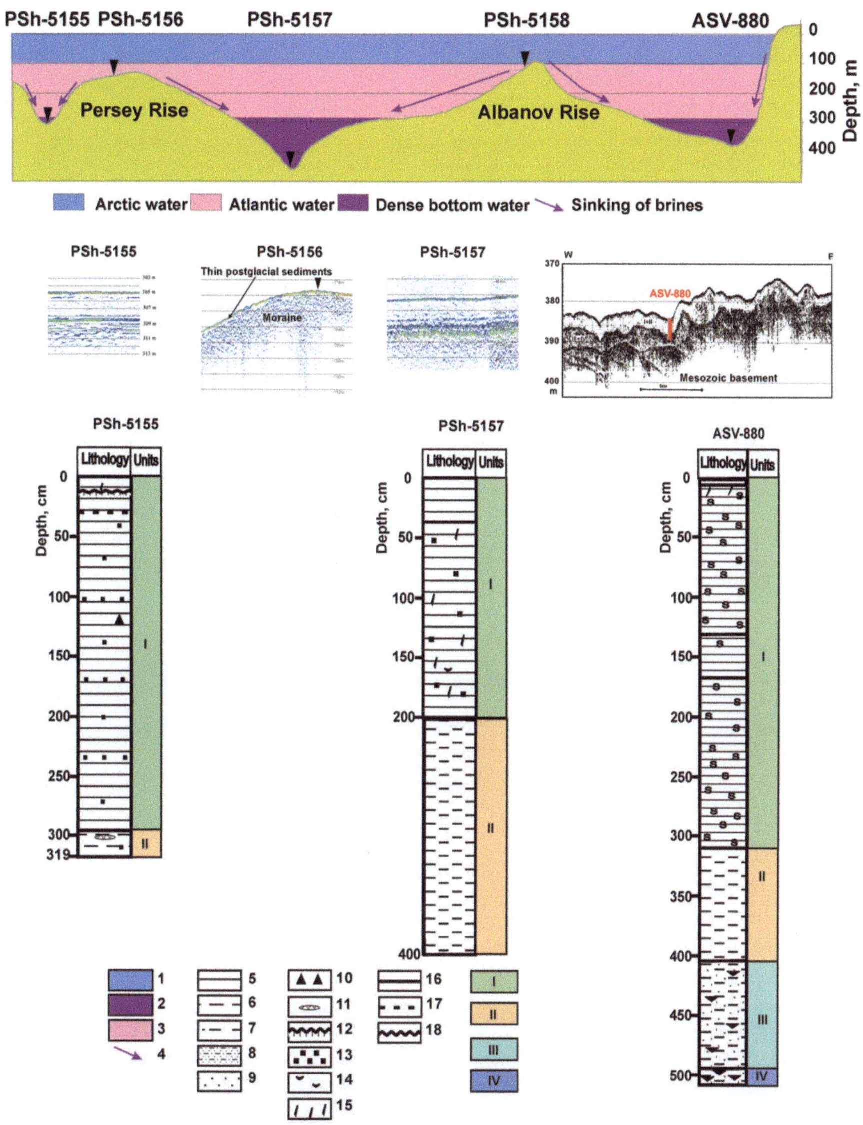

Plate 4.3 Sediment cores from three southern branches of the Franz Victoria Trough. Rise summits in between are practically barren of Holocene sediments (upper acoustically transparent unit in seismic records). Main water masses of the stratified water column filling the trough branches are shown in the profile above (for the location see Fig. 4.1). *Arrows* indicate assumed sinking of brines which carry the near-bottom nepheloid layer from banks to local deeps, thus maintaining a rapid sedimentation in those. Stratigraphic units in cores: (I) Holocene; (II) Late Deglaciation; (III) Early Deglaciation; (IV) Last (Late Weichselian) glaciation (LGM?). The figure is compiled using materials obtained during Cruise 11 of the R.V. *Akademik Sergei Vavilov* (1997) and Cruise 63 of the R.V. *Professor Shtokman* (2004) (modified after Ivanova 2004)

Plate 4.4 Relationship between two groups of benthic foraminifera in cores BP00-07/5, BP01-62/6, and PS2718/6 from the Kara Sea: (1) species preferring normal marine salinity, (2) species tolerant to decreased salinity. The upper part of core PS2718/6 was lost during the coring, and the middle part is barren

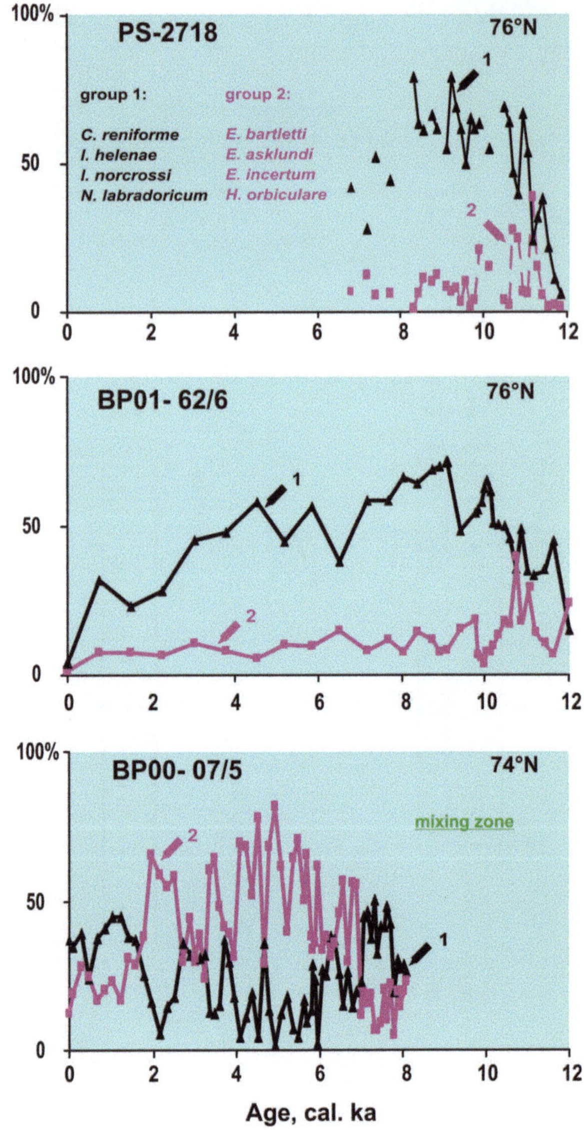

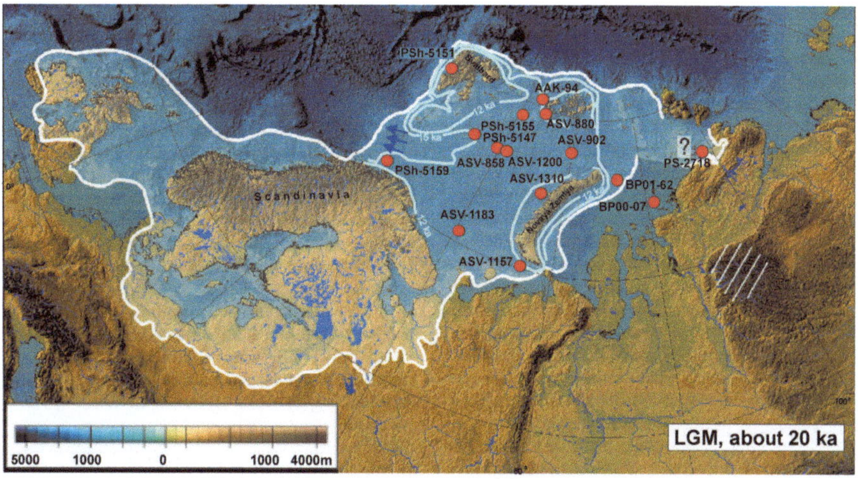

Plate 4.5 Reconstruction of the Eurasian ice sheet at the Last Glacial Maximum (LGM), about 20 ka BP (after Svendsen et al. 2004 with additions). *Dark blue arrows* show direction of the major outlet glacier via the Bear Island Trough to the Norwegian Sea; and *blue lines* in Barents and Kara seas are isochrones of the glacier front retreat during deglaciation, 15 ka and 12 ka BP (from Landvik et al. 1998). *Red circles* indicate location of reference cores used in Chapter 4

Plate 4.6 Maximum and minimum ice sheet boundary in the Kara Sea at the LGM according to different authors. Submerged Ob and Yenisei paleoriver channels are shown. Position of the ice sheet edge according to data obtained during cruises 35 and 36 of the R.V. *Boris Petrov* is marked by *solid red line* (modified from Stein et al. 2002). *Red filled circles* indicate location of two cores studied by the author

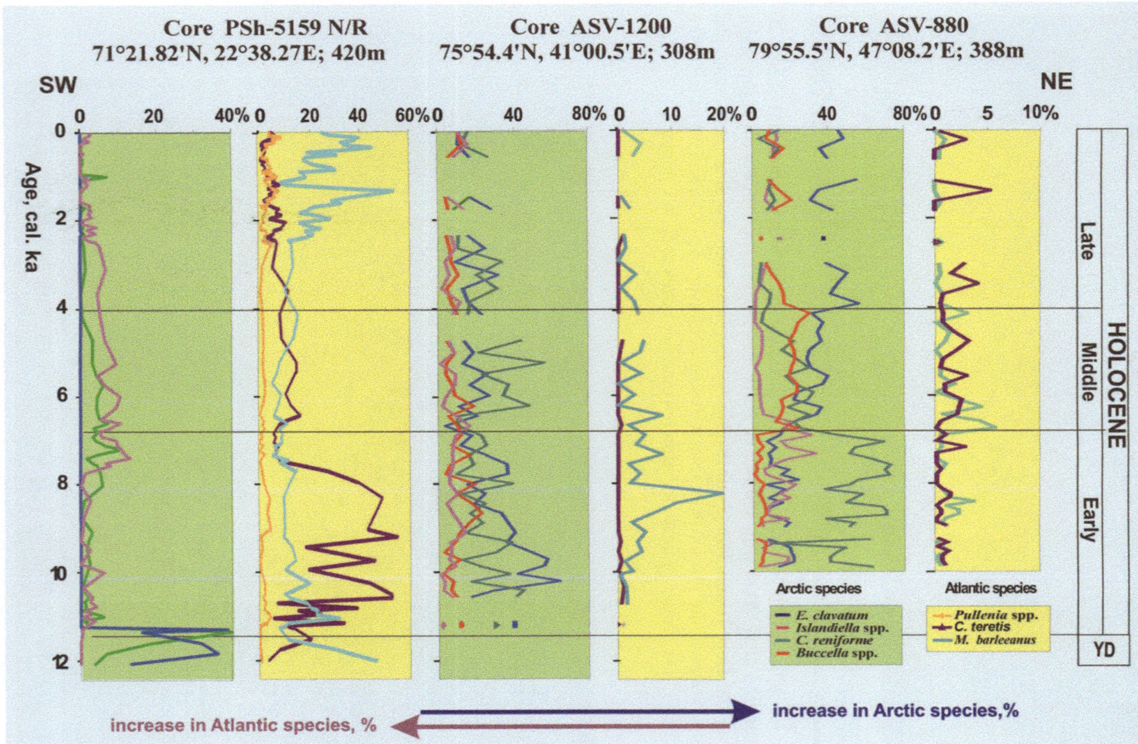

Plate 4.7 Distribution of indicative benthic foraminiferal species versus calendar age BP throughout the Holocene section in three reference cores along the SW–NE transect in the Barents Sea. The foraminiferal data for cores ASV 880 and ASV 1200 from (Ivanova 2003), age models from (Duplessy et al. 2001, 2005), foraminiferal data and age model for core PSh 5159N/R from (Chistyakova et al. in press, Risebrobbaken et al. submitted). The gaps in the species records result from foraminiferal dissolution

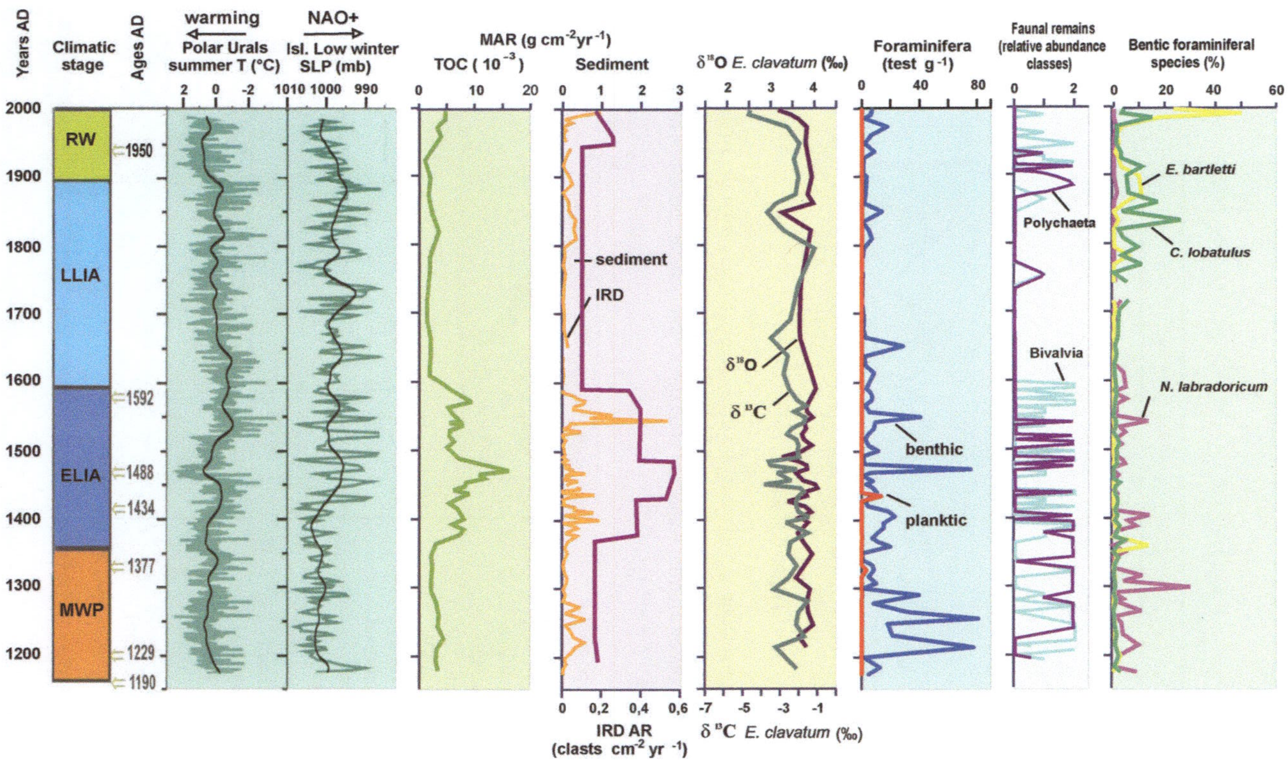

Plate 4.8 Correlation of multi-proxy records of Core ASV-987 from the Russian Gavan' Fjord, NE Barents Sea, with the records of summer temperature anomaly based on tree-ring data from Polar Urals (Briffa et al. 1995; Briffa 2000), and reconstructed Icelandic Low intensity based on GISP-2 sea-salt Na data (Meeker and Mayewski, 2002) over the last 800 years (modified after Ivanova et al. 2003a, Polyak et al. 2004)

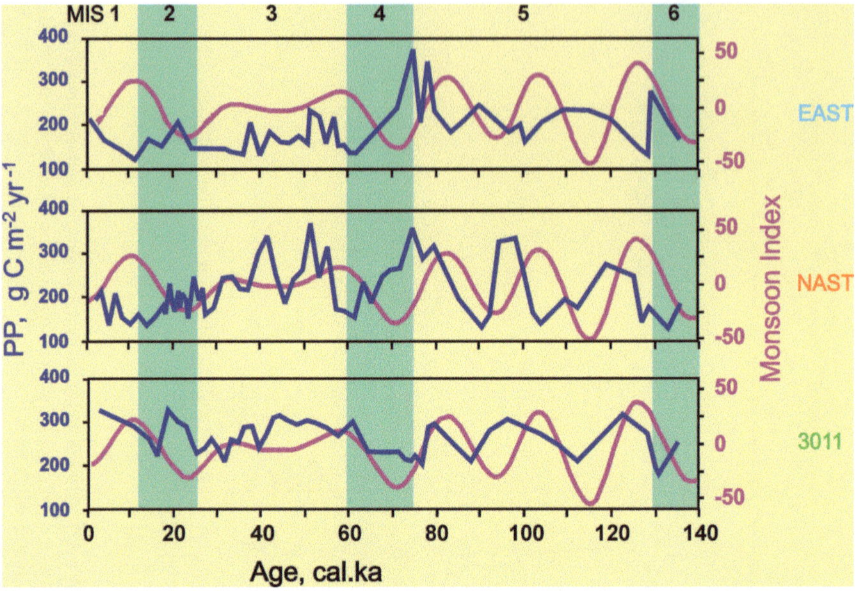

Plate 5.1 Primary production estimated in the three deep-sea cores from the Arabian Sea and compared with the summer Monsoon Index (MI) by Rossignol-Strick (1983). Glacial stages are marked by stippling (modified from Ivanova et al. 2003b). Note general correlation of PP maxima to MI maxima with a few thousand years lag

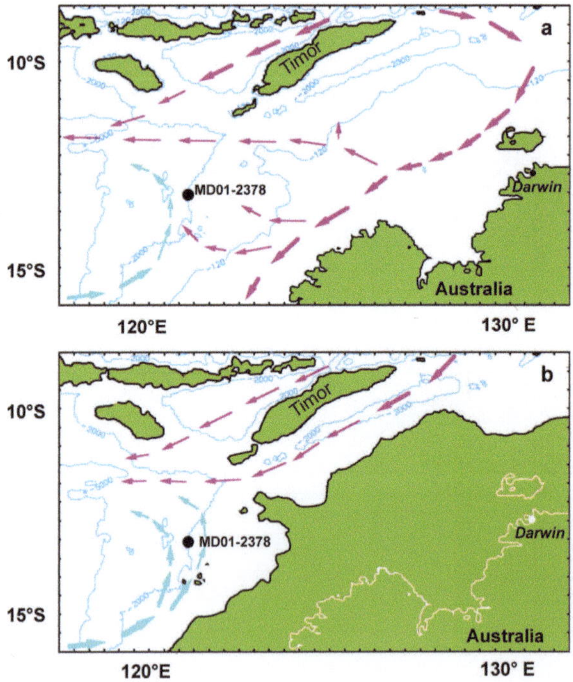

Plate 5.2 Glacial topography and surface current systems in the Indonesian Throughflow (ITF) outflow area of the Timor passage. Today, warm ITF water (lilac) is transported over the NW Australian shelf and forms the predominant water mass at the location of Core MD01-2378. During the Last Glacial Maximum the ITF was strongly reduced and cold current systems (*blue*) originating from the south probably reached the Timor Sea (after Holbourn et al. 2005 with AGU permission)

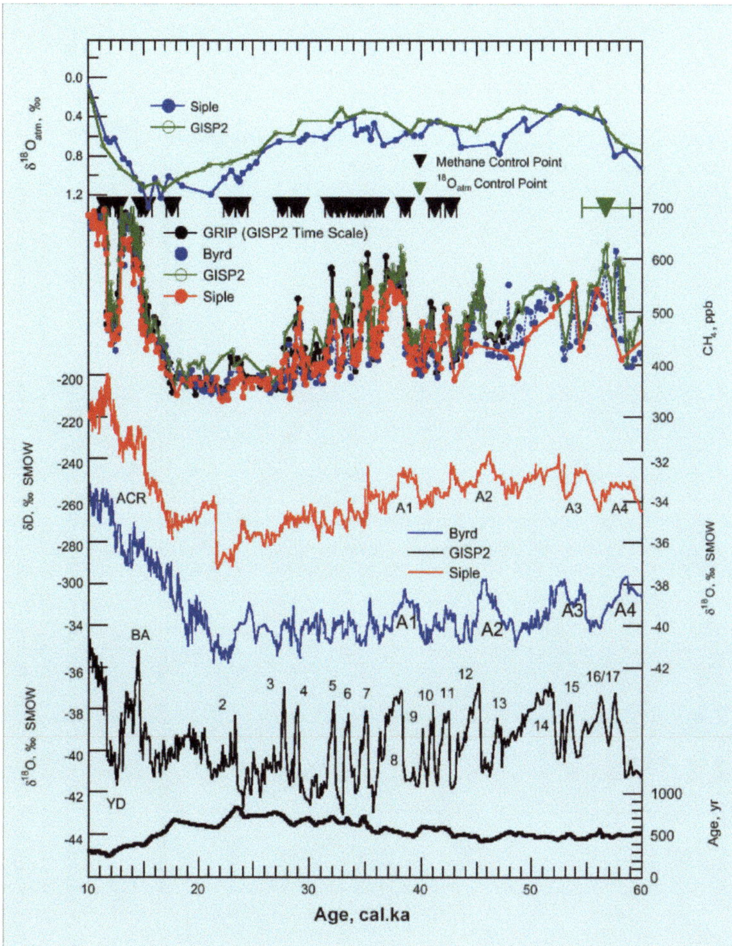

Plate 7.1 The correlation of the proxy time series records from the Antarctic ice cores (Brook et al. 2005 with permission from Elsevier). Age control points plotted at the top of figure (*solid triangles*). The youngest control point used (8.33 ka at 514.78 m based on CH_4) is not shown. Numbers indicate DO events, BA = Bølling/Allerød, YD = Younger Dryas, ACR = Antarctic Cold Reversal, A1 - A4 = Antarctic warmings

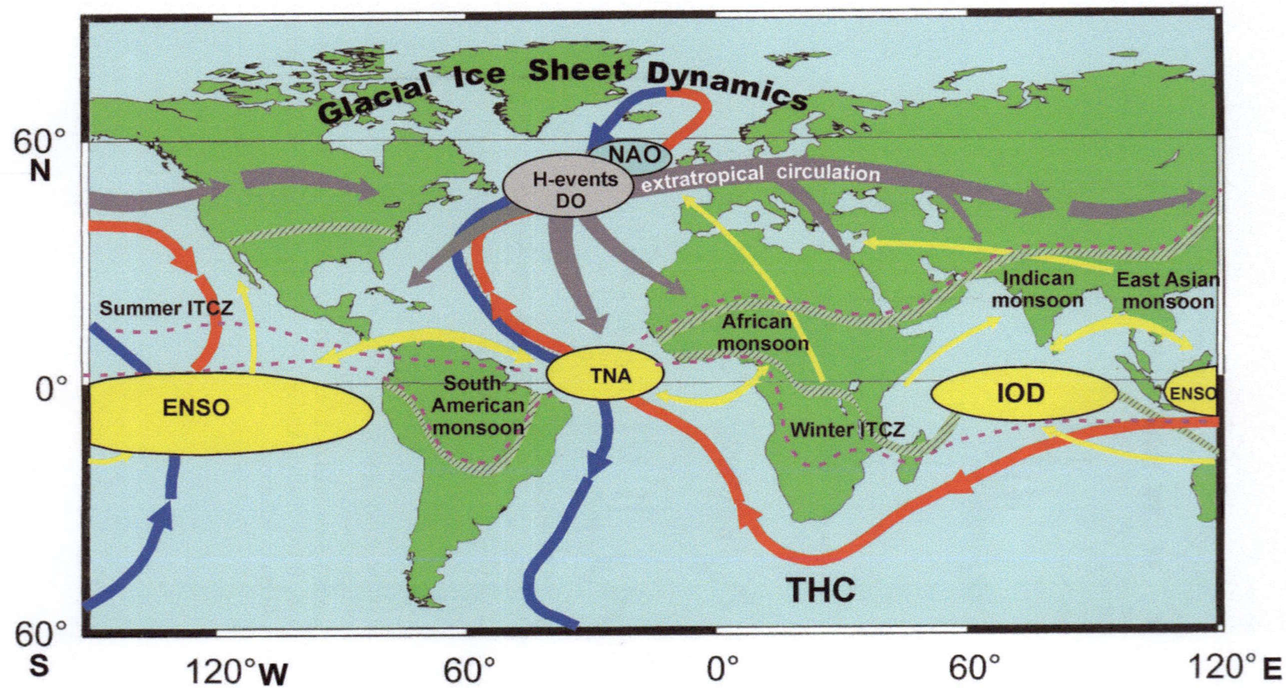

Plate 7.2 Linkages between agents of the climatic teleconnections in the atmosphere and ocean operating at different time scales (after Zahn 2002; Vidal and Arz 2004). *Arrows* indicate linkages between different agents of the system. Surface and deep limbs of the global thermohaline circulation (THC) are shown by *red* and *blue colors*, respectively. Winter and summer position of the ITCZ is shown by *pink dashed line*. H-events = Heinrich events, DO = Dansgaard-Oeschger cycles. NAO = North Atlantic Oscillation; ENSO = El Niño - Southern Oscillation; IOD = Indian Ocean dipole; TNA = tropical North Atlantic. *Shaded belts* mark boundaries of the monsoon area

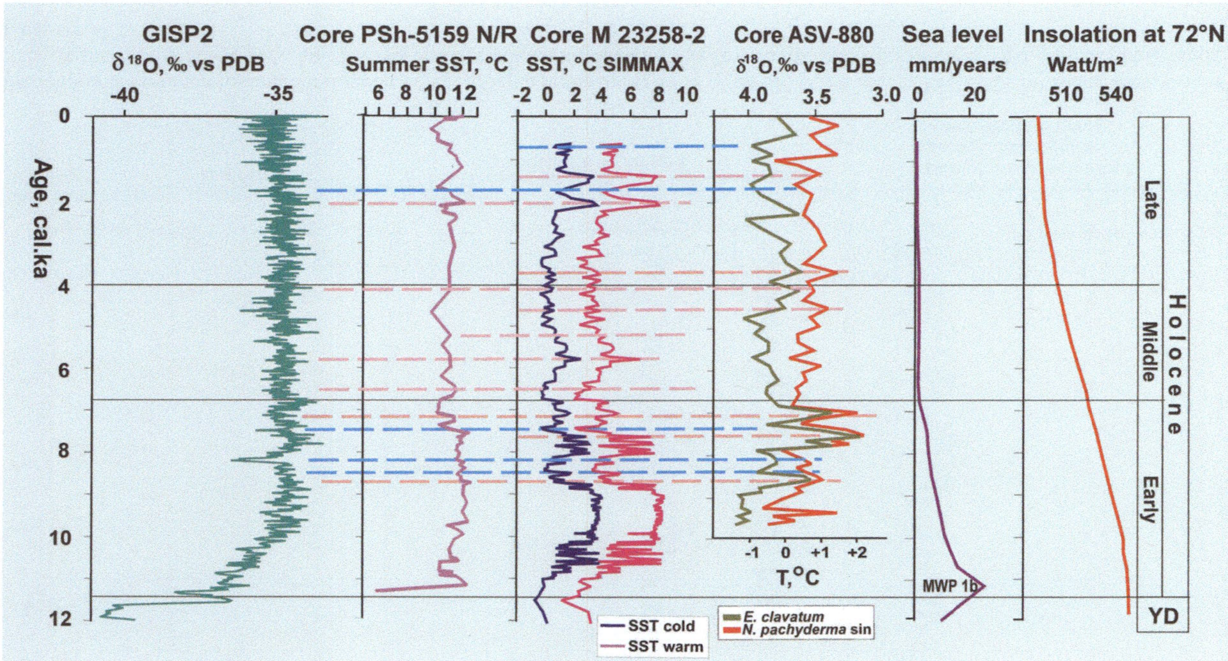

Plate 7.3 Correlation of paleotemperature time series in three Barents Sea cores discussed in section 4.5 with GISP2 oxygen isotope record (Grootes and Stuiver 1997), June insolation at 72°N (Laskar 1990) and sea-level record (Fairbanks 1989) over the last 12 ka. For location see Col. Fig. 4.1. Paleotemperature records from core ASV 880 (Duplessy et al. 2001) correspond to subsurface and bottom water layers, for core M 23258-2 (Sarnthein et al. 2003a) winter (cold) and summer (warm) sea-surface temperatures (SST) are shown, for core PSh 5159N/R (Chistyakova et al. in press) summer SST is shown. Short-term warming and cooling episodes are indicated by pink and blue hatch lines respectively. YD = Younger Dryas. MWP 1b = melt water pulse 1b at the end of the last deglaciation. Note that the Holocene warming in the Barents Sea starts after the insolation maximum, and is coeval to the MWP 1b event in core PSh 5159N/R from the south-western part of the sea

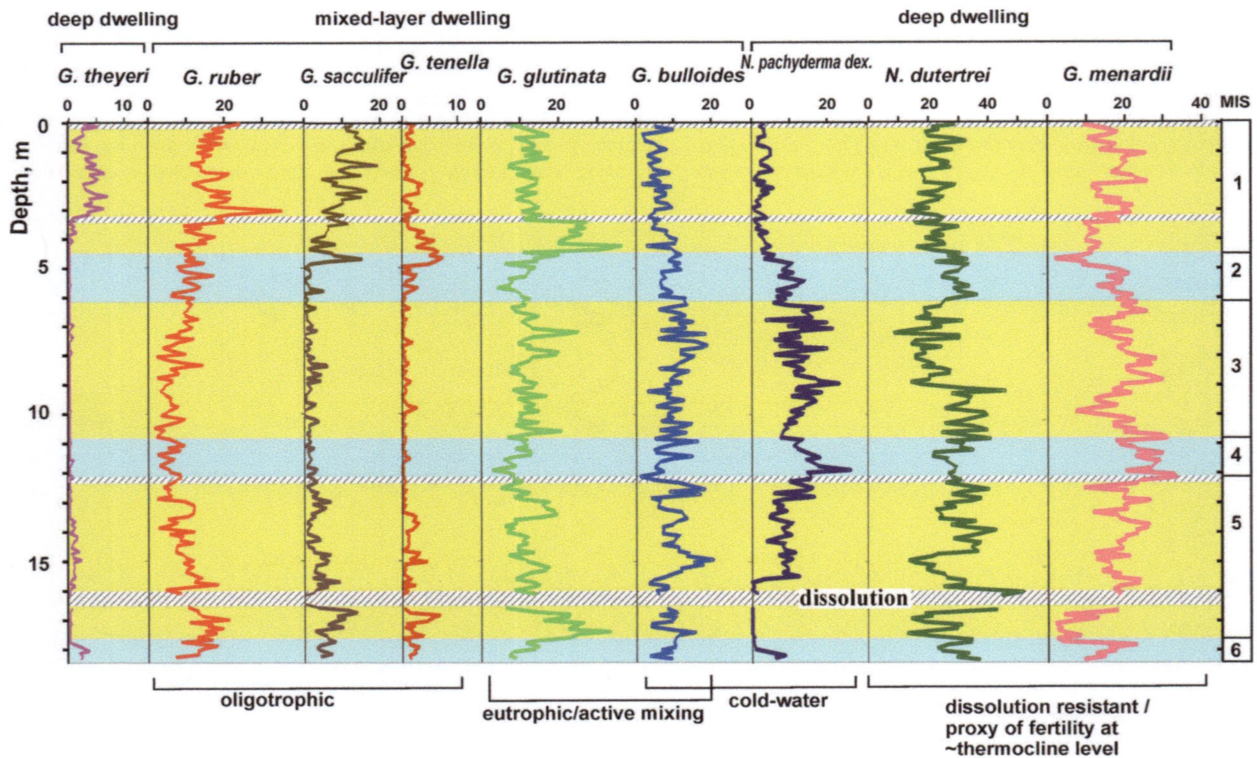

Plate 7.4 Distribution of common foraminiferal species (%) vs. depth in the upper 18 m of the IMAGES core MD02-2529 off Costa Rica, eastern equatorial Pacific. Glacials are marked by *blue bands*, interglacials and interstadial are shown by *yellow* bands according to the oxygen isotope stratigraphy (data from Ivanova et al. in prep.)

Index

H

O

P

Q

T

MIX
Papier aus verantwortungsvollen Quellen
Paper from responsible sources
FSC® C105338

If you have any concerns about our products,
you can contact us on
ProductSafety@springernature.com

In case Publisher is established outside the EU,
the EU authorized representative is:
Springer Nature Customer Service Center GmbH
Europaplatz 3, 69115 Heidelberg, Germany

Printed by Libri Plureos GmbH
in Hamburg, Germany